Lecture Notes in Computer Science 12489

More information about this series at http://www.springer.com/series/7409

Cesar Analide · Paulo Novais ·
David Camacho · Hujun Yin (Eds.)

Intelligent Data Engineering and Automated Learning – IDEAL 2020

21st International Conference
Guimaraes, Portugal, November 4–6, 2020
Proceedings, Part I

 Springer

Editors
Cesar Analide (iD)
University of Minho
Braga, Portugal

Paulo Novais (iD)
University of Minho
Braga, Portugal

David Camacho (iD)
Technical University of Madrid
Madrid, Spain

Hujun Yin (iD)
University of Manchester
Manchester, UK

ISSN 0302-9743   ISSN 1611-3349   (electronic)
Lecture Notes in Computer Science
ISBN 978-3-030-62361-6   ISBN 978-3-030-62362-3   (eBook)
https://doi.org/10.1007/978-3-030-62362-3

LNCS Sublibrary: SL3 – Information Systems and Applications, incl. Internet/Web, and HCI

This Springer imprint is published by the registered company Springer Nature Switzerland AG
The registered company address is: Gewerbestrasse 11, 6330 Cham, Switzerland

Preface

The International Conference on Intelligent Data Engineering and Automated Learning (IDEAL) is an annual international conference dedicated to emerging and challenging topics in intelligent data analytics and associated machine learning systems and paradigms. After the hugely successful IDEAL 2018 in Madrid, Spain, and its 20th edition in Manchester, UK, last year, IDEAL 2020 was hosted by the Synthetic Intelligence Lab (ISLab) from the ALGORITMI Center at the University of Minho, in Guimarães, the birthplace of Portugal, during November 4–6, 2020. It was also technically co-sponsored by the Portuguese Artificial Intelligence Association (APPIA) and the IEEE Computational Intelligence Society Portuguese Chapter. Due to the COVID-19 pandemic, the Organization Committee decided to hold the IDEAL 2020 completely online, to ensure the safety of the organization and the delegates of the conference.

This year marked the 21st edition of IDEAL, a conference series which has been serving an important role in data analytics and machine learning communities. The conference aims to bring researchers and practitioners together, share the latest knowledge, disseminate cutting-edge results, and forge alliances on tackling many real-world challenging problems. The core themes of the IDEAL 2020, as usual, included big data challenges, machine learning, data mining, information retrieval and management, bio-/neuro-informatics, bio-inspired models, agents and hybrid intelligent systems, real-world applications of intelligent techniques, and AI.

The IDEAL 2020 event consisted of the main track, nine special sessions, and one colocated workshop. These special sessions were:

- Special Session 1:   Data Generation and Data Pre-processing in Machine Learning
- Special Session 2:   Optimization and Machine Learning for Industry 4.0
- Special Session 3:   Practical Applications of Deep Learning
- Special Session 4:   New trends and challenges on Social Networks Analysis
- Special Session 5:   Machine Learning in Automatic Control
- Special Session 6:   Emerging Trends in Machine Learning
- Special Session 7:   Machine Learning, Law and Legal Industry
- Special Session 8:   Data Recovery Approach to Clustering and Interpretability
- Special Session 9:   Automated learning for industrial applications

The colocated event was the Workshop on Machine Learning in Smart Mobility (MLSM 2020).

A total of 122 submissions were received and were then subjected to rigorous peer reviews by the members and experts of the Program Committee. Only those papers that were found to be of the highest quality and novelty were accepted and included in the proceedings. These proceedings contain 93 papers, in particular, 34 for the main track with an acceptance rate of 59%, 54 for special sessions, and 5 for the colocated workshop, were accepted and presented at IDEAL 2020.

We deeply appreciate the efforts of our invited speakers Paulo Lisboa of Liverpool John Moores University, UK; João Gama of the University of Porto, Portugal; and George Baciu of Hong Kong Polytechnic University, China; and thank them for their interesting and inspiring lectures.

We would like to thank our sponsors for their technical support. We would also like to thank everyone who invested so much time and effort in making the conference a success, particularly the Program Committee members and reviewers, the organizers of the special sessions, as well as all the authors who contributed to the conference.

A big thank you to the special sessions and workshop chairs Antonio J. Tallón-Ballesteros and Susana Nascimento for their fantastic work.

Special thanks also go to the MLSM 2020 organizers Sara Ferreira, Henrique Lopes Cardoso, and Rosaldo Rossetti.

Finally, we are also very grateful to the hard work by the local Organizing Committee at the University of Minho, Portugal, especially Pedro Oliveira, António Silva, and Bruno Fernandes for checking all the camera-ready files. The continued support, sponsorship, and collaboration from Springer LNCS are also greatly appreciated.

November 2020

Cesar Analide
Paulo Novais
David Camacho
Hujun Yin

Organization

General Chairs

Paulo Novais	University of Minho, Portugal
Cesar Analide	University of Minho, Portugal
David Camacho	Universidad Politecnica de Madrid, Spain

Honorary Chair

Hujun Yin	The University of Manchester, UK

Program Chairs

Cesar Analide	University of Minho, Portugal
Paulo Novais	University of Minho, Portugal
David Camacho	Universidad Politecnica de Madrid, Spain
Hujun Yin	The University of Manchester, UK

Steering Committee

Hujun Yin	The University of Manchester, UK
Colin Fyfe	University of the West of Scotland, UK
Guilherme Barreto	Federal University of Ceará, Brazil
Jimmy Lee	The Chinese University of Hong Kong, Hong Kong, China
John Keane	The University of Manchester, UK
Jose A. Costa	Federal University of Rio Grande do Norte, Brazil
Juan Manuel Corchado	University of Salamanca, Spain
Laiwan Chan	The Chinese University of Hong Kong, Hong Kong, China
Malik Magdon-Ismail	Rensselaer Polytechnic Institute, USA
Marc van Hulle	KU Leuven, Belgium
Ning Zhong	Maebashi Institute of Technology, Japan
Peter Tino	University of Birmingham, UK
Samuel Kaski	Aalto University, Finland
Vic Rayward-Smith	University of East Anglia, UK
Yiu-ming Cheung	Hong Kong Baptist University, Hong Kong, China
Zheng Rong Yang	University of Exeter, UK

Special Session and Workshop Chairs

Antonio J. Tallón-Ballesteros	University of Seville, Spain
Susana Nascimento	Universidade Nova de Lisboa, Portugal

Publicity and Liaisons Chairs

Bin Li	University of Science and Technology of China, China
Guilherme Barreto	Federal University of Ceará, Brazil
Jose A. Costa	Federal University of Rio Grande do Norte, Brazil
Yimin Wen	Guilin University of Electronic Technology, China

Local Organizing Committee

Paulo Novais
Paulo Moura Oliveira
Cesar Analide
José Luís Calvo-Rolle
José Machado
Ana Silva
António Silva
Héctor Alaiz Moretón
Bruno Fernandes

Leandro Freitas
Dalila Durães
Leonardo Nogueira Matos
Filipe Gonçalves
Marco Gomes
Francisco Marcondes
Pedro Oliveira
Fábio Silva

Program Committee

Álvaro Herrero
Ajalmar Rêgo Da Rocha Neto
Ajith Abraham
Alejandro Martín
Alexandros Tzanetos
Alfredo Cuzzocrea
Alfredo Vellido
Ana Madureira
Anabela Simões
Andre de Carvalho
Angel Arcos
Angel Arroyo
Ângelo Costa
Anna Gorawska
Antonio Fernández-Caballero
Antonio Gonzalez-Pardo
Antonio J. Tallón-Ballesteros
Antonio Neme

Armando Mendes
Bibiana Clara
Bin Li
Bogusław Cyganek
Boris Delibašić
Boris Mirkin
Bruno Baruque
Carlos A. Iglesias
Carlos Cambra
Carlos Carrascosa
Carlos Coello Coello
Carlos M. Travieso-Gonzalez
Carlos Pereira
Carlos Ramos
Carmelo J. A. Bastos Filho
Cristian Mihaescu
Cristian Ramírez-Atencia
Dalila Duraes

Daniel Urda
Dariusz Frejlichowski
Dariusz Jankowski
Diana Manjarres
Davide Carneiro
Dinu Dragan
Dongqing Wei
Dragan Simic
Edyta Brzychczy
Eiji Uchino
Emilio Carrizosa
Eneko Osaba
Ernesto Damiani
Esteban Jove
Eva Onaindia
Federico Mata
Felipe M. G. França
Fernando De La Prieta
Fernando Diaz
Fernando Nuñez
Fionn Murtagh
Florentino Fdez-Riverola
Francesco Corona
Fábio Silva
Gema Bello Orgaz
Gerard Dreyfus
Giancarlo Fortino
Gianni Vercelli
Goreti Marreiros
Grzegorz J. Nalepa
Hamido Fujita
Héctor Alaiz Moretón
Héctor Quintián
Ignacio Hidalgo
Ioannis Hatzilygeroudis
Irene Pulido
Ireneusz Czarnowski
Isaias Garcia
Ivan Silva
Izabela Rejer
J. Michael Herrmann
Jaakko Hollmén
Jason Jung
Javier Bajo
Javier Del Ser
Jean-Michel Ilie

Jerzy Grzymala-Busse
Jesus Alcala-Fdez
Jesus López
Jesús Sánchez-Oro
João Carneiro
Joaquim Filipe
Jochen Einbeck
Jose Andrades
Jose Alfredo Ferreira Costa
Jose Carlos Montoya
Jose Dorronsoro
Jose Luis Calvo-Rolle
Jose M. Molina
José Maia Neves
Jose Palma
Jose Santos
Josep Carmona
José Fco. Martínez-Trinidad
José Alberto Benítez-Andrades
José Machado
José Luis Casteleiro-Roca
José Ramón Villar
José Valente de Oliveira
João Ferreira
João Pires
Juan G. Victores
Juan Jose Flores
Juan Manuel Dodero
Juan Pavón
Leandro Coelho
Lino Figueiredo
Lourdes Borrajo
Luis Javier García Villalba
Luís Cavique
Luís Correia
Maciej Grzenda
Manuel Grana
Manuel Jesus Cobo Martin
Marcin Gorawski
Marcin Szpyrka
Marcus Gallagher
Margarida Cardoso
Martin Atzmueller
Maria Teresa Garcia-Ordas
María José Ginzo Villamayor
Matilde Santos

Mercedes Carnero
Michal Wozniak
Miguel J. Hornos
Murat Caner Testik
Ngoc-Thanh Nguyen
Pablo Chamoso
Pablo García Sánchez
Paulo Cortez
Paulo Moura Oliveira
Paulo Quaresma
Paulo Urbano
Pawel Forczmanski
Pedro Antonio Gutierrez
Pedro Castillo
Pedro Freitas
Peter Tino
Qing Tian
Radu-Emil Precup
Rafael Corchuelo
Raquel Redondo
Raul Lara-Cabrera
Raymond Wong
Raúl Cruz-Barbosa
Renato Amorim
Ricardo Aler
Ricardo Santos

Richard Allmendinger
Richar Chbeir
Robert Burduk
Roberto Carballedo
Roberto Casado-Vara
Roberto Confalonieri
Romis Attux
Rui Neves Madeira
Rushed Kanawati
Songcan Chen
Stefania Tomasiello
Stelvio Cimato
Sung-Bae Cho
Susana Nascimento
Tatsuo Nakajima
Tzai-Der Wang
Valery Naranjo
Vasile Palade
Vicent Botti
Vicente Julián
Víctor Fernández
Wei-Chiang Hong
Wenjian Luo
Xin-She Yang
Ying Tan

Additional Reviewers

Adrián Colomer
Aldo Cipriano
Alejandro Álvarez Ayllón
Alfonso González Briones
Alfonso Hernandez
Angel Panizo Lledot
Beatriz Ruiz Reina
David Garcia-Retuerta
Eloy Irigoyen
Francisco Andrade
Gabriel Garcia
Javier Huertas-Tato
Jesus Fernandez-Lozano
Joana Silva
Juan Albino Mendez
Juan José Gamboa-Montero

Julio Silva-Rodríguez
Kevin Sebastian Luck
Lorena Gonzalez Juarez
Manuel Masseno
Marcos Maroto Gómez
Mariliana Rico Carrillo
Mario Martín
Marta Arias
Pablo Falcón Oubiña
Patrícia Jerónimo
Paweł Martynowicz
Szymon Bobek
Vanessa Jiménez Serranía
Wenbin Pei
Ying Bi

Special Session on Data Generation and Data Pre-processing in Machine Learning

Organizers

Antonio J. Tallón-Ballesteros	University of Huelva, Spain
Bing Xue	Victoria University of Wellington, New Zealand
Luis Cavique	Universidade Aberta, Portugal

Special Session on Optimization and Machine Learning for Industry 4.0

Organizers

Eneko Osaba	Basque Research and Technology Alliance, Spain
Diana Manjarres	Basque Research and Technology Alliance, Spain
Javier Del Ser	University of the Basque Country, Spain

Special Session on Practical Applications of Deep Learning

Organizers

Alejandro Martín	Universidad Politécnica de Madrid, Spain
Víctor Fernández	Universidad Politécnica de Madrid, Spain

Special Session on New Trends and Challenges on Social Networks Analysis

Organizers

Gema Bello Orgaz	Universidad Politécnica de Madrid, Spain
David Camacho	Universidad Politécnica de Madrid, Spain

Special Session on Machine Learning in Automatic Control

Organizers

Matilde Santos	University Complutense de Madrid, Spain
Juan G. Victores	University Carlos III of Madrid, Spain

Special Session on Emerging Trends in Machine Learning

Organizers

Davide Carneiro	Polytechnic Institute of Porto, School of Technology and Management, Portugal
Fábio Silva	Polytechnic Institute of Porto, School of Technology and Management, Portugal

José Carlos Castillo   University Carlos III of Madrid, Spain
Héctor Alaiz Moretón   University of León, Spain

Special Session on Machine Learning, Law and Legal Industry

Organizers

Pedro Miguel Freitas   Portuguese Catholic University, Portugal
Federico Bueno de Mata   University of Salamanca, Spain

Special Session on Data Recovery Approach to Clustering and Interpretability

Organizers

Susana Nascimento   NOVA University of Lisbon, Portugal
José Valente de Oliveira   University of Algarve, Portugal
Boris Mirkin   National Research University, Russia

Special Session on Automated Learning for Industrial Applications

Organizers

José Luis Calvo-Rolle   University of A Coruña, Spain
Álvaro Herrero   University of Burgos, Spain
Roberto Casado Vara   University of Salamanca, Spain
Dragan Simić   University of Novi Sad, Serbia

Workshop on Machine Learning in Smart Mobility

Organizers

Sara Ferreira   University of Porto, Portugal
Henrique Lopes Cardoso   University of Porto, Portugal
Rosaldo Rossetti   University of Porto, Portugal

Program Committee

Achille Fonzone   Edinburgh Napier University, UK
Alberto Fernandez   Rey Juan Carlos University, Spain
Ana L. C. Bazzan   Federal University of Rio Grande do Sul, Brazil
Ana Paula Rocha   University of Porto, LIACC Portugal
Carlos A. Iglesias   Polytechnic University of Madrid, Spain
Cristina Olaverri-Monreal   Johannes Kepler University Linz, Austria
Daniel Castro Silva   University of Porto, LIACC, Portugal
Dewan Farid   United International University, Bangladesh
Eduardo Camponogara   Federal University of Santa Catarina, Brazil
Eftihia Nathanail   University of Thessaly, Greece
Fenghua Zhu   Chinese Academy of Sciences, China

Contents – Part I

Contents – Part II

Special Session on Practical Applications of Deep Learning

Special Session on New Trends and Challenges on Social Networks Analysis

Special Session on Machine Learning in Automatic Control

Special Session on Emerging Trends in Machine Learning

Special Session on Machine Learning, Law and Legal Industry

Special Session on Machine Learning Algorithms for Hard Problems

Workshop on Machine Learning in Smart Mobility

Prostate Gland Segmentation in Histology Images via Residual and Multi-resolution U-NET

Julio Silva-Rodríguez[1]([✉]), Elena Payá-Bosch[2], Gabriel García[2], Adrián Colomer[2], and Valery Naranjo[2]

[1] Institute of Transport and Territory, Universitat Politècnica de València, Valencia, Spain
jjsilva@upv.es
[2] Institute of Research and Innovation in Bioengineering, Universitat Politècnica de València, Valencia, Spain

Abstract. Prostate cancer is one of the most prevalent cancers worldwide. One of the key factors in reducing its mortality is based on early detection. The computer-aided diagnosis systems for this task are based on the glandular structural analysis in histology images. Hence, accurate gland detection and segmentation is crucial for a successful prediction. The methodological basis of this work is a prostate gland segmentation based on *U-Net* convolutional neural network architectures modified with residual and multi-resolution blocks, trained using data augmentation techniques. The residual configuration outperforms in the test subset the previous state-of-the-art approaches in an image-level comparison, reaching an average *Dice Index* of 0.77.

Keywords: Prostate cancer · Histology · Gland segmentation · *U-Net* · Residual

1 Introduction

Prostate cancer was the second most prevalent cancer worldwide in 2018, according to the Global Cancer Observatory [15]. The final diagnosis of prostate cancer is based on the visual inspection of histological biopsies performed by expert pathologists. Morphological patterns and the distribution of glands in the tissue are analyzed and classified according to the Gleason scale [4]. The Gleason patterns range from 3 to 5, inversely correlating with the degree of glandular differentiation. In recent years, the development of computer-assisted diagnostic systems has increased in order to raise the level of objectivity and support the work of pathologists.

One of the ways to reduce mortality in prostate cancer is through its early detection [14]. For this reason, several works have focused on the first stage of prostate cancer detection by differentiating between benign and Gleason pattern 3 glands [1,3,9]. The benign glands differentiate from Gleason pattern 3

C. Analide et al. (Eds.): IDEAL 2020, LNCS 12489, pp. 1–8, 2020.
https://doi.org/10.1007/978-3-030-62362-3_1

structures in the size, morphology, and density in the tissue (see Fig. 1). In order to automatically detect the early stage of prostate cancer, the main methodology used in the literature is based on detecting and segmenting glands and then, classifying each individual gland. For the classification of cancerous glands, both classic approaches based on hand-driven feature extraction [11] and modern deep-learning techniques [1] have been used. Nevertheless, those results are limited by a correct detection and delimitation of the glands in the image. This encourages the development of accurate systems able to detect and segment the glandular regions.

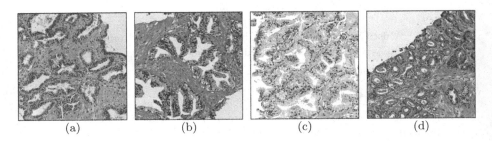

(a)   (b)   (c)   (d)

Fig. 1. Histology regions of prostate biopsies. Examples (a) and (b) present benign glands, including dilated and fusiform patterns. Images (c) and (d) contain patterns of Gleason grade 3, with small sized and atrophic glands.

For the prostate gland segmentation, different approaches have been carried out. In the work of Nguyen et al. [8–11] this procedure is based on the unsupervised clustering of the elements in the tissue, i.e. lumen, nuclei, cytoplasm, and stroma. Then, for each detected lumen, a gland is associated if enough nuclei are found in a region surrounding the lumen's contour. In the research carried out by García et al. [1–3] the components in the image are clustered by working in different color spaces, and then a Local Constrained Watershed Transform algorithm is fitted using the lumens and nuclei as internal and external markers respectively. As final step, the both aforementioned methodologies require a supervised model to differentiate between artifacts or glands. To the best of the authors' knowledge, the best performing state-of-the-art techniques for semantic segmentation, based on convolutional neural networks, have not been studied yet for prostate gland segmentation. In particular, one of the most used techniques in recent years for semantic segmentation is the *U-Net* architecture, proposed for medical applications in [12].

In this work, we present an *U-Net*-based model that aims to segment the glandular structures in histology prostate images. To the extent of our knowledge, this is the first time that an automatic feature-learning method is used for this task. One of the main contributions of this work is an extensive validation about different convolutional block modifications and regularization techniques on the basic *U-Net* architecture. The proposed convolutional block configurations are based on well-known CNN architectures of the literature (i.e. residual and Inception-based blocks). Furthermore, we study the impact of regularization approaches during the training stage based on data augmentation and using the gland contour as an independent class. Finally, we perform, as novelty, an image-level comparison of the most relevant methods in the literature for prostate gland segmentation under the same database. Using our proposed modified *U-Net* with residual blocks, we outperform in the test subset previous approaches.

2 Materials and Methods

2.1 Materials

The database used in this work consists of 47 whole slice images (WSIs, histology prostate tissue slices digitised in high-resolution images) from 27 different patients. The ground truth used in this work was prepared by means of a pixel-level annotation of the glandular structures in the tissue. In order to work with the high dimensional WSIs, they were sampled to 10× resolution and divided into patches with size 1024^2 and overlap of 50% among them. For each image, a mask was extracted from the annotations containing the glandular tissue. The resulting database includes 982 patches with its respective glandular masks.

2.2 *U-Net* architecture

The gland segmentation in the prostate histology images process is carried out by means of the *U-Net* convolutional neural network architecture [12] (see Fig. 2). As input, the images of dimensions 1024^2 are resized to 256^2 to avoid memory problems during the training stage. The *U-Net* configuration is based on a symmetric encoder-decoder path. In the encoder part, a feature extraction process is carried out based on convolutional blocks and dimensional reduction through max-pooling layers. Each block increases the number of filters in a factor of 2×, starting from 64 filters up to 1024. After each block, the max-pooling operation reduces the activation maps dimension in a factor of $2x$. The basic convolutional block (hereafter referred to as *basic*) consist of two stacked convolutional layers with filters of size 3×3 and ReLU activation. Then, the decoder path builds the segmentation maps, recovering the original dimensions of the image. The reconstruction process is based on deconvolutional layers with filters of size 3×3 and ReLU activation. These increase the spatial dimensions of the activation volume in a factor of 2× while reducing the number of filters in a half. The encoder features from a specific level are joined with the resulting activation maps of the same decoder level by a concatenation operation, feeding a convolutional block that combines them. Finally, once the original image dimensions are recovered, a convolutional layer with as many filters as classes to segment and soft-max activation creates the segmentation probability maps.

2.3 Loss Function

The loss function defined for the training process is the categorical *Dice*. This measure takes as input the reference glands and background masks (y) and the predicted probability maps ($\widehat{y}$) and is defined as follows:

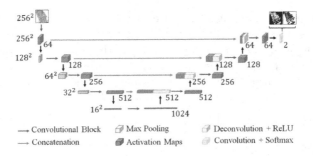

Fig. 2. *U-Net* architecture for prostate gland segmentation.

$$Dice(y, \widehat{y}) = \frac{1}{C} \sum_{c=1}^{C} \frac{2 \sum \widehat{y}_c \circ y_c}{\sum \widehat{y}_c^2 + y_c^2} \qquad (1)$$

where y is the pixel-level one hot encoding of the reference mask for each class c and $\widehat{y}$ is the predicted probability map volume.

Using a categorical average among the different classes brings robustness against class imbalance during the training process.

2.4   Introducing Residual and Multi-resolution Blocks to the *U-Net*

To increase the performance of the *U-Net* model, different convolutional blocks are used to substitute the *basic* configuration. In particular, residual and multi-resolution Inception-based blocks are used during the encoder and decoder stages.

The residual block [5] (from now on *RB*) is a configuration of convolutional layers that have shown good performance in deep neural networks optimisation. The residual block proposed to modify the basic U-Net consist of three convolutional layers with size 3×3 and ReLU activation. The first layer is in charge of normalizing the activation maps to the output' amount of filters for that block. The resultant activation maps from this layer are combined in a shortcut connection with the results of two-stacked convolutional layers via an adding operation.

Regarding the multi-resolution block (referred to in this work as *MRB*), it was recently introduced in [6] as a modification of the *U-Net* with gains of accuracy on different medical applications. This configuration, based on Inception blocks [13], combines features at different resolutions by concatenating the output of three consecutive convolutional layers of size 3×3 (see Fig. 3). The number of activation maps in the output volume (F_{out}) is progressively distributed in the three blocks ($\frac{F_{out}}{4}$, $\frac{F_{out}}{4}$, and $\frac{F_{out}}{2}$ respectively). Furthermore, a residual connection is established with the input activation maps, normalizing the number of maps with a convolutional layer of size 1×1.

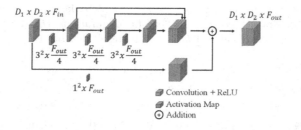

Fig. 3. Multi-resolution block. F_{in}: activation maps in the input volume. F_{out}: number of activation maps in the output volume. D: activation map dimensions.

2.5 Regularization Techniques

To improve the training process, two regularization techniques are proposed: data augmentation and the addition of the gland border as an independent class. Data augmentation (DA) is applied during the training process making random translations, rotations, and mirroring are applied to the input images. Regarding the use of the gland contour as an independent class (BC), this strategy has shown to increase the performance in other histology image challenges such as nuclei segmentation [7]. The idea is to highlight more (during the training stage) the most important region for obtaining an accurate segmentation: the boundary between the object and the background. Thus, the reference masks and the output of the *U-Net* are modified with an additional class in this approach.

3 Experiments and Results

To validate the different *U-Net* configurations and the regularization techniques, the database was divided in a patient-based 4 groups cross-validation strategy. As figure of merit, the image-level average *Dice Index* (DI) for both gland and background was computed. The *Dice Index* for certain class c is obtained from the *Dice* function (see Eq. 1) such that: $DI = 1 - Dice$. The metric ranges 0 to 1, from null to perfect agreement.

The different *U-Net* architectures, composed of basic (*basic*), residual (*RB*) and multi-resolution (*MRB*) blocks were trained with the proposed regularisation techniques, data augmentation (DA) and the inclusion of the border class (BC), in the cross-validation groups. The training was performed in mini-batches of 8 images, using NADAM optimiser. Regarding the learning rates, those were empirically optimised at values $5 * 10^{-4}$ for the *basic* and *RB* configurations and to $1 * 10^{-4}$ for the *MRB* one. All models were trained during 250 epochs. The results obtained in the cross-validation groups are presented in Table 1.

Analysing the use of different *U-Net* configurations, the best performing method was the *U-Net* modified with residual blocks ($RB + DA$), reaching an average DI of 0.75, and outperforming the basic architecture in 0.06 points. Regarding the use of the multi-resolution blocks (MRB), an increase in the performance was not registered. Concerning the use of the gland profile as an

Table 1. Results in the validation set for gland and background segmentation. The average *Dice Index* is presented for the different configurations. *DA*: data augmentation, *BC*: border class, *RB*: residual and *MRB*: multi-resolution blocks.

Method	DI_{gland}	$DI_{background}$
basic	0.5809(0.2377)	0.9766(0.0240)
basic + DA	0.6941(0.2515)	0.9845(0.01664)
basic + DA + BC	0.6945(0.2615)	0.9842(0.0168)
RB	0.5633(0.2340)	0.9759(0.0255)
RB + DA	**0.7527(0.2075)**	**0.9862(0.0148)**
RB + DA + BC	0.7292(0.2395)	0.9854(0.0161)
MRB	0.5710(0.2378)	0.9765(0.0253)
MRB + DA	0.7294(0.2173)	0.9843(0.0161)
MRB + DA + BC	0.7305(0.2247)	0.9846(0.0163)

independent class (*BC*), the results obtained were similar to the ones with just two classes.

The best performing strategy, *RB + DA*, was evaluated in the test subset. A comparison of the results obtained in this cohort with the previous methods presented in the literature is challenging. While previous works report object-based metrics, we consider more important to perform an image-level comparison of the predicted segmentation maps with the ground truth, in order to take into account false negatives in object detection. For this reason, and in order to establish fair comparisons, we computed the segmentation results in our test cohort applying the main two approaches found in the literature: the work of Nguyen et al. [9] and the research developed by García et al. [2]. This is, to the best of the authors' knowledge, the first time in the literature that the main methods for prostate gland segmentation are compared at image level in the same database. The figures of merit obtained in the test set are presented in the Table 2. Representative examples of segmented images for all the approaches are shown in Fig. 4.

Table 2. Results in the test set for gland and background segmentation. The average *Dice Index* for both classes is presented for the state-of-the-art methods and our proposed model. *DA*: data augmentation and *RB*: residual blocks.

Method	DI_{gland}	$DI_{background}$
Nguyen et al. [9]	0.5152(0.2201)	0.9661(0.0168)
García et al. [2]	0.5953(0.2052)	0.9845(0.01664)
U-Net + RB + DA	**0.7708(0.2093)**	**0.9918(0.0075)**

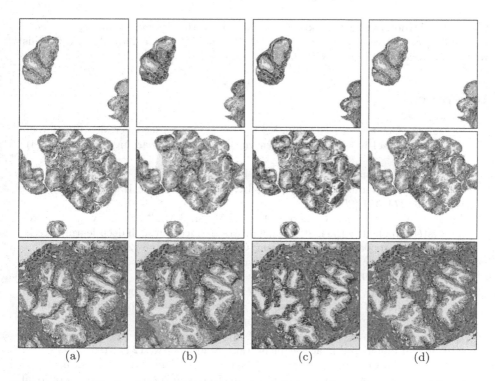

(a)   (b)   (c)   (d)

Fig. 4. Semantic gland segmentation in regions of images from the test set. (a): reference, (b): Nguyen et al., (c): García et al., and (d): proposed *U-Net*.

Our model outperformed previous methods in the test cohort, with an average *DI* of 0.77 for the gland class. The method proposed by Nguyen et al. and the one of García et al. obtained 0.51 and 0.59 respectively. The main differences were observed in glands with closed lumens (see first and second row in Fig. 4). The previous methods, based on lumen detection, did not segment properly those glands, while our proposed *U-Net* obtains promising results. Our approach also shows a better similarity in the contour of the glands with respect the reference annotations (see third row in Fig. 4).

4   Conclusions

In this work, we have presented modified *U-Net* models with residual and multi-resolution blocks able to segment glandular structures in histology prostate images. The *U-Net* with residual blocks outperforms in an image-level comparison previous approaches in the literature, reaching an average *Dice Index* of 0.77 in the test subset. Our proposed model shows better performance in both glands with closed lumens and in its shape definition. Further research will focus on studying the gains in accuracy in the first-stage cancer identification with a better gland segmentation based on our proposed *U-Net*.

Acknowledgments. This work was supported by the Spanish Ministry of Economy and Competitiveness through project DPI2016–77869. The Titan V used for this research was donated by the NVIDIA Corporation.

References

1. García, G., Colomer, A., Naranjo, V.: First-stage prostate cancer identification on histopathological images: hand-driven versus automatic learning. Entropy **21**(4), 356 (2019)
2. García, J.G., Colomer, A., Naranjo, V., Peñaranda, F., Sales, M.A.: Identification of Individual glandular regions Using LCWT and machine learning techniques. IDEAL **3**, 374–384 (2018)
3. García, J.G., Colomer, A., López-Mir, F., Mossi, J.M., Naranjo, V.: Computer aid-system to identify the first stage of prostate cancer through deep-learning techniques. In: European Signal Processing Conference, pp. 1–5 (2019)
4. Gleason, D.: Histologic grading of prostate cancer: a perspective. Hum. Pathol. **23**(3), 273–279 (1992)
5. He, K., Zhang, X., Ren, S., Sun, J.: Deep residual learning for image recognition. In: Proceedings of the IEEE Computer Society Conference on Computer Vision and Pattern Recognition 2016-Decem, pp. 770–778 (2016)
6. Ibtehaz, N., Rahman, M.S.: MultiResUNet: rethinking the U-Net architecture for multimodal biomedical image segmentation. Neural Netw. **121**, 74–87 (2020)
7. Kumar, N., Verma, R., Sharma, S., Bhargava, S., Vahadane, A., Sethi, A.: A dataset and a technique for generalized nuclear segmentation for computational pathology. IEEE Trans. Med. Imaging **36**(7), 1550–1560 (2017)
8. Nguyen, K., Jain, A.K., Allen, R.L.: Automated gland segmentation and classification for gleason grading of prostate tissue images. In: International Conference on Pattern Recognition, pp. 1497–1500 (2010)
9. Nguyen, K., Sabata, B., Jain, A.K.: Prostate cancer grading: gland segmentation and structural features. Pattern Recogn. Lett. **33**(7), 951–961 (2012). https://doi.org/10.1016/j.patrec.2011.10.001
10. Nguyen, K., Sarkar, A., Jain, A.K.: Structure and context in prostatic gland segmentation and classification. In: Ayache, N., Delingette, H., Golland, P., Mori, K. (eds.) MICCAI 2012. LNCS, vol. 7510, pp. 115–123. Springer, Heidelberg (2012). https://doi.org/10.1007/978-3-642-33415-3_15
11. Nguyen, K., Sarkar, A., Jain, A.K.: Prostate cancer grading: use of graph cut and spatial arrangement of nuclei. IEEE Trans. Med. Imaging **33**(12), 2254–2270 (2014)
12. Ronneberger, O., Fischer, P., Brox, T.: U-Net: convolutional networks for biomedical image segmentation. In: Navab, N., Hornegger, J., Wells, W.M., Frangi, A.F. (eds.) MICCAI 2015. LNCS, vol. 9351, pp. 234–241. Springer, Cham (2015). https://doi.org/10.1007/978-3-319-24574-4_28
13. Szegedy, C., et al.: Going deeper with convolutions. In: Proceedings of the IEEE Computer Society Conference on Computer Vision and Pattern Recognition 07–12-June, pp. 1–9 (2015)
14. Vickers, A.J., Lilja, H.: Predicting prostate cancer many years before diagnosis: how and why? World J. Urol. **30**(2), 131–135 (2012)
15. World Health Organization: Global Cancer Observatory (2019). http://gco.iarc.fr

Evaluation of Machine Learning Algorithms for Automated Management of Wireless Links

Francisca Frias[1,2], André R. S. Marcal[2], Rui Prior[2,3],
Waldir Moreira[1(✉)], and Antonio Oliveira-Jr[1,4]

[1] Fraunhofer Portugal AICOS, Porto, Portugal
{francisca.rodrigues,waldir.junior}@fraunhofer.pt, antoniojr@ufg.br
[2] Faculdade de Ciências da Universidade do Porto, Porto, Portugal
andre.marcal@fc.up.pt, rprior@dcc.fc.up.pt
[3] Instituto de Telecomunicações, Porto, Portugal
[4] Institute of Informatics - Federal University of Goiás, Goiás, Brazil

Abstract. Machine learning, a subfield of artificial intelligence, has been widely used to automate tasks usually performed by humans. Some applications of these techniques are understanding network traffic behavior, predicting it, classifying it, fixing its faults, identifying malware applications, and preventing deliberate attacks. The goal of this work is to use machine learning algorithms to classify, in separate procedures, the errors of the network, their causes, and possible fixes. Our application case considers the WiBACK wireless system, from which we also obtained the data logs used to produce this paper. WiBACK is a collection of software and hardware with auto-configuration and self-management capabilities, designed to reduce CAPEX and OPEX costs. A principal components analysis is performed, followed by the application of decision trees, k nearest neighbors, and support vector machines. A comparison between the results obtained by the algorithms trained with the original data sets, balanced data sets, and the principal components data is performed. We achieve weighted F1-score between 0.93 and 0.99 with the balanced data, 0.88 and 0.96 with the original unbalanced data, and 0.81 and 0.89 with the Principal Components Analysis.

Keywords: Machine learning algorithms · Principal components analysis · Traffic classification · Wireless links management

1 Introduction

The design and implementation of autonomous networks are one of the top challenges of networking research. The emerging networking technologies and software render networks more and more flexible. Hence, including capabilities of Machine Learning (ML) techniques allows for the prediction of changes in network and user behavior, providing mechanisms to configure networking equipment

© Springer Nature Switzerland AG 2020
C. Analide et al. (Eds.): IDEAL 2020, LNCS 12489, pp. 9–16, 2020.
https://doi.org/10.1007/978-3-030-62362-3_2

without the intervention of an administrator, allowing network management and control tasks to be performed automatically and autonomously. Our motivation for this work is that intelligent autonomous networks clearly benefit from ML in order to classify network behavior and adapt themselves according to the demands of network operation to improve performance. We aim to bridge networking and ML, to take decisions online and in an automated fashion. Our application case is the WiBACK [7], a wireless software and hardware solution with directed radio technology that provides a communication backhaul in challenged areas (i.e., rural, sparse population, without expert readily available, digital illiteracy). It is designed to be self-configurable and self-managed, taking care of its own setup and operation of the network, increasing reliability and performance. With WiBACK, there is less need for technical skills to set up and operate a network than in alternative technologies. It is important to note that WiBACK is very often installed in places where there are no people with the needed abilities to perform technical interventions. Thus, a ML-based approach for the identification of network faults, their causes, and solutions is of great interest. With that in mind, we can state that the main goal of this work is to further automate WiBACK's management. We defined a set of specific objectives. We use the WiBACK data logs that contain the numerical variables, and apply Decision Trees (DT) [4], K Nearest Neighbors (KNN) [8] and Support Vector Machines (SVM) [12] to classify those faults, causes and fixes. Another specific objective relates to the visual representation of data in the different classes of each categorical variable (i.e., errors, causes, and fixes). We want to obtain a compact representation of data, but also to conclude how the description of the data according to the labeled classes varies with the different components. That is also why we apply a Principal Components Analysis PCA [10].

The rest of the paper is organized as follows: Sect. 2 presents related work. Section 3 introduces the data set description, PCA, and the classification algorithms. Section 4 discusses the evaluation of the models with the weighted F1-score. Section 5 presents conclusions and future work.

2   Related Work

Boutaba et al. [3] surveyed the ML approaches applied to network management, which includes traffic prediction, routing and classification, congestion control, resource and fault management, quality of end-user and quality of experience management, and network security. Traffic classification, congestion control, and fault management have a greater relevance for our work.

Daniels [6] focused on wireless link adaptations, which leads to a practical concern of our work, since our networks' problems are highly related to the quality of the links. He developed online ML link adaptation procedures to adjust to changes that occur and affect transmitter or receiver. Online learning was performed with KNN and SVM.

Alzamly [1] work consists in understanding the WiBACK behavior, correlating the features of its logs, discovering its faults, identifying their causes and how to solve them. Then, based on the acquired knowledge, a set of numerical variables were selected. Also based on that knowledge, each instance of the data set was labeled with one error, one cause, and one fix. These are the three categorical variables in the data set [2]. Initially, errors were classified with a Neural Network (NN), considering just the numerical variables as input. Then, also with a NN, the causes of the errors were classified, considering the categorical variable *errors* as input. Finally, the solutions were classified with a DT, using the categorical variables *errors* and *causes* as inputs. The obtained accuracy figures were 99% with NN and 100% with the DT. There are three main differences between Alzamly's work and ours: (i) besides the DT, we consider SVM and KNN, and we do not consider NN; (ii) we do not consider categorical variables *errors* and *causes* as inputs, so we do three classification procedures separately, and (iii) we have a data set with bidirectional features, instead of unidirectional ones, which reflects the different behavior of the networks in the two directions of the channels. We perform the classification of errors, causes and solutions separately, because there is a high correspondence between errors, causes, and solutions (e.g., when the error is a link overload, that is always because of self cell overload, and the solution is always to adjust the shaping), which could mean to introduce some redundant information in the models. Besides, particularly to classify fixes, we do not want to limit the list of errors and causes, because we believe that the listed solutions could fix non-labeled problems.

3   Machine Learning Proposal

This section presents the data set description, PCA, and the classification algorithms.

3.1   Data Set Description

The data set was extracted from the WiBACK logs. Initially, there were 68 numerical variables, the 3 categorical variables we want to classify, and 3680 instances. All instances of the data set are labeled. We are considering two interfaces, S and R, and the bidirectionality of the channels. Both interfaces have the sender and the receiver mode. However, the direction from S to R plays a more important role because S is the interface suffering from transmitter-overrun errors (tx-overrun). The first step is to exclude the instances with incoherent values. The next step is variable selection. Initially, we exclude those that concern time and node identification, those used just to synchronize data, and the constant ones. Then, we exclude variables with high Spearman's correlation. We end up working with 20 numerical variables and 3250 instances. The numerical variables are related to the i) *time* spent sending the packets and used by external sources to the interfaces; ii) *signal quality*, important to inspect the links faults (though, as we can see by the following categorization, other factors are also affecting the network's performance); iii) *Packet Error Indicator (PEI)*, which does not tell us what the problem is, but how the network behavior is from a global perspective, so it is an appropriate measure for us to evaluate continuously; iv) *transmission and reception rates*, which allows concluding if there is some issue with the links in a prior analysis; v) *link capacity* estimated by WiBACK (based, for example, on the distance between transmitter and receiver); vi) randomly and forced *dropped packets* (due to several reasons, such as the large distance between the interfaces, packets queue exceeding the buffer capacity, or weak signal link); vii) *transmission power*, which helps understand whether the number of dropped packets is due to low transmission power or other interference; and viii) *retransmission rate*, useful to compare with the number of packets sent that could be misleading if individually analyzed.

Regarding the variables to predict, there are errors, causes, and solutions. Each error that occur in the network is extracted from a log related to the S interface. There are 4 classes for the errors: link overload (62.5%), occupied channel (25.0%), weak link (12.2%), and unknown (0.3%), and 4 classes for the causes: self cell overload (62.5%), non-WLAN interference on the sender (24.0%), bad signal (12.5%), and non-WLAN interference on the receiver (1%). In general, there are some relations between the numerical variables and the causes: when a self cell is overloaded, there is a low retransmission rate; when there is non-WLAN interference on the sender, PEI is high; when there is non-WLAN interference on the receiver, there are many packets retransmissions (higher than 10.0%), high PEI, and the transmission bit rate of S is much lower than the estimated link capacity. Finally, there are 5 classes for the solutions: adjust the shaping (62.5%), change the channel (24.0%), recalibrate the link (12.2%), increase the power (1.0%), and unknown (0.3%). Note that some classes of the different categorical variables have the same absolute frequency. Those occurrences are the same: when there is a link overload, that is because of self cell overload, and the solution is to adjust the shaping; when there is a weak link, the proposed solution is to recalibrate the link; when the cause is non-WLAN interference on the sender, the proposed solution is to change the channel; when the cause is non-WLAN interference on the receiver, the proposed solution is to increase the power; and when the error is unknown, the solution is also unknown. This suggests that there is some redundancy across the variables, and that is the main reason for us not to consider categorical variables as input to the classification. The relations mentioned between the numerical variables and the causes also suggest some redundancy.

Our original data is unbalanced, which could lead to the suboptimal performance of the classifiers [5]. Thus, we oversample the data with the Synthetic Minority Over Sampling TEchnique (SMOTE) [9], and compare the results with the original and the PCA data. For each categorical variable, the percentage between the number of instances of the most represented class and the sum of the instances of the other classes is 60%.

3.2   Principal Components Analysis

Data is centered and normalized (standard deviation equal to 0, and mean equal to 1) before the application of PCA, and it is performed with the unbalanced data (the original data set). Our goal by performing this PCA analysis is twofold: (i) to carry out a visual exploratory study of the data, understanding how well the PCA describes the data in each class, and (ii) to compare the performance of the classification algorithms with and without the PCA transformation. We work with 10 principal components, that correspond to 91% of the data variance.

For each categorical variable, to understand how well the projection of the points onto the principal components describe the points of each class, for each class, we take each component separately, and compute the mean of the Euclidean distance between all those points. The distance increases as we move to the next component. The data projection into the pairs of components two to ten, and the evolution of the Euclidean distance suggest that the first two principal components show a more clear visual representation of the data in each class than the other pairs of components. The less represented classes achieve higher distances, and the distance in those classes (particularly, the error and the solution 'unknown') have a greater growth in the last components. This suggests a greater dispersion in the less represented classes.

3.3   Classification Algorithms

We apply DT, KNN, and SVM with grid search and 10-cross validation for the classification. SVM are implemented for multi-classification to increase the accuracy of traffic classification [12]. We chose this method because, thanks to its non-linear transformation, it is adaptable to several structures in the data. SVM and KNN are also applied to a flow feature-based data set, and the traffic classification accuracy achieved was equal to 95.6% [8]. We also chose DT because of their great advantage of being interpretable, giving us somehow information about the numerical variables of the wireless context that the other algorithms do not, when we look the node splits.

4   Performance Evaluation

Table 1 shows the F1-weighted score with DT, KNN, and SVM for the balanced, unbalanced, and PCA data for the *errors, causes,* and *solutions* classification. KNN and DT have always lower performance, but the differences between the three algorithms are not very significant. However, we perform tests with other data sets, and the difference between the performance of the two first algorithms and SVM increases, being that SVM still present better results. We believe the lower results could be due to noise, which can be provenient from the classes or the numerical variables [13]. Noisy points might affect particularly KNN because, having a value set for k, the classification of each label is made always based on the k nearest neighbors. The data set might have some noise due to variables like the link capacity, estimated by WiBACK and not 100% accurate (which is normal in the problem context). There are some points in the PCA data appearing to be outliers, and they might also affect the classification made by KNN. The number of neighbors, k, is a very sensitive parameter, and the best choice could vary for different test sets. It is 2 (balanced data) or between 4 and 8 (unbalanced and PCA data). With the larger set, data may be overlapping more, which could lead different classes to overlap and, consequently, to a smaller valuer of k. We amplify the range we have set initially for k in grid search, as the range for the complexity parameter of DT pruning algorithm, but the results remained the same.

Undersampling the data can decrease the performance because it removes samples with valuable information, while oversampling can increase the performance [9]. Regarding the results with our data sets, it seems to be and advantage in oversampling, while the PCA transformation does not increase the performance.

In different locations of WiBACK networks and in different time periods, we might observe different behaviors of the data, and even the necessity to create new classes. In some cases, we may have less unbalanced data sets. We need a model that behaves well in both cases. Besides the reduced performance of DT and KNN, KNN does not seem to be the most efficient model to run online since every time that a new log event occurs, the k nearest neighbors of all data points are constantly changing, and KNN would have to do all distance computations reacting to these changes [11]. After the first test, we have actually retrained the models with more data, and SVM continue to present the best results. Besides, with the right kernel and training set, SVM could reduce the online implementation complexity [6].

Table 1. Weighted F1-score of all classifiers

	Data set	DT	KNN	SVM
Errors	Unbalanced	0.89	0.90	0.96
	Balanced	0.93	0.97	0.99
	PCA	0.82	0.88	0.89
Causes	Unbalanced	0.88	0.88	0.95
	Balanced	0.93	0.97	0.98
	PCA	0.81	0.86	0.88
Fixes	Unbalanced	0.89	0.90	0.96
	Balanced	0.93	0.97	0.99
	PCA	0.82	0.88	0.89

Regarding the numerical variables, it is also very important to understand which are the right ones to consider, not just regarding the results, but also the problem context. The way that every numerical variable is calculated or obtained makes some of them more accurate than the others. Note that, in the initial set, we had 68 variables, and it is not clear which ones we should consider. Table 2 shows the variables considered in the nodes of the DT trained with the balanced data (the ones with the higher weighted F1-score). We see that there are considered just 6 of the 20, which could mean that we do not need so many variables as input. Note also that the variables in the three classifiers are almost the same, which is understandable since there is a high correspondence between errors, causes and solutions, as aforementioned. The variables in the nodes of the DT are particularly important, because they might change some of our assumptions about the problem context.

Table 2. Numerical variables in the nodes of the DT

Numerical variables	Errors classifier	Causes classifier	Fixes classifier
AT_out	x	x	x
AT_in	x	x	x
Cap_of_MCS_down	x		x
Rx_br_down	x	x	x
Rx_br_up	x	x	x
RED		x	x

5   Conclusions and Future Work

The results presented above are the product of a first approach to the automated management of wireless links. Our choice for an algorithm to deploy on the WiBACK system would favor SVM. However, we know that the performance can significantly change when we run the algorithms online with more data, because there might be a different behavior. So, it is necessary to test the algorithms with further data.

Future work consists of applying ML algorithms on the WiBACK system. If the performance is satisfactory online, which means classifying correctly the events of the software logs, we aim to move to an implementation phase where the machine is capable of deciding whether the proposed solution to a specific problem is the best solution for the network as a whole. For this purpose, we have to take into account the network behavior, a list of parameters that interfere with the network operation, and the notion that the usage patterns could be not entirely stable, i.e. change dynamically over time. Then, the labeled solutions might have weights defined according not just the problem itself, but also these parameters, and the changes with negative impact in the network that each solution may originate. This would be a possibility for how the network shall decide which solution is the best for each error. Future work also includes a process of re-fixing the errors, even if with the weighted solutions, the network does not respond satisfactorily.

Acknowledgment. We are grateful to Hossam Alzamly, Mathias Kretschmer and Jens Moedeker from Fraunhofer FIT, and Osianoh Glenn Aliu from Defutech that provided us with data sets and knowledge about the WiBACK's features, as well as fruitful discussions about the current work. This work was funded under the scope of Project AFRICA: On-site air-to-fertilizer mini-plants relegated by sensor-based ICT technology to foster African agriculture (LEAP-Agri-146) co-funded by the European Union's Horizon 2020 research and innovation programme under grant agreement No 727715 and Fundação para a Ciência e a Tecnologia (FCT) under reference LEAPA-gri/0004/2017.

References

1. Alzamly, H.: Evaluation of errors on wiback wireless links. Tech. rep., Cologne University of Applied Sciences, Faculty of Information, Media and Electrical Engineering (2018)
2. Alzamly, H.: Automated Error Detection and Classification using Machine Learning Techniques for WiBACK Wireless Links. Master's thesis, Faculty of Information, Media and Electrical Engineering (2019)
3. Boutaba, R., et al.: A comprehensive survey on machine learning for networking: evolution, applications and research opportunities. J. Internet Serv. Appl. **9**(1), 1–99 (2018). https://doi.org/10.1186/s13174-018-0087-2
4. Buntine, W.: Learning classification trees. Stat. Comput. **2**(2), 63–73 (1992)
5. Chen, X.W., Wasikowski, M.: Fast: a roc-based feature selection metric for small samples and imbalanced data classification problems. In: Proceedings of the 14th ACM SIGKDD international conference on Knowledge discovery and data mining, pp. 124–132 (2008)
6. Daniels, R.C.: Machine learning for link adaptation in wireless networks. Ph.D. thesis (2011)
7. Fraunhofer FOKUS: Introducing Fraunhofer's Wireless Backhaul Technology (2014). https://www.wiback.org/content/dam/wiback/en/documents/Whitepaper_Introducing_WiBACK.pdf
8. He, L., Xu, C., Luo, Y.: VTC: machine learning based traffic classification as a virtual network function. In: Proceedings of the 2016 ACM International Workshop on Security in Software Defined Networks & Network Function Virtualization, pp. 53–56 (2016)
9. Kaur, P., Gosain, A.: Comparing the behavior of oversampling and undersampling approach of class imbalance learning by combining class imbalance problem with noise. In: Saini, A.K., Nayak, A.K., Vyas, R.K. (eds.) ICT Based Innovations. AISC, vol. 653, pp. 23–30. Springer, Singapore (2018). https://doi.org/10.1007/978-981-10-6602-3_3
10. Le Borgne, Y., Bontempi, G.: Unsupervised and supervised compression with principal component analysis in wireless sensor networks. In: Proceedings of the Workshop on Knowledge Discovery from Data, 13th ACM International Conference on Knowledge Discovery and Data Mining, pp. 94–103 (2007)
11. Pärkkä, J., Cluitmans, L., Ermes, M.: Personalization algorithm for real-time activity recognition using pda, wireless motion bands, and binary decision tree. IEEE Trans. Inf Technol. Biomed. **14**(5), 1211–1215 (2010)
12. Wang, R., Liu, Y., Yang, Y., Zhou, X.: Solving the app-level classification problem of p2p traffic via optimized support vector machines. In: Sixth International Conference on Intelligent Systems Design and Applications, vol. 2, pp. 534–539. IEEE (2006)
13. Zhu, X., Wu, X.: Class noise vs. attribute noise: a quantitative study. Artif. Intell. Rev. **22**(3), 177–210 (2004)

Imputation of Missing Boarding Stop Information in Smart Card Data with Machine Learning Methods

Nadav Shalit[1]($\boxtimes$), Michael Fire[1], and Eran Ben-Elia[2]

[1] Department of Software and Information Systems Engineering,
Ben-Gurion University of the Negev, Beersheba, Israel
{shanad,mickyfi}@post.bgu.ac.il
[2] GAMESLab, Department of Geography and Environmental Development,
Ben-Gurion University of the Negev, Beersheba, Israel
benelia@bgu.ac.il

Abstract. With the increase in population densities and environmental awareness, public transport has become an important aspect of urban life. Consequently, large quantities of transportation data are generated, and mining data from smart card use has become a standardized method to understand the travel habits of passengers.

Increase in available data and computation power demands more sophisticated methods to analyze big data. Public transport datasets, however, often lack data integrity. Boarding stop information may be missing either due to imperfect acquirement processes or inadequate reporting. As a result, large quantities of observations and even complete sections of cities might be absent from the smart card database. We have developed a machine (supervised) learning method to impute missing boarding stops based on ordinal classification. In addition, we present a new metric, Pareto Accuracy, to evaluate algorithms where classes have an ordinal nature. Results are based on a case study in the city of Beer Sheva utilizing one month of data. We show that our proposed method significantly outperforms schedule-based imputation methods and can improve the accuracy and usefulness of large-scale transportation data. The implications for data imputation of smart card information is further discussed.

Keywords: Machine learning · Smart card · Boarding stop imputation

1 Introduction

Smart card technology is a popular automatic fare collection (AFC) due to it's efficiency and cost saving [9]. Nowadays most of the monetary transactions in public transport are based solely on smart cards [6]. Recent developments in the field of data science, such as the increase in data volume, new data mining tools [13], and cloud computing [12], have created new opportunities to analyze

© Springer Nature Switzerland AG 2020
C. Analide et al. (Eds.): IDEAL 2020, LNCS 12489, pp. 17–27, 2020.
https://doi.org/10.1007/978-3-030-62362-3_3

and determine travel behavior at the individual level over long periods and large areas [16].

Similar to datasets in other domains, large-scale transportation datasets may be affected by issues of data integrity, such as incorrect or missing values. For example, operators may have only partial data on travelers' boarding stops. This can result in severe bias as complete sections of cities can me unrepresented. Solving this problem demands replacing the missing or erroneous data by utilizing public accessible data [15] or discarding the aforementioned data [19]. However, these two solutions are not optimal.

In this study, we establish a new machine learning based methodology for improving the quality and integrity of transportation datasets by predicting missing or corrupted travelers' records. Namely, we propose a novel algorithm for predicting passengers' boarding stops. Our proposed algorithm is based on features extracted by fusing multiple big data sources, such as planned General Transit Feed Specification (GTFS), defined as a common format for public transportation schedules and associated geographic information [15]. schedule data, smart card data, and supplementary geospatial data (see Table 1). Then, we apply a machine learning model on these features to predict boarding stops (see Sect. 3).

We utilized a real-world smart card dataset, which consists of over a million trips taken by more than 85,000 people. Specifically, our supervised model's accuracy was almost two times better than the schedule-based model constructed using GTFS data (see Sect. 4). Moreover, we show that our constructed prediction model is generic and can be used to predict boarding stops across different cities (see Sect. 3.4). Furthermore, our proposed model predicted considerably more boarding stops with over 50% accuracy than the schedule-based method (see Fig. 1).

(a) Schedule-based predictions   (b) Proposed model predictions

Fig. 1. Boarding stations with predicted accuracy of over 50%.

The problem we address is ordinal classification because after embedding the boarding stops (see Sect. 3.4) they are ordered. Ordinal classification is a classification task where the classes have an inherent order between them. Existing methods for evaluation of this problem are not suitable in our case that also requires a high level of interpretability of the results. For this purpose, we propose a new method of evaluation which shows the percentage of each error dimension, Pareto Accuracy (see Sect. 3.6).

Our study presents three main contributions:

1. We present a novel algorithm for completing missing boarding stops using supervised learning.
2. We develop a new method for evaluating this problem that can be applied to other domains of ordinal classification.
3. We introduce a new method for embedding boarding stations. This method reduces imbalance significantly and increases transferability. This enables a model generated from data of one city to be utilized for prediction boarding stations in other cities.

In addition to the above contributions, we show which features are most important in predicting the imputed boarding stops using state-of-the-art SHAP values [14].

The rest of the paper is organized as follows: In Sect. 2, we give a brief overview of related works. Section 3 describes the experimental framework and methods used to develop the model and the extraction of the features. In Sect. 4, we present the results of this study. In Sect. 5, we discuss the implications of the findings. Lastly, in Sect. 6, we present our conclusions and future research directions.

2   Related Work

The smart card system was introduced as a smart and efficient automatic fare collection (AFC) system in the early 2000s [7] and has become an increasingly popular payment method [20]. In recent years, smart cards have not only become an effective payment tool, but also an increasingly popular big data source for research [1,9]. A comprehensive review of smart card usage literature is provided by Pelletier et al. [17].

Researchers have now come to realize that the traditional analysis methods are sub par when used in combination with big data. In recent years, there has been a shift towards harvesting the prognostic nature of machine learning to yield predictive analytics. Welch and Widita [21] presented a number of machine learning algorithms in the field of public transportation.

The current belief is that the proper use of smart card data will yield insights that were not previously available from traditional methods [21]. A literature survey on this topic can be found in the paper by Li et al. [13].

One of the well-known problems in data mining is that data is often incomplete, and a significant amount of data could be missing or incorrect, as evident in the study by Lakshminarayan et al. [11]. Missing data imputation should be carefully handled or bias might be introduced [2], and as shown in their paper, common methods are not always optimal.

In recent years, techniques to optimize missing data imputation have been explored further [3], even using state-of-the art deep learning methods to impute

[4], showing the importance of this area. This field of study, however, has not been optimized in the field of public transport, and therefore it is compelling to further examine.

Classification - the domain of this paper, is a form of supervised machine learning which tries to generalize a hypothesis from a given set of records. It learns to create $h(x_i) \rightarrow y_i$ where y has a finite number of classes [10]. Ordinal classification is a form of multi-class classification where there is a natural ordinal ordering between the classes but not necessarily numerical traits for each class. For this case, a classifier will not necessarily be chosen based on traditional metrics, such as accuracy, but rather on the severity of its errors [8].

3   Methods and Experiments

Our study's main goal is to improve public transportation data integrity using machine learning algorithms. Namely, we developed a supervised learning-based model to impute missing boarding stops given a smart card dataset. Moreover, we created a generic model that is fully transferable to other datasets and can be used to impute missing data in similar transportation datasets without any adjustments.

First, given a smart card dataset, we pre-processed and cleaned the dataset (see Sect. 3.2). Then, we extracted a variety of features from the geospatial and GTFS datasets (see Sect. 3.3). In addition, we converted boarding stops from their original identifiers to a numerical representation by utilizing GTFS data (see Sect. 3.4). Next, we applied machine learning algorithms to generate a prediction model that can predict boarding stops (see Sect. 3.5). Lastly, we evaluated our model compared to schedule-based methods with common metrics. Moreover, we evaluated our supervised algorithm performance using a novel Pareto Accuracy metric (see Sect. 3.6). In the following subsections, we describe in detail each step in our methodology.

3.1   Datasets

To apply our methodology for predicting missing boarding stops, we needed to fuse three types of datasets:

1. *Smart card dataset* - contains data of smart card unique IDs; traveler types, such as student or senior travelers; boarding stops; timestamps of boarding; and unique trip identifiers for the line at that time.
2. *GTFS dataset* - aligns with the smart card dataset and consists of a detailed timetable of every trip made by public transport. The GTFS dataset was used both to enrich the features space as well as transform boarding stops into a numerical value (see Sect. 3.4).
3. *Geospatial dataset*- contains a variety of geospatial attributes, such as number of traffic lights, population density, and more, derived from a geospatial database.

3.2   Data Preprocessing

To make the dataset suitable for constructing prediction models, we needed to remove any record that did not have a boarding stop or a trip ID (a unique identifier of a trip provided by a specific and unique public transport operator) from the smart card dataset. Next, we joined the smart card dataset with the GTFS dataset by matching the trip ID attribute. Furthermore, we joined the geospatial dataset with the smart card dataset using the GTFS dataset, which contains all the geographic coordinates of each route.

Table 1. Extracted features

Dataset	Feature	Explanation
City geospatial records	Addresses_average	Calculate the amount of addresses listed along the route
	Street_light_average	Calculate the amount of street lights along the route
	Light_traffics_average	Calculate the amount of trafic lights along the route
GTFS	Number_of_points	Calculate the number of points in shape file in GTFS per route
	Average_distance_per stop	Calculate total length of route divided by number of points
	Average_time_per_stop	Calculate total expected travel time of route divided by number of points
	Average_points_to stops	Calculate the number of points in shape file in GTFS per route divided by number of points
	Time_diff_of_trip	Total travel time
GTFS & smart card	Time_from_boarding to_last_stop	Time from boarding time to expected last stop of the route
	Time_from_departue to_boarding	Time from route departue to boarding time
	Predicted_sequence	GTFS prediction sequence of the most likely stop
	Hourly_expected_lateness	Average lateness per hour
Smart card	Boardingtime_Seconds from_midnight	Time stamp of boarding to numerical value- second from midnight
	Boardingtime_weekday	Day of the week in which the boarding accord
	Is_weekend	Is it a weekend

3.3 Features Extraction

While the smart card dataset contains valuable data regarding passengers' trip, this data provides only a partial picture of the world.

In addition, there were physical characteristics that could improve the prediction model. These characteristics included such attributes as the number of traffic lights along the line, which increases the probability of traffic congestion forming and consequently additional delays. In the feature extraction process[1] we extracted three features by utilizing the smart card dataset, eight features by utilizing the GTFS dataset, and three features by utilizing the geospatial dataset (see Table 1).

3.4 Embedding Boarding Stops

To construct the prediction model, we used the GTFS dataset to create a naive schedule-based prediction. This prediction is the transit vehicle's position along a line according to the GTFS schedule. Namely, let S_i be the sequence number of the boarding stop based on the GTFS schedule and let A_i be the actual boarding stop sequence number. Then, we define D_i as $D_i = A_i - S_i$. Our prediction model goal was to predict D_i by utilizing the variety of features presented in the previous section (see Table 1).

For instance, consider a passenger who boarded a line at the third stop, i.e., $A_i = 3$, at the time the transit vehicle was scheduled to arrive at the second stop. The schedule-based prediction would be 2, i.e., $S_i = 2$, the stop where it was supposed to be at that time. Then, the difference is $D_i = A_i - S_i = 3 - 2 = 1$, and this is the class the algorithm will predict.

3.5 Constructing Prediction Model

To construct a model which can predict boarding stops, we performed the following steps: First, we selected several well-known classification algorithms. Namely, we used Random Forest [18], Logistic Regression [18], and XGBoost [5]. Second, we split our dataset into training dataset, which consisted of the first three weeks of data, and test dataset, which consisted of the last week of data. The algorithm classifies the embedded stop (see Sect. 3.4). Third, for both training and test datasets, we extracted all 15 features (see Sect. 3.3). Fourth, we constructed the prediction models using each one of the selected algorithms. Lastly, we evaluate each model (see Sect. 4).

3.6 Model Evaluation

We evaluated each model and compared it to the schedule-based method on the test dataset using common metrics: *Accuracy, recall, precision, F1,* and *Pareto Accuracy.*

---

[1] In Initial experiments we tested 41 features, using stepwise feature selection we reduced the feature space to 15 (see Table 1).

We used the following variables for our novel *Pareto Accuracy* metric: Let p_i be the predicted sequence of $stop_i$, a_i be the actual sequence and d_i be the absolute difference between them. Let l be the limit of acceptable difference for imputation, i.e. if an error of one stop is tolerated, such as for neighborhood segmentation, then $l = 1$.

Let indicator $X_i = \begin{cases} 1 & \text{if } di \leq l \\ 0 & \text{otherwise} \end{cases}$. We defined Pareto Accuracy as follows:

$$PA_l = \frac{\sum_{i=1}^{n} X_i}{n}.$$

The PA metric is a generalization of the accuracy metric. Namely, PA_0 is the well-known accuracy metric. The primary advantage of using the PA metric is for evaluating the true dimension of error while being extremely robust to outliers (by setting parameter l), unlike other ordinal classification methods. Moreover, this metric is highly informative since its outcome value can be interpreted easily; for example, 0.6 means that 60% of predictions had at most l difference from true labels.

For example, let us consider a set of eight observations of embedded boarding stops $\{-2, 0, 3, 20, -3, 4, 3, 2\}$, where each observation is a simulated boarding by a passenger where each number (D_i) in the set represents the difference between expected (S_i) and actual boarding stops (A_i), (see Sect. 3.4). The fourth observation, with a value of 20, is an outlier, which might occur due to some fault in the decoder device of the public transport operator. We do not want to predict it, as it is naturally unpredictable. We seek a metric that will both be resilient to outliers, as they are unpredictable, and still account for the true dimension of the errors (see Sect. 2). Let us compare two classifiers, A and B. Classifier A predicted the following boarding stops $\{-2, 0, 4, 3, -2, 3, 2, 2\}$, while Classifier B predicted $\{3, 0, 3, 7, 1, 1, 3, 2\}$. Classifier A is a more useful classifier since, in general, its predicted values are closer to the actual values; i.e., its variance is very small, which makes it more reliable. However, when using the classical accuracy and RMSE metrics, Classifier B has a higher accuracy and lower RMSE values than Classifier A, with accuracy values of 50% vs. 37.5%, and RMSE values of 5.2 vs. 6. By using the Pareto Accuracy (PA_1), we obtain a more accurate picture in which Classifier A clearly outperforms Classifier B (87.5% vs. 50%). Here, we see a case were metrics used for both classical classification (accuracy) and ordinal classification (RMSE) do not reflect the actual performance of each classifier.

Lastly, since we wanted to enrich our understanding on the nature and patterns of public transport, we produced and analysed feature importance (see Fig. 2) by utilizing the state-of-the-art method SHAP values [14].

3.7   Experiment Settings

To evaluate the above methodology, we chose to focus on the city of Beer Sheva, Israel. We utilized a smart card dataset, consisting of over a million records from over 85,000 distinct travelers for one month during November-December 2018. We chose this city since the boarding stop information was complete and road traffic is not prone to congestion.

We also used a GTFS dataset, containing over 27,000 stops with over 200,000 trips, and including all the agencies (operators) in Israel.[2] The dataset includes a detailed timetable for every transit trip taken by public transport. We utilize the GTFS and geospatial datasets to enrich our feature space (see Sect. 3.3) and to transform our boarding stops with the detailed timetable (see Sect. 3.4). We extracted the 15 features from the above datasets as mentioned in Sect. 3.3. Next, we embedded the stops to a numerical value as described in Sect. 3.4. Lastly, we developed a machine learning algorithm to classify the above boarding stops (see Sect. 3.5) and evaluated the classifier's performance (see Sect. 3.6).

Additionally we used the above methodology on a smaller dataset from the city of Kiryat Gat, Israel. Evaluation was done on one day with about 4 K observations. Boarding stops were embedded and features were created exactly as mentioned above. In the future we would like to validate this on larger dataset.

4   Results

Using the aforementioned datasets, we trained several classifiers and evaluated their performance. Among all trained classifiers, the XGBoost classifier presented the highest performance (see Table 2). Comparison was done with common metrics as stated in Sect. 3.6. In addition, we evaluated our new proposed metric, Pareto Accuracy. We chose to evaluate on error size of 1 i.e. PA_1. This is due to the fact that a larger gap would be usually unacceptable and PA_1 is highly correlated to PA_i for $i > 1$.

Table 2. Classifiers performances

Algorithm	Accuracy	Recall	Precision	AUC	*PA1*
Schedule based	0.209	0.209	0.212	0.59	0.47
Logistic Regression	0.205	0.205	0.097	0.562	0.474
Random Forest	0.368	*0.368*	0.348	0.65	0.672
XGBoost	**0.41**	**0.41**	**0.393**	**0.675**	**0.712**

---

[2] https://transitfeeds.com/p/ministry-of-transport-and-road-safety/820.

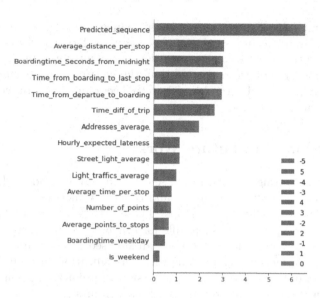

Fig. 2. Feature importance using shap values

5 Discussion

From the above results we can conclude the following: First, our methodology for embedding boarding stops demonstrates several advantages: (a) the proposed algorithm is a *generic model* that can be used with different smart card datasets since the labels (numeric representations) are always aligned (see Sect. 3.4); (b) by encoding the boarding stops, our proposed method ensures that the number of distinct labels will be very small and can *reduce the computation time significantly* for generating a prediction model since classification computation time is correlated to the number of labels; and (c) boarding stops inherently have an extremely imbalanced nature. This poses a considerable challenge to classify many classes accurately with such imbalance; however, we managed *to reduce the imbalanced* nature of the problem significantly.

Second, from the SHAP values presented in Fig. 2, the following can be noted: (a) most important feature is the prediction created by the scheduling-based method, i.e., Predicted sequence, which shows it is highly correlated to actual patterns and is very useful for classifying; (b) other than the first two features, the next four features are temporal, which is logical in that different time periods have an impact on traffic (such as morning commute); (c) while geospatial features do not have the highest values of importance, they are not insignificant, and thus we can conclude that physical attributes affect the nature of our problem, such as denser areas can lead to more congestion; and (d) the two least significant pertain to day of the week, and we can conclude that daily public transport routines remained quite stable in our case study.

Lastly, we introduced a new metric in this paper, a generalized accuracy metric we named Pareto Accuracy, which helps compare classifiers for ordinal classification. This metric is more robust to outliers, more interpretable, and accounts for the true dimension of errors. In addition, the metric is easy to implement. In the future we hope to understand how Pareto Accuracy can improve additional ordinal classification use cases.

6   Conclusions and Future Work

We assert that the straightforward used schedule-based methods suffer from sub par performance in terms of accuracy and other key metrics, as well as being highly dependent on centrality of stops. These attributes reduce the imputed data's integrity and make it less suitable for usage. In contrast, we showed that the proposed model outperformed the schedule-based method in all metrics.

In addition, our method is based on generic classification and thus can be used in a wide variety of transportation use cases, such as prediction of alighting stops, imputing other attributes of interest, etc. Initial results on another city (Kiryat Gat, Israel) looks promising compared to the current schedue-based method, yielding 30% accuracy vs 20%. In the future, we would like to test our model in other cities to verify its generalization. In addition, we would like to test the influence of transfer learning on new datasets.

There are a few limitations to the study worth noting. Our method requires several constraints to succeed, such as the availability of timestamp, trip IDs, and existing trip timetable. In addition, the generality of our method can increase bias as it ignores features that cannot be transferred from one dataset to another.

In conclusion, by mining smart card data we were able to construct a passenger boarding stop prediction model that clearly surpasses the schedule-based method. Our study revealed that applying machine learning techniques improves the integrity of public transport data, which can significantly benefit the field of transportation planning and operations.

Acknowledgments. This research was supported by a grant from the Ministry of Science & Technology, Israel and The Ministry of Science & Technology of the People's Republic of China (No. 3-15741). Special thanks to Prof. Itzhak Benenson of Tel Aviv University for helpful advice and to Carol Teegarden for editing and proofreading.

References

1. Agard, B., Morency, C., Trépanier, M.: Mining public transport user behaviour from smart card data (2007)
2. Batista, G.E., Monard, M.C.: An analysis of four missing data treatment methods for supervised learning. Appl. Artif. Intell. **17**(5–6), 519–533 (2003)
3. Bertsimas, D., Pawlowski, C., Zhuo, Y.D.: From predictive methods to missing data imputation: an optimization approach. J. Mach. Learn. Res. **18**(1), 7133–7171 (2017)

4. Camino, R.D., Hammerschmidt, C.A., State, R.: Improving missing data imputation with deep generative models (2019). arXiv preprint: arXiv:1902.10666
5. Chen, T., Guestrin, C.: Xgboost: a scalable tree boosting system. In: Proceedings of the 22nd ACM sigkdd international conference on knowledge discovery and data mining, pp. 785–794 (2016)
6. Chen, Z., Fan, W.: Extracting bus transit boarding stop information using smart card transaction data. J. Mod. Transp. **26**(3), 209–219 (2018). https://doi.org/10.1007/s40534-018-0165-y
7. Chien, H.Y., Jan, J.K., Tseng, Y.M.: An efficient and practical solution to remote authentication: smart card. Comput. Secur. **21**(4), 372–375 (2002)
8. Gaudette, L., Japkowicz, N.: Evaluation methods for ordinal classification. In: Gao, Y., Japkowicz, N. (eds.) AI 2009. LNCS (LNAI), vol. 5549, pp. 207–210. Springer, Heidelberg (2009). https://doi.org/10.1007/978-3-642-01818-3_25
9. Jang, W.: Travel time and transfer analysis using transit smart card data. Transp. Res. Rec. **2144**(1), 142–149 (2010)
10. Kotsiantis, S.B., Zaharakis, I., Pintelas, P.: Supervised machine learning: a review of classification techniques. Emerg. Artif. Intell. Appl. Comput. Eng. **160**, 3–24 (2007)
11. Lakshminarayan, K., et al.: Imputation of missing data using machine learning techniques. In: KDD, pp. 140–145 (1996)
12. Li, H., Li, F., Song, C., Yan, Y.: Towards smart card based mutual authentication schemes in cloud computing. KSII Trans. Internet Inf. Syst. (TIIS) **9**(7), 2719–2735 (2015)
13. Li, T., Sun, D., Jing, P., Yang, K.: Smart card data mining of public transport destination: a literature review. Information **9**(1), 18 (2018)
14. Lundberg, S.M., Lee, S.I.: A unified approach to interpreting model predictions. In: Advances in neural information processing systems, pp. 4765–4774 (2017)
15. Ma, X.L., Wang, Y.H., Chen, F., Liu, J.F.: Transit smart card data mining for passenger origin information extraction. J. Zhejiang Univ. Sci. C **13**(10), 750–760 (2012)
16. Ma, X., Liu, C., Wen, H., Wang, Y., Wu, Y.J.: Understanding commuting patterns using transit smart card data. J. Transp. Geogr. **58**, 135–145 (2017)
17. Pelletier, M.P., Trépanier, M., Morency, C.: Smart card data use in public transit: a literature review. Transp. Res. Part C: Emerg. Technol. **19**(4), 557–568 (2011)
18. Singh, A., Thakur, N., Sharma, A.: A review of supervised machine learning algorithms. In: 2016 3rd International Conference on Computing for Sustainable Global Development (INDIACom), pp. 1310–1315. IEEE (2016)
19. Tao, S., Rohde, D., Corcoran, J.: Examining the spatial-temporal dynamics of bus passenger travel behaviour using smart card data and the flow-comap. J. Transp. Geogr. **41**, 21–36 (2014)
20. Trépanier, M., Tranchant, N., Chapleau, R.: Individual trip destination estimation in a transit smart card automated fare collection system. J. Intell. Transp. Syst. **11**(1), 1–14 (2007)
21. Welch, T.F., Widita, A.: Big data in public transportation: a review of sources and methods. Transp. Rev. **39**(6), 795–818 (2019)

HEAT: Hyperbolic Embedding of Attributed Networks

David McDonald$^{(\boxtimes)}$ (iD) and Shan He (iD)

School of Computer Science, University of Birmingham,
Edgbaston, Birmingham B15 2TT, UK
{office,dxm237,s.he}@cs.bham.ac.uk
https://www.cs.bham.ac.uk

Abstract. Finding a low dimensional representation of hierarchical, structured data described by a network remains a challenging problem in the machine learning community. An emerging approach is embedding networks into hyperbolic space since it can naturally represent a network's hierarchical structure. However, existing hyperbolic embedding approaches cannot deal with attributed networks, where nodes are annotated with additional attributes. To overcome this, we propose HEAT (for Hyperbolic Embedding of Attributed Networks). HEAT first extracts training samples from the network that captures both topological and attribute node similarity and then learns a low-dimensional hyperboloid embedding using full Riemannian Stochastic Gradient Descent. We show that HEAT can outperform other network embedding algorithms on several common downstream tasks. As a general network embedding method, HEAT opens the door to hyperbolic manifold learning on a wide range of both attributed and unattributed networks.

Keywords: Network embedding · Hyperbolic embedding · Random walk

1 Introduction

The success of machine learning algorithms often depends upon data representation. Unsupervised representation learning – learning alternative (low dimensional) representations of data – has become common for processing information on non-Euclidean domains, such as complex networks. Prediction over nodes and edges requires careful feature engineering and representation learning leads to the extraction of features from a graph that are most useful for downstream tasks, without careful design or a-priori knowledge [1].

An emerging representation learning approach for complex networks is hyperbolic embedding. This is based on compelling evidence that the underlying metric space of many complex networks is hyperbolic [2]. A hyperbolic space can be interpreted as a continuous representation of a discrete tree structure that captures the hierarchical organisation of elements within a complex system [2]. Furthermore, hyperbolic metric spaces have been shown to explain other characteristics typical to complex networks, characteristics such as clustering [2] and

© Springer Nature Switzerland AG 2020
C. Analide et al. (Eds.): IDEAL 2020, LNCS 12489, pp. 28–40, 2020.
https://doi.org/10.1007/978-3-030-62362-3_4

the "small world" phenomenon [3]. Hyperbolic spaces therefore offer a natural continuous representations of hierarchical complex networks [2].

However, existing hyperbolic embedding approaches cannot deal with attributed networks, of which nodes (entities) are richly annotated with attributes [4]. For example, a paper within a citation network may be annotated with the presence of keywords, and the people in a social network might have additional information such as interests, hobbies, and place of work. These attributes might provide additional proximity information to constrain the representations of the nodes. Therefore, incorporating node attributes can improve the quality of the final embedding, with respect to many different downstream tasks [4].

This paper proposes the first hyperbolic embedding method for attributed networks called HEAT. The intuition behind HEAT is to extract training samples from the original graph, which can capture both topological and attribute similarities, and then learn a hyperbolic embedding based on these samples. To extract training samples, a novel random walk algorithm with a teleport procedure is developed. The purpose of this walk is to capture phantom links between nodes that do not necessarily share a topological link, but have highly similar attributes. To learn the embeddings from these extracted samples, HEAT employs a novel learning objective that is optimized using full Riemannian stochastic gradient descent in hyperbolic space.

Thorough experimentation shows that HEAT can achieve better performance on several downstream tasks compared with several state-of-the-art embedding algorithms. As a general framework, HEAT can embed both unattributed and attributed networks with continuous and discrete attributes, which opens the door to hyperbolic manifold learning for a wide range of complex networks.

1.1 Related Work

Recently, graph convolution has been generalised to hyperbolic space to allow for non-Euclidean representation learning on attributed networks [5]. The algorithm in [6] embeds networks to the Poincaré ball using retraction updates to optimize an objective that aims to maximize the likelihood of observing true node pairs versus arbitrary pairs of nodes in the network. This approach was improved by the same authors and applied to the hyperboloid model of hyperbolic space [7]. Additionally, trees can be embedded in hyperbolic space without distortion [8] and so, some works by embed general graphs to trees and then compute an exact distortion-free embedding of the resulting tree [9].

For attributed network embedding in Euclidean space, several algorithms have been proposed. Text-assisted Deepwalk (TADW) [10] generalizes Deepwalk [11] to nodes with text attributes. By generalizing convolution from regular pixel lattices to arbitrary graphs, it is possible embed and classify entire graphs [12]. Furthermore, the popular Graph Convolutional Network (GCN) extend this approach to simplify graph convolution in the semi-supervised setting [13]. Graph-SAGE introduces an inductive framework for online learning of node embeddings capable of generalizing to unseen nodes [4].

2   Hyperboloid Model of Hyperbolic Geometry

Due to the fundamental difficulties in representing spaces of constant negative curvature as subsets of Euclidean spaces, there are not one but many equivalent models of hyperbolic spaces [14]. The models are equivalent because all models of hyperbolic geometry can be freely mapped to each other by a distance preserving mapping called an *isometry*. Each model emphasizes different aspects of hyperbolic geometry, but no model simultaneously represents all of its properties.

For HEAT, the hyperboloid model is selected for the simple form of Riemannian operations. The main advantage of this is that the simple forms allow for the inexpensive computation of exact gradients and, therefore, HEAT is optimized using full Riemannian Stochastic Gradient Descent (RSGD) [7], rather than approximating gradient descent with retraction updates [6]. Ultimately, this has the advantage of faster convergence [7]. Unlike disk models, that sit in an ambient Euclidean space of dimension n, the hyperboloid model of n-dimensional hyperbolic geometry sits in $n+1$-dimensional Minkowski space-time. Minkowski space-time is denoted $\mathbb{R}^{n:1}$, and a point $\mathbf{x} \in \mathbb{R}^{n:1}$ has spacial coordinates $\mathbf{x}^i$ for $i = 1, 2, ..., n$ and time co-ordinate $\mathbf{x}^{n+1}$.

HEAT requires the *Minkowski bilinear form* for both its learning objective as well as for parameter optimization. The Minkowski bilinear form is defined as $\langle \mathbf{u}, \mathbf{v} \rangle_{\mathbb{R}^{n:1}} = \sum_{i=1}^{n} \mathbf{u}^i \mathbf{v}^i - \psi^2 \mathbf{u}^{n+1} \mathbf{v}^{n+1}$ where ψ is the speed of information flow in our system (here set to 1 for simplified calculations). This bilinear form is an inner product and allows the computation of the Minkowski norm: $||\mathbf{u}||_{\mathbb{R}^{n:1}} := \sqrt{\langle \mathbf{u}, \mathbf{u} \rangle_{\mathbb{R}^{n:1}}}$. Using the bilinear form, the hyperboloid is defined as $\mathbb{H}^n = \{\mathbf{u} \in \mathbb{R}^{n:1} \mid \langle \mathbf{u}, \mathbf{u} \rangle_{\mathbb{R}^{n:1}} = -1, \mathbf{u}^{n+1} > 0\}$. The first condition defines a hyperbola of two sheets, and the second one selects the top sheet.

In addition, the bilinear form is required to define the distance between two points on the hyperboloid, which is incorporated as part of HEAT's training objective function. The distance between two points $\mathbf{u}, \mathbf{v} \in \mathbb{H}^n$ is given by the length of the geodesic between them $D_{\mathbb{H}^n}(\mathbf{u}, \mathbf{v}) = \text{arccosh}(\gamma)$ where $\gamma = -\langle \mathbf{u}, \mathbf{v} \rangle_{\mathbb{R}^{n:1}}$ [7].

To update parameters, HEAT performs full RSGD. This requires defining the tangent space of a point $\mathbf{u} \in \mathbb{H}^n$, that is denoted $T_{\mathbf{u}}\mathbb{H}^n$. It is the collection of all points in $\mathbb{R}^{n:1}$ that are orthogonal to $\mathbf{u}$, and is defined as $T_{\mathbf{u}}\mathbb{H}^n = \{\mathbf{x} \in \mathbb{R}^{n:1} \mid \langle \mathbf{u}, \mathbf{x} \rangle_{\mathbb{R}^{n:1}} = 0\}$. It can be shown that $\langle \mathbf{x}, \mathbf{x} \rangle_{\mathbb{R}^{n:1}} > 0 \; \forall \mathbf{x} \in T_{\mathbf{u}}\mathbb{H}^n \; \forall \mathbf{u} \in \mathbb{H}^n$, and so the tangent space of every point the hyperboloid is positive-definite, and so $\mathbb{H}^n$ is a Riemannian manifold [7].

3   Hyperbolic Embedding of Attributed Networks

We consider a network of N nodes given by the set V with $|V| = N$. We use E to denote the set of all interactions between the nodes in our network. $E = \{(u, v)\} \subseteq V \times V$. We use the matrix $\mathbf{W} \in \mathbb{R}^{N \times N}$ to encode the weights of these interactions, where $\mathbf{W}_{uv}$ is the weight of the interaction between node u and node v. We have that $\mathbf{W}_{uv} \neq 0 \iff (u,v) \in E$. If the network is unweighted then $\mathbf{W}_{uv} = 1$ for all $(u,v) \in E$. Furthermore, the matrix $\mathbf{X} \in \mathbb{R}^{N \times d}$ describes the attributes of each node in the network. These attributes may be discrete or continuous. Edge attributes could be handled by transforming them into node attributes shared by both nodes connected by the edge. We consider the problem of representing a graph given as $\mathbb{G} = (V, E, \mathbf{W}, \mathbf{X})$ as set of low-dimensional vectors in the n-dimensional hyperboloid $\{\mathbf{x}_v \in \mathbb{H}^n \mid v \in V\}$, with $n \ll N$. The described problem is unsupervised.

3.1 HEAT Overview

Our proposed HEAT consists of two main components:

1. A novel network sampling algorithm based on random walks to extract samples than can capture both topological and attribute similarity.
2. A novel learning algorithm that can learn hyperbolic embeddings from training samples using Riemannian stochastic gradient descent (RSGD) in hyperbolic space.

3.2 Sample the Network Using Random Walks with Jump

We propose a novel random-walk procedure to obtain training samples that capture both topological and attribute similarity. Random walks have been proposed in the past as a robust sampling method of elements from structured data, such as graphs, since they provide an efficient, flexible and parallelizable sampling method [11]. For every node in the network, several 'walks' with a fixed length l are performed [1]. We note that random walks traditionally take into account only first-order topological similarity, that is: nodes are similar if they are connected in the network. However, additional topological similarity can be considered. For example, second-order similarity between nodes (that is: the similarity of neighbourhoods) could be incorporated into the topological similarity matrix using a weighted sum [15]. We leave this as future work.

We propose that, in addition to standard random walks which capture topological similarity, we use attribute similarity to 'jump' the random walker to the nodes with similar attributes. To this end, we define the attribute similarity $\mathbf{Y}$ as cosine similarity of the attribute vectors of the nodes. We assign a value of 0 similarity to any negative similarity values. We select cosine similarity as it can readily handle high dimensional data well without making a strong assumption about the data. We propose HEAT as a general framework and, so, can change the cosine similarity to a more sophisticated and problem-dependant measure of pairwise node attribute similarity.

To define the probability of moving from a node to another based on both topological and attribute similarity, we then additionally define $\bar{\mathbf{W}}$ and $\bar{\mathbf{Y}}$ to be the row-normalized versions of the weight matrix $\mathbf{W}$ and attribute similarity matrix $\mathbf{Y}$ respectively. Each row in $\bar{\mathbf{W}}$ and $\bar{\mathbf{Y}}$ describes a discrete probability distribution corresponding to the likelihood of jumping from one node to the next according to either topological similarity or attribute similarity.

To control the trade-off between topology and attributes, we define the hyper-parameter $0 \leq \alpha \leq 1$. Formally, we use i to denote the ith node in the walk $(x_0 = s)$, and for each step $i = 1, 2, ..., l$ in the walk, we sample $\pi_i \sim U(0, 1)$ and determine the ith node as follows: if $\pi_i < \alpha$, then $P(x_i = v \mid x_{i-1} = u) = \hat{\mathbf{Y}}_{uv}$, else $P(x_i = v \mid x_{i-1} = u) = \hat{\mathbf{W}}_{uv}$.

We follow previous works and consider nodes that appear within a maximum distance of each other in the same walk to be "context pairs" [1,11]. We call this maximum distance the "context-size" and it is a hyper-parameter that controls the size of a local neighbourhood of a node. Previous works show that increasing context size typically improves performance, at some computational cost [1]. All of the source-context pairs are added into a set $\mathcal{D}$.

3.3 Setting α

For practical applications, we suggest the following approach to set the value of the hyper-parameter α that controls the trade-off between topology and similarity in the sampling process: First, randomly sample a proportion of edges from the network and remove them. Next, select an equal number of 'non-edge' node pairs $(u, v) \in V \times V \setminus E$. These two edge sets will form a validation set. Then perform HEAT to generate hyperboloid embeddings for a range of values of α. Removed edges can be ranked against the sampled non-edges and a global ranking measure, such as AUROC or AP, could be used to select a value for α. This procedure is performed in Sect. 4.3, where we evaluate link prediction. As we show in Sect. 4.4, HEAT is robust to the setting of α for three common downstream tasks.

3.4 Hyperboloid Embedding Learning

For the hyperboloid embedding learning procedure of HEAT, we aim to maximise the probability of observing all of the pairs in $\mathcal{D}$ in the low-dimensional embedded space. We define the probability of two nodes sharing a connection to be a function of their distance in the embedding space. This is motivated by the intuition of network embedding that nodes separated by a small distances share a high degree of similarity and should, therefore share a high probability of connection, and nodes very far apart in the embedding space should share a low probability of connection. This principle forms the basis of an objective function that is optimized by HEAT to learn node embeddings.

We make the common simplifying assumption that the distance between a source node and neighbourhood is symmetric (i.e. $P((u, v)) = P((v, u))$ for all $u, v \in V$) [1]. To this end, we define the symmetric function $\hat{P}((u, v)) := -D^2_{\mathbb{H}^n}(\mathbf{u}, \mathbf{v})$ to be the unnormalized probability of observing a link between source node u and context node v, where $\mathbf{u}$ and $\mathbf{v}$ are their respective hyperbolic positions. We square the distance as this leads to stable gradients [9].

We normalize the probability using: $P((u, v)) := \exp(\hat{P}((u, v)))/Z(u)$, where $Z(u) := \sum_{v' \in V} \exp(\hat{P}((u, v')))$.

However, computing the gradient of the partition function $Z(u)$ involves a summation over all nodes $v \in V$, which for large networks, is prohibitively computationally expensive [1]. Following previous works, we overcome this limitation through *negative sampling*. We define the set of negative samples for u, as the set of v for we observe no relation with u: $\text{Neg}(u) := \{v \in V \mid (u, v) \notin \mathcal{D}\}$. We further define $\text{Neg}_K(u, v) := \{x_i \sim^{P_n} \text{Neg}(u) \mid i = 1, 2, ..., K\} \cup \{v\}$ to be a random sample with replacement of size K from the set of negative samples of u, according to a noise distribution P_n including v. Following [1], we set $P_n = U^{\frac{3}{4}}$, the unigram distribution raised to the 3/4 power.

We aim to represent the obtained distribution of pairs in a low-dimensional hyperbolic space. To this end, we formulate a loss function L that encourages maximising the probability of observing all positive sample pairs $P(v \mid u)$ for all $(u, v) \in \mathcal{D}$ and minimising the probability of observing all other pairs. To this end, we define the loss function L for an embedding $\Theta = \{\mathbf{u} \in \mathbb{H}^n \mid u \in V\}$ to be the mean of negative log-likelihood of observing all the source-context pairs in $\mathcal{D}$, against the negative sample noise:

$$L(\Theta) = -\frac{1}{|\mathcal{D}|} \sum_{(u,v) \in \mathcal{D}} \log \left[\frac{\exp\left(-D_{\mathbb{H}^n}^2(\mathbf{u}, \mathbf{v})\right)}{\sum_{v' \in \text{Neg}_K(u,v)} \exp\left(-D_{\mathbb{H}^n}^2(\mathbf{u}, \mathbf{v}')\right)} \right] \tag{1}$$

The numerator of Eq. (1), $\exp\left(-D_{\mathbb{H}^n}^2(\mathbf{u}, \mathbf{v})\right)$, is concerned with the hyperbolic distance between nodes u and v in the positive sample set $\mathcal{D}$. Minimising L involves minimising the distance between $\mathbf{u}$ and $\mathbf{v}$ in the embedding space. The denominator is a sum over all v' in a given sample of size m of the negative samples for node u. Minimising L involves maximising this term, thereby pushing $\mathbf{u}$ and $\mathbf{v}'$ far apart in the embedding space. Overall, we observe that minimising L involves maximising $P((u, v))$ for all $(u, v) \in \mathcal{D}$ as required. This encourages source-context pairs to be close together in the embedding space, and u to be embedded far from the noise nodes v' [6].

3.5 Optimization

Since we use hyperboloid model, unlike some previous works [6,9] that use the *Poincaré ball* model and approximate gradients with retraction updates, we are able to use full Riemannian optimization and so our gradient computation is exact and possesses a simple form [7]. We follow a three step procedure in to compute gradients and then update hyperbolic coordinates [7].

To compute the gradient of L for vector $\mathbf{u} \in \mathbb{H}^n$ with respect to the hyperboloid $\nabla_{\mathbf{u}}^{\mathbb{H}^n} L$, we first compute the gradient with respect to the Minkowski ambient space $\mathbb{R}^{n:1}$ as

$$\nabla_{\mathbf{u}}^{\mathbb{R}^{n:1}} L = \left(\frac{\partial L}{\partial u^1}\bigg|_{\mathbf{u}}, ..., \frac{\partial L}{\partial u^n}\bigg|_{\mathbf{u}}, -\frac{\partial L}{\partial u^{n+1}}\bigg|_{\mathbf{u}} \right) \tag{2}$$

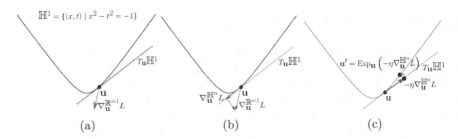

Fig. 1. Three step optimization on $\mathbb{H}^n$. (a) provides a representation of $\mathbb{H}^1$ a one dimensional manifold in two dimensional Minkowski space $\mathbb{R}^{1:1}$. One point on $\mathbb{H}^1$ is highlighted: $\mathbf{u}$. The red arrow is an example $\nabla_{\mathbf{x}_u}^{\mathbb{R}^{n:1}} L$ vector. Finally, $T_u\mathbb{H}^n$ is given by the dotted black line. (b) highlights the component of $\nabla_{\mathbf{x}_u}^{\mathbb{R}^{n:1}} L$ lying on $T_u\mathbb{H}^n$. Finally, (c) plots the mapping from $-\eta\nabla_{\mathbf{u}}^{\mathbb{H}^n} L$ back to the hyperboloid using the exponential map $\mathrm{Exp}_{\mathbf{u}}$.

Let $o_{uv} := -D_{\mathbb{H}^n}^2(\mathbf{u}, \mathbf{v})$ and $\mathrm{Neg}_K(u) := \bigcup_{\{v|(u,v)\in\mathcal{D}\}} \mathrm{Neg}_K(u, v)$. Then

$$\nabla_{\mathbf{u}}^{\mathbb{R}^{n:1}} L = \frac{1}{|\mathcal{D}|} \underset{v\in\mathrm{Neg}_K(u)}{\Sigma} (\delta_{vv'} - P((u, v))) \cdot 2 \cdot \frac{\mathrm{arccosh}\left(-\langle\mathbf{u}, \mathbf{v}\rangle_{\mathbb{R}^{n:1}}\right)}{\sqrt{\langle\mathbf{u}, \mathbf{v}\rangle_{\mathbb{R}^{n:1}}^2 - 1}} \cdot \mathbf{v} \quad (3)$$

where $\delta_{vv'}$ is the Kronecker delta function. We then use the vector projection formula to compute the projection of the ambient gradient to its component in the tangent space:

$$\nabla_{\mathbf{u}}^{\mathbb{H}^n} L = \nabla_{\mathbf{u}}^{\mathbb{R}^{n:1}} L + \langle\mathbf{u}, \nabla_{\mathbf{u}}^{\mathbb{R}^{n:1}} L\rangle_{\mathbb{R}^{n:1}} \cdot \mathbf{u} \quad (4)$$

Having computed the gradient component in the tangent space of $\mathbf{u}$, we define the exponential map to take a vector $\mathbf{x} \in T_u\mathbb{H}^n$ to its corresponding point on the hyperboloid:

$$\mathrm{Exp}_{\mathbf{u}}(\mathbf{x}) = \cosh(r) \cdot \mathbf{u} + \sinh(r) \cdot \mathbf{x}/r \quad (5)$$

where $r = ||\mathbf{x}||_{\mathbb{R}^{n:1}} = \sqrt{\langle\mathbf{x}, \mathbf{x}\rangle}_{\mathbb{R}^{n:1}}$ denotes the Minkowski norm of $\mathbf{x}$.

Altogether, we have that the three-step procedure for computing the new position of $\mathbf{u}$, with learning rate η is: First, calculate ambient gradient $\nabla_{\mathbf{u}}^{\mathbb{R}^{n:1}} L$ (Eq. (3)). Next, project $\nabla_{\mathbf{u}}^{\mathbb{R}^{n:1}} L$ to tangent $\nabla_{\mathbf{u}}^{\mathbb{H}^n} L$(Eq. (4)). Finally, set $\mathbf{u} = \mathrm{Exp}_{\mathbf{u}}\left(-\eta\nabla_{\mathbf{u}}^{\mathbb{H}^n} L\right)$ (Eq. (5)). Figure 1 provides an example of this procedure operating on $\mathbb{H}^1$.

4   Experimental Validation

4.1   Datasets and Benchmarks

We evaluate HEAT on three citation networks: Cora_ML ($N = 2995$, $|E| = 8416$, $d = 2879$, $y = 7$), Citeseer ($N = 4230$, $|E| = 5358$, $d = 2701$, $y = 6$), and

Pubmed ($N = 18230$, $|E| = 79612$, $d = 500$, $y = 3$) [16]; one PPI network ($N = 3480$, $|E| = 54806$, $d = 50$, $y = 121$) [4]; and one social network for MIT university ($N = 6402$, $|E| = 251230$, $d = 2804$, $y = 32$) [17]. Here we use N, $|E|$, d, and y to denote the number of nodes, edges, attributes and labels respectively. Features for all networks were scaled to have a mean of 0 and a standard deviation of 1.

We compare against the following benchmark algorithms from literature: DEEPWALK [11], TADW [10], AANE [18], the GraphSAGE [4] implementation of the GCN (SAGEGCN) [13], and N&K, an unattributed embedding approach to the Poincaré ball [6]. Furthermore, we use a method based purely on SVD of the attribute similarity matrix (ATTRPURE) [17].

For all benchmark methods we adopt the original source code and follow the suggestions of the original papers for parameter settings. For Deepwalk: we set walks per node to 10, walk length to 80, context size to 10 and top-k value to 30. For TADW, and AANE, we set the balancing factors to 0.2, 0.05, and 0.8 respectively. For SAGEGCN, we set learning rate, dropout rate, batch size, normalization, weight decay rate and epochs to 0.001, 0.5, 512, true, 1e−4 and 100 respectively. For N&K, we set learning rate, epochs, negative samples, batch size, and burn-in to 1.0, 1500, 10, 512 and 20 respectively. For HEAT, we set learning rate, epochs, negative samples, batch size, context-size, walks per node, and walk length to 1.0, 5, 10, 512, 10, 10, and 80 respectively. We fix $\alpha = 0.2$ for all experiments, which we determined by applying the procedure described in Sect. 3.3 on the Cora_ML network. We train all methods in an unsupervised manner. For each experiment, we perform 30 independent runs of all algorithms and report the mean result.

4.2 Network Reconstruction and Link Prediction

Following common practice, we use network reconstruction to evaluate the capacity of the learned embeddings to reflect the original data [6]. After training our model to convergence upon the complete information, we compute distances in the embedding space between all pairs of nodes according to both models. For link prediction, we randomly select 10% of the edges in the network and remove them [6]. We then randomly select also an equal number of non-edges in the network. An embedding is learned for each incomplete network and pairs of nodes are ranked by distance.

Table 1 provides a summary of the network reconstruction and link prediction results for embedding dimension 10. For reconstruction, we observe that HEAT has a high capacity for learning network structure, even at low dimensions. Further, by incorporating attributes, performance increased further on two out of the five networks studied. The link prediction results demonstrate that HEAT is capable of highly competitive link prediction ability with and without attributes. We observe that the inclusion of attributes improves performance on three out of five networks and suggest that this is because of the high level of homophily in citation networks.

Table 1. Summary of network reconstruction and link prediction results for embedding dimension 10. Bold text indicates the best score.

	Reconstruction								Link Prediction			
Cora_ML												
	Rank	AUROC	AP	mAP	p@1	p@3	p@5	p@10	Rank	AUROC	AP	mAP
N&K	209.8	0.987	0.986	0.674	0.791	0.736	0.723	0.713	96.6	0.922	0.935	0.248
AANE	3945.7	0.758	0.775	0.145	0.182	0.192	0.206	0.238	315.3	0.743	0.761	0.098
TADW	539.9	0.967	0.959	0.401	0.458	0.425	0.438	0.444	72.8	0.941	0.933	0.180
ATTRPURE	4768.0	0.708	0.735	0.124	0.156	0.164	0.174	0.203	356.5	0.710	0.736	0.093
DEEPWALK	185.6	0.989	0.986	0.712	0.757	0.723	0.722	0.720	178.9	0.855	0.896	0.224
SAGEGCN	913.9	0.944	0.935	0.289	0.304	0.341	0.363	0.392	160.5	0.870	0.873	0.114
HEAT$_{\alpha=0.00}$	**40.9**	**0.998**	**0.997**	**0.853**	0.874	**0.869**	**0.858**	**0.850**	131.8	0.893	0.928	0.276
HEAT$_{\alpha=0.20}$	58.6	0.996	0.995	0.838	**0.895**	0.838	0.823	0.813	**49.0**	**0.961**	**0.963**	**0.288**
Citeseer												
N&K	67.5	0.994	0.994	0.799	0.774	0.772	0.800	0.837	145.5	0.820	0.853	0.204
AANE	3741.9	0.650	0.645	0.097	0.088	0.112	0.132	0.196	284.8	0.646	0.643	0.086
TADW	297.1	0.972	0.964	0.376	0.314	0.352	0.399	0.469	55.9	0.931	0.917	0.149
ATTRPURE	3922.2	0.633	0.630	0.089	0.083	0.102	0.126	0.189	297.1	0.630	0.630	0.083
DEEPWALK	23.5	0.998	0.997	0.798	0.721	0.805	0.853	0.899	259.9	0.677	0.775	0.213
SAGEGCN	618.9	0.942	0.939	0.216	0.160	0.291	0.364	0.470	138.1	0.829	0.848	0.132
HEAT$_{\alpha=0.00}$	9.6	0.999	0.999	0.830	0.740	**0.898**	**0.932**	**0.967**	16.5	0.981	0.979	0.791
HEAT$_{\alpha=0.20}$	**7.6**	**0.999**	**0.999**	**0.926**	**0.899**	0.894	0.908	0.933	**6.3**	**0.993**	**0.994**	**0.810**
Pubmed												
N&K	451.5	0.995	0.994	0.737	0.743	0.803	0.835	0.835	801.7	0.880	0.900	0.210
AANE	19746.8	0.777	0.784	0.104	0.106	0.187	0.219	0.253	1503.8	0.774	0.782	0.088
TADW	2155.6	0.976	0.973	0.514	0.502	0.550	0.600	0.617	465.5	0.930	0.923	0.157
ATTRPURE	25982.8	0.707	0.707	0.097	0.098	0.175	0.207	0.241	1954.5	0.706	0.706	0.082
DEEPWALK	565.1	0.994	0.993	0.816	0.794	0.808	0.845	0.856	1742.8	0.738	0.833	0.203
SAGEGCN	4118.9	0.954	0.945	0.261	0.215	0.341	0.405	0.467	635.5	0.905	0.901	0.114
HEAT$_{\alpha=0.00}$	**115.8**	**0.999**	0.998	0.823	0.753	0.863	**0.896**	**0.905**	463.6	0.930	0.931	0.202
HEAT$_{\alpha=0.20}$	161.8	0.998	**0.998**	**0.908**	**0.913**	**0.876**	0.892	0.893	**322.8**	**0.952**	**0.954**	**0.256**
PPI												
N&K	6757.7	0.938	0.940	0.421	**0.801**	**0.699**	**0.664**	0.640	722.1	0.912	0.918	0.185
AANE	50719.7	0.537	0.588	0.089	0.470	0.251	0.201	0.168	3822.2	0.535	0.582	0.070
TADW	23408.8	0.786	0.767	0.142	0.502	0.298	0.258	0.242	2237.5	0.728	0.722	0.080
ATTRPURE	51304.6	0.511	0.512	0.038	0.164	0.084	0.065	0.051	3854.5	0.511	0.511	0.018
DEEPWALK	9962.9	0.909	0.903	0.388	0.721	0.582	0.543	0.523	1066.4	0.870	0.873	0.144
SAGEGCN	44688.5	0.592	0.607	0.095	0.455	0.218	0.169	0.137	3622.2	0.560	0.574	0.067
HEAT$_{\alpha=0.00}$	**4926.0**	**0.955**	**0.952**	**0.468**	0.782	0.696	0.664	**0.643**	431.6	**0.948**	**0.947**	**0.255**
HEAT$_{\alpha=0.20}$	5628.5	0.949	0.946	0.457	0.786	0.667	0.627	0.604	493.0	0.940	0.940	0.241
MIT												
N&K	32506.7	0.927	0.930	0.565	1.000	0.884	0.855	0.827	2727.1	0.918	0.922	0.256
AANE	157174.0	0.647	0.639	0.189	1.000	0.528	0.414	0.321	11816.8	0.646	0.638	0.098
TADW	79981.9	0.820	0.814	0.397	1.000	0.760	0.700	0.639	6133.4	0.816	0.815	0.180
ATTRPURE	171241.0	0.615	0.617	0.184	0.993	0.543	0.428	0.323	12823.6	0.616	0.617	0.098
DEEPWALK	33037.6	0.926	0.919	0.578	1.000	0.857	0.819	0.782	2834.4	0.915	0.909	0.241
SAGEGCN	89637.4	0.798	0.797	0.380	1.000	0.670	0.599	0.545	7142.0	0.786	0.784	0.153
HEAT$_{\alpha=0.00}$	**24776.7**	**0.944**	**0.940**	**0.639**	1.000	**0.902**	**0.877**	**0.850**	**2246.9**	**0.933**	**0.930**	**0.279**
HEAT$_{\alpha=0.20}$	28621.0	0.936	0.934	0.633	1.000	0.901	0.870	0.839	2510.4	0.925	0.925	0.273

4.3   Node Classification

To evaluate node classification, we learn an embedding, using complete topological and attribute information, in an unsupervised manner. We then use an out-of-the-box Support Vector Classifier (SVC) to evaluate the separation of classes in the embedding space For hyperbolic embeddings (HEAT and N&K), we first project to the Klein model of hyperbolic space, which preserves straight lines [14]. For the PPI network, each protein has multiple labels from the Gene Ontology [4]. To evaluate our model in this multi-label case, we adopt a one-vs-all setting, where we train a separate classifier for each class.

Table 2. Summary of node classification results for embedding dimension 10. All measures are micro-averaged. Bold text indicates best performance.

	Cora_ML				PPI			
	F1	Precision	Recall	AUROC	F1	Precision	Recall	AUROC
N&K	0.761	0.827	0.704	0.953	0.387	**0.695**	0.268	0.704
AANE	0.706	0.814	0.624	0.944	0.395	0.688	0.277	0.705
TADW	0.837	0.869	0.808	**0.983**	0.393	0.688	0.275	0.705
ATTRPURE	0.675	0.782	0.594	0.937	**0.398**	0.688	**0.281**	**0.705**
DEEPWALK	**0.855**	**0.886**	**0.827**	0.977	0.387	0.694	0.269	0.704
SAGEGCN	0.679	0.758	0.616	0.933	0.388	0.694	0.269	0.704
HEAT$_{\alpha=0.0}$	0.804	0.858	0.756	0.965	0.388	0.694	0.269	0.704
HEAT$_{\alpha=0.2}$	0.849	0.884	0.817	0.980	0.388	0.694	0.269	0.704
	Citeseer				MIT			
	F1	Precision	Recall	AUROC	F1	Precision	Recall	AUROC
N&K	0.516	0.856	0.370	0.861	0.508	**0.858**	0.361	0.937
AANE	0.609	0.829	0.482	0.897	0.036	0.414	0.019	0.866
TADW	0.878	0.899	0.857	0.980	0.544	0.739	0.430	0.949
ATTRPURE	0.597	0.837	0.465	0.892	0.056	0.409	0.030	0.869
DEEPWALK	**0.927**	**0.939**	**0.916**	**0.987**	**0.661**	0.841	**0.544**	**0.965**
SAGEGCN	0.466	0.727	0.347	0.843	0.480	0.832	0.337	0.929
HEAT$_{\alpha=0.0}$	0.594	0.857	0.455	0.879	0.634	0.849	0.507	0.959
HEAT$_{\alpha=0.2}$	0.887	0.914	0.861	0.981	0.637	0.819	0.522	0.960
	Pubmed				Mean Rank on All Datasets			
	F1	Precision	Recall	AUROC	F1	Precision	Recall	AUROC
N&K	0.797	0.815	0.780	0.933	5.4	3	5.6	5.6
AANE	**0.830**	**0.840**	**0.819**	**0.947**	4.2	5.8	4.2	4.2
TADW	0.778	0.797	0.761	0.921	3.8	4.8	3.6	3.2
ATTRPURE	0.781	0.806	0.758	0.912	5.2	6.6	5.4	5.2
DEEPWALK	0.765	0.792	0.740	0.907	3.8	3.2	3.6	3.2
SAGEGCN	0.768	0.791	0.746	0.906	6.8	6.6	6.8	7.4
HEAT$_{\alpha=0.0}$	0.807	0.823	0.791	0.933	4	3	4	4
HEAT$_{\alpha=0.2}$	0.797	0.815	0.780	0.933	**2.6**	**2.8**	**2.6**	**3**

Table 2 proves a summary of the node classification results. While HEAT never ranked first for any network, it consistently ranked highly on all networks unlike all other benchmark algorithms – with HEAT$_{\alpha=0.2}$ the best overall rank averaged across all the datasets. Further, we observe that, when considering node attributes, HEAT outperformed the other hyperbolic embedding algorithm N&K on all networks. We also note that, even without node attributes, HEAT obtained better results than N&K on all networks. Comparing with the Euclidean benchmark algorithms, we observe competitive results – especially when incorporating attributes.

4.4   Sensitivity to the Setting of α

We carried out preliminary experiments to evaluate HEAT's robustness to the setting of the control parameters. Our results indicated that the most sensitive parameter is α, which controls the trade off between considering topology and attributes. We run HEAT over a range of values $\alpha \in [0, 1]$ in steps of 0.05. Figure 2 plots AUROC scores for network reconstruction, link prediction and node classification obtained from a 5 dimensional embedding. From the three plots, we observe that the performance of HEAT on the three tasks on all of the networks is robust to a wide range of values of α (especially $\alpha \in [0, 0.5]$).

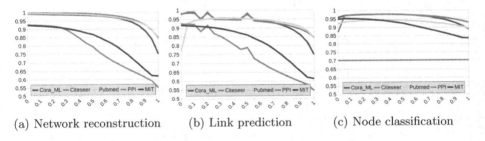

(a) Network reconstruction   (b) Link prediction   (c) Node classification

Fig. 2. The effect of the setting of α (x-axis) on the AUROC scores (y-axis) on three downstream machine learning tasks: (a) network reconstruction, (b) link prediction, and (c) node classification.

5   Conclusion

This paper presents HEAT to fill the gap of embedding attributed networks in hyperbolic space. We have designed a random walk algorithm to obtain the training samples that capture both network topological and attribute similarity. We have also derived an algorithm that learns hyperboloid embeddings from the training samples. Our results on five benchmark attributed networks show that, by including attributes, HEAT can improve the quality of a learned hyperbolic embedding in a number of downstream machine learning tasks. We find that, in general, HEAT does not perform as well in the node classification task compared with the Euclidean benchmark algorithms (Table 2) than in the network reconstruction and link prediction tasks (Table 1). While it is the most consistent across all datasets, it does not rank first on any particular dataset as it does for reconstruction and link prediction. This could be attributed to the SVC learning sub-optimal decision boundaries, since it is using a Euclidean optimization procedure. Logistic regression has been generalized to the Poincaré ball [19], and this may provide superior results. However, we leave this as future work.

Overall, we find that HEAT provides a general hyperbolic embedding method for both unattributed and attributed networks, which opens the door to hyperbolic manifold learning on a wide ranges of networks.

References

1. Grover, A., Leskovec, J.: node2vec: scalable feature learning for networks. In: ACM SIGKDD, pp. 855–864. ACM (2016)
2. Krioukov, D., Papadopoulos, F., Kitsak, M., Vahdat, A., Boguná, M.: Hyperbolic geometry of complex networks. Phys. Rev. E **82**(3), 036106 (2010)
3. Bianconi, G., Rahmede, C.: Emergent hyperbolic network geometry. Sci. Rep. **7**, 41974 (2017)
4. Hamilton, W., Ying, Z., Leskovec, J.: Inductive representation learning on large graphs. In: Advances in Neural Information Processing Systems, pp. 1024–1034 (2017)
5. Chami, I., Ying, Z., Ré, C., Leskovec, J.: Hyperbolic graph convolutional neural networks. In: Advances in Neural Information Processing Systems, pp. 4869–4880 (2019)
6. Nickel, M., Kiela, D.: Poincaré embeddings for learning hierarchical representations. In: Advances in Neural Information Processing Systems, pp. 6338–6347 (2017)
7. Nickel, M., Kiela, D.: Learning continuous hierarchies in the lorentz model of hyperbolic geometry. In: International Conference on Machine Learning, pp. 3776–3785 (2018)
8. Sarkar, R.: Low distortion delaunay embedding of trees in hyperbolic plane. In: van Kreveld, M., Speckmann, B. (eds.) GD 2011. LNCS, vol. 7034, pp. 355–366. Springer, Heidelberg (2012). https://doi.org/10.1007/978-3-642-25878-7_34
9. De Sa, C., Albert, G., Ré, C., Sala, F.: Representation tradeoffs for hyperbolic embeddings. Proc. Mach. Learn. Res. **80**, 4460 (2018)
10. Yang, C., Liu, Z., Zhao, D., Sun, M., Chang, E.: Network representation learning with rich text information. In: Twenty-Fourth International Joint Conference on Artificial Intelligence (2015)
11. Perozzi, B., Al-Rfou, R., Skiena, S.: Deepwalk: online learning of social representations. In: ACM SIGKDD, pp. 701–710. ACM (2014)
12. Defferrard, M., Bresson, X., Vandergheynst, P.: Convolutional neural networks on graphs with fast localized spectral filtering. In: Advances in Neural Information Processing Systems, pp. 3844–3852 (2016)
13. Kipf, T.N., Welling, M.: Semi-supervised classification with graph convolutional networks (2016). arXiv:1609.02907
14. Cannon, J.W., et al.: Hyperbolic geometry. Flavors Geom. **31**, 59–115 (1997)

15. Wang, D., Cui, P., Zhu, W.: Structural deep network embedding. In: Proceedings of the 22nd ACM SIGKDD International Conference on Knowledge Discovery and Data Mining, pp. 1225–1234 (2016)
16. Bojchevski, A., Günnemann, S.: Deep gaussian embedding of graphs: unsupervised inductive learning via ranking. In: International Conference on Learning Representations, pp. 1–13 (2018)
17. Hou, C., He, S., Tang, K.: RoSANE: robust and scalable attributed network embedding for sparse networks. Neurocomputing **409**, 231–243 (2020)
18. Huang, X., Li, J., Hu, X.: Accelerated attributed network embedding. In: Proceedings of the 2017 SIAM International Conference on Data Mining, pp. 633–641. SIAM (2017)
19. Ganea, O.-E., Bécigneul, G., Hofmann, T.: Hyperbolic neural networks (2018). arXiv:1805.09112

Least Squares Approach for Multivariate Split Selection in Regression Trees

Marvin Schöne[✉] and Martin Kohlhase

Center for Applied Data Science Gütersloh, Faculty of Engineering and Mathematics,
Bielefeld University of Applied Sciences, Bielefeld, Germany
{marvin.schoene,martin.kohlhase}@fh-bielefeld.de

Abstract. In the context of Industry 4.0, an increasing number of data-driven models is used in order to improve industrial processes. These models need to be accurate and interpretable. *Regression Trees* are able to fulfill these requirements, especially if their model flexibility is increased by multivariate splits that adapt to the process function. In this paper, a novel approach for multivariate split selection is presented. The direction of the split is determined by a first-order *Least Squares* model, that adapts to process function gradient in a local area. By using a forward selection method, the curse of dimensionality is weakened, interpretability is maintained and a generalized split is created. The approach is implemented in CART as an extension to the existing algorithm for constructing the *Least Squares Regression Tree* (LSRT). For evaluation, an extensive experimental analysis is performed in which LSRT leads to much smaller trees and a higher prediction accuracy than univariate CART. Furthermore, low sensitivity to noise and performance improvements for high dimensional input spaces and small data sets are achieved.

Keywords: Industry 4.0 · Interpretability · Curse of dimensionality · *Oblique Regression Trees* · *Mixed Regression Trees* · CART

1 Introduction

Within the framework of Industry 4.0, industrial processes become more efficient, sustainable and secure. A key role in these improvements is played by data-driven models, which have to fulfill specific requirements to be applicable. To provide reliable process control using data-driven models, they need to be accurate. Furthermore, to gain process knowledge and confidence towards the operators, the decisions of data-driven models must be interpretable [13].

Data-driven models that are able to fulfill these requirements, especially the last one, are *Decision Trees*. Due to the decision rule-based structure, *Decision Trees* are easy to interpret. In addition, they are able to deal with missing values, measure the importance of the input variables and handle both numerical and categorical input variables [10]. *Decision Trees* can predict either a categorical (*Classification Trees*) or a numerical (*Regression Trees*) output variable.

© Springer Nature Switzerland AG 2020
C. Analide et al. (Eds.): IDEAL 2020, LNCS 12489, pp. 41–50, 2020.
https://doi.org/10.1007/978-3-030-62362-3_5

The structure of the *Decision Tree* is a hierarchical and it is represented as a directed acyclic graph that starts with a single root and ends in multiple leaves containing a local model. Each local model is only valid in a certain partition of the input space and is connected to the root by an unique decision rule. Most *Decision Trees* are binary trees and to generate them, a set of labeled samples is recursively split into partitions until a stopping criterion is reached. The tree's size is controlled by a pruning method. Depending on the M input variables $\mathbf{x} = [x_1, x_2, \ldots, x_M]$ of a sample, the set of samples is split by a divide-and-conquer strategy, which is specific for the algorithm that generates the tree. Typically, univariate splits $x_m \leq c$ for $c \in \mathbb{R}$ or $x_m \in \mathcal{B}$ for $\mathcal{B} \subset \mathcal{A}$ are used to divide the input space axis orthogonally by a single input variable x_m with $m \in \mathbb{N} \mid 1 \leq m \leq M$. For calculating a suitable threshold value c or subset of categorical values $\mathcal{B}$ from all possible categorical characteristics $\mathcal{A}$ of x_m, different impurity measurements are used [7].

Nevertheless, there is potential for improvement. Especially for *Regression Trees*, an approach to increase the model flexibility and to overcome the curse of dimensionality is necessary [10]. Because of the axis orthogonal splits, the model flexibility is limited. Depending on the process function that is modeled, this leads to lower prediction accuracy and due to larger trees to less interpretability [2]. To increase the model flexibility and reduce the tree size, multivariate splits $\sum_{m=1}^{M} \beta_m x_m \leq c$ can be used to construct an *Oblique Regression Tree* or, if both multi- and univariate splits are used, a *Mixed Regression Tree* [7]. The efficiency of multivariate split selection depends on the method used to determine the direction of the split. For partitioning the input space into suitable local areas, split directions orthogonally to the function gradient should be chosen [7,11]. Especially for high dimensional or noisy data sets, this is a challenging task.

A well-known approach for multivariate split selection is the *Linear Combination Search Algorithm* that is implemented in *Classification And Regression Tree* (CART) [1]. The algorithm uses a heuristic-based selection method that leads to high computational costs. Another approach uses *Partial Least Squares Regression* (PLSR) for generating clusters and finding patterns [8]. For split selection, the first principal component of PLSR is used, which is a linear combination of M input variables. A third method uses the *Principal Hessian Directions* [4], which leads to a similar splitting criterion as PLSR.

In the following, a novel method for multivariate split selection in *Regression Trees* is presented. The aim is to improve the prediction accuracy of *Oblique Regression Trees* and *Mixed Regression Trees* while maintaining as much interpretability as possible. To achieve this, a *Least Squares* (LS) approach is used, which is implemented into CART. Compared to the previously described approaches, a generalized and more interpretable split can be generated by limiting the LS approach to significant input variables. The resulting tree is called *Least Squares Regression Tree* (LSRT), which is further described in Sect. 2. In Sect. 3, LSRT is tested and analyzed with both synthetic and real-world data sets and compared with the conventional univariate CART. Finally, Sect. 4 summarizes the results and an overview on further research is given.

2   Method

To determine a multivariate split whose direction adapts to the function gradient $\nabla f(\mathbf{x})$ in a partition, the function $f(\mathbf{x})$ is linearly approximated in this area by a first-order LS model $\hat{y}(\mathbf{x})$. For splitting the partition, a constant output value of $\hat{y}(\mathbf{x})$ is chosen. In this way, a splitting direction orthogonal to the gradient $\nabla \hat{y}(\mathbf{x})$ is achieved. If the nonlinearity in a considered area is not excessive, $\nabla \hat{y}(\mathbf{x})$ is similar to $\nabla f(\mathbf{x})$ and a suitable splitting direction for the process function is given. A more detailed overview is presented in Sect. 2.1.

To construct the *Oblique* or *Mixed Regression Tree* LSRT, the generalization and interpretability of the multivariate split is improved by a forward selection method and implemented into CART, which is further explained in Sect. 2.2.

2.1   Least Squares Approach

The aim of the LS approach is to determine a multivariate split direction that should be as orthogonal as possible to the function gradient $\nabla f(\mathbf{x})$. To achieve this, the linear affine LS model

$$\hat{y}(\mathbf{x}) = \beta_0 + \sum_{m=1}^{M} \beta_m x_m \tag{1}$$

is estimated. The coefficients are calculated by using

$$\boldsymbol{\beta} = [\beta_0, \beta_1, \ldots \beta_M] = (\mathbf{X}^T\mathbf{X})^{-1}\mathbf{X}^T\mathbf{y} \tag{2}$$

with the extended $N \times (1 + M)$ predictor matrix

$$\mathbf{X} = \begin{pmatrix} 1 & \mathbf{x}_1 \\ 1 & \mathbf{x}_2 \\ \vdots & \vdots \\ 1 & \mathbf{x}_N \end{pmatrix} = \begin{pmatrix} 1 & x_{1,1} & x_{1,2} & \cdots & x_{1,M} \\ 1 & x_{2,1} & x_{2,2} & \cdots & x_{2,M} \\ \vdots & \vdots & \vdots & \ddots & \vdots \\ 1 & x_{N,1} & x_{N,2} & \cdots & x_{N,M} \end{pmatrix}, \tag{3}$$

that consists of N samples $\mathbf{x}_n = [x_{n,1}, x_{n,2}, \ldots, x_{n,M}]$ and an additional column of ones for the constant part β_0 in the linear affine model [11]. The vector $\mathbf{y} = [y_1, y_2, \ldots, y_N]^T \in \mathbb{R}^N$ contains the related labels. For a more detailed explanation, the LS approach is applied to an example function.

Figure 1a shows a linear LS model $\hat{y}(\mathbf{x})$ trained on the 20 white samples. The samples are generated by the function $f(\mathbf{x}) = x_1 x_2$ and to cover the whole input space, an optimized *Latin Hypercube Design* [3] was used. To get a constant model output $\hat{y}(\mathbf{x}) = \alpha$ of the value α, various input combinations are available. These input combinations result in a contour line, contour plane or contour hyperplane (depending on M), that runs along (1) through the input space and can be used as an oblique border of a multivariate split. The direction of the oblique border depends on the coefficients $[\beta_1, \beta_2, \ldots, \beta_M]$ and is orthogonal to $\nabla \hat{y}(\mathbf{x})$ that is passed through in the input space. This is illustrated in Fig. 1b,

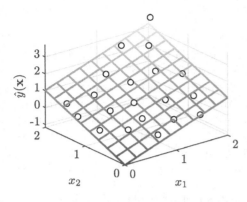
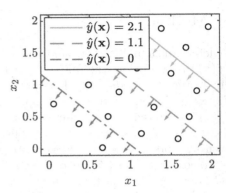

(a) Model output $\hat{y}(\mathbf{x})$ of a first-order LS model (colored grid) trained on the white samples generated from $f(\mathbf{x}) = x_1 x_2$.

(b) Three possible multivariate splits that result from the LS model and are orthogonal to the gradient $\nabla \hat{y}(\mathbf{x})$.

Fig. 1. Example of the LS approach in multivariate split selection.

which shows three possible oblique borders that result from $\hat{y}(\mathbf{x})$ in Fig. 1a. The oblique borders are contour lines of three different constant model outputs $\hat{y}(\mathbf{x}) = \alpha$ and divide the input space into two partitions. By adjusting α, the border is shifted parallel trough the input space. In this way, a suitable threshold value for splitting can be determined. Finally, based on the oblique border that results from (1) and by subtracting β_0 from the LS model, the multivariate split

$$\sum_{m=1}^{M} \beta_m x_m \leq \alpha - \beta_0 \qquad (4)$$

is created. To determine a suitable threshold value $c = \alpha - \beta_0$, the LS approach is implemented into CART, which is described next.

2.2 Multivariate Split Selection

In addition to the direction of the multivariate split, it is necessary to adjusting the threshold value c so that the impurity in the resulting partitions is reduced as much as possible. To determine c and construct a whole tree using multivariate split selection, the LS approach is implemented as an extension of CART.

The implementation leads to LSRT, which is binary structured and has constant output values in its leaves. Figure 2a shows the model structure of an *Oblique Regression Tree* LSRT, trained on the 20 samples of Fig. 1. The tree consists of three nodes t_1, t_3 and t_5 for splitting the input space axis oblique by the multivariate criterion next to the nodes. Furthermore, the tree consists of four leaves $\tilde{t}_2$, $\tilde{t}_4$, $\tilde{t}_6$ and $\tilde{t}_7$ containing a constant output value listed below the leaves. This value is only valid in a certain partition of the input space, which is shown in Fig. 2b. Each partition is modeled by a constant output value represented by the corresponding leaf. The gray contour lines of $f(\mathbf{x})$ indicate that

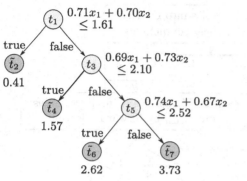

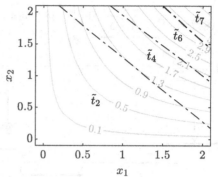

(a) Model structure of LSRT, in which the nodes t contain a multivariate splitting criterion (next to t) and the leaves $\tilde{t}$ contain a constant output value $\hat{y}(\mathbf{x})$ (below $\tilde{t}$).

(b) Partitioning of the input space by LSRT. Each partition has adapted to the gray contour lines of $f(\mathbf{x})$ and contains a locally valid model (represented by $\tilde{t}$).

Fig. 2. *Oblique Regression Tree* LSRT resulting from the samples shown in Fig. 1.

the partition borders (dot-dashed lines) are adapted to $\nabla f(\mathbf{x})$. In addition, areas with a higher gradient are modeled more detailed by using smaller partitions.

Figure 3 illustrates how LSRT is splitting a set of samples $\mathcal{D}_t = \{\mathbf{X}, \mathbf{y}\}$ in node t into two partitions. First, in step a) of Fig. 3 the predictor matrix is divided into a matrix $\mathbf{X}_{\text{num}}$ with numerical input variables and a matrix $\mathbf{X}_{\text{cat}}$ with categorical input variables. In order to overcome the curse of dimensionality and create a generalized and interpretable multivariate split, at step b) $\tilde{M} \leq M$ significant input variables are selected from $\mathbf{X}_{\text{num}}$. These variables are selected by a forward selection method and a chosen information criterion. In this paper, two different criteria are used. The bias-corrected *Akaike's Information Criterion*

$$\text{AIC}_C = N \log \frac{\text{RSS}}{N} + 2\tilde{M} + N + N \log(2\pi) + \frac{2(\tilde{M} + 2)(\tilde{M} + 3)}{N - (\tilde{M} + 2) - 1} \tag{5}$$

and the *Bayesian Information Criterion*

$$\text{BIC} = N \log \frac{\text{RSS}}{N} + \tilde{M} \log N + N + N \log(2\pi) \tag{6}$$

with the residual sum of squares RSS. These criteria are well established in model selection and take both the model accuracy and complexity into account. In (5) and (6), model accuracy is considered by the first term and complexity by the remaining terms. AIC_C differs from the uncorrected AIC and BIC by the additional bias-correction results from the last fraction, which should improve the performance for small data sets or high dimensional input spaces. In addition, by the second term of (5) and (6), BIC penalizes model complexity more than AIC_C, resulting in less complex models [9, 11].

If more than two input variables are selected, a multivariate split is performed. In this case, the selected variables are combined in $\tilde{\mathbf{X}}$ and used in

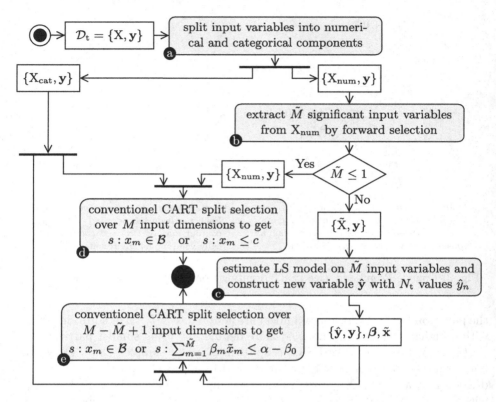

Fig. 3. Illustration of splitting a node t into two partitions using the algorithm that generates LSRT.

step c) to estimate $\hat{y}(\tilde{\mathbf{x}})$. Based on the set of samples $\tilde{X}$, the model is used to predict N_t new output values that construct a new variable for splitting $\hat{\mathbf{y}} = [\hat{y}(\tilde{\mathbf{x}}_1), \hat{y}(\tilde{\mathbf{x}}_2), \ldots, \hat{y}(\tilde{\mathbf{x}}_{N_t})]^T = [\hat{y}_1, \hat{y}_2, \ldots, \hat{y}_{N_t}]^T$. Finally, depending on the size of $\tilde{M}$, in step d) and e) the best multi- or univariate split s is chosen.

To determine s, both a suitable input variable and the corresponding c or $\mathcal{B}$ must be selected. Therefore, CART uses a brute force method in which all possible threshold values and subsets are analyzed over the entire input space. In case of multivariate splitting, $\hat{\mathbf{y}}$ is ordered and all discrete threshold values $c = \hat{y}_n + 0.5(\hat{y}_{n+1} - \hat{y}_n)$ between two values are checked. The brute force method leads to $N_t - 1$ possible splits for numerical variables and $2^{|\mathcal{A}|-1} - 1$ for categorical variables [10]. For all possible splits, the mean squared error MSE in each of the two partitions t_{left} and t_{right} is calculated. The split that maximizes

$$\Delta\text{MSE} = \text{MSE}(\mathbf{y}) - \frac{N_{\text{tleft}}}{N_{\text{t}}}\text{MSE}(\mathbf{y}_{\text{tleft}}) - \frac{N_{\text{tright}}}{N_{\text{t}}}\text{MSE}(\mathbf{y}_{\text{tright}}) \qquad (7)$$

is performed, where N_{tleft} and N_{tright} denote the number of samples in the corresponding partition [1]. In the following, LSRT is tested in an extensive experimental analysis.

3   Experimental Analysis

In order to analyze the performance of LSRT in comparison to the conventional univariate CART, in Sect. 3.1 synthetic data sets and in Sect. 3.2 real-world data sets are used. Furthermore, by using $\text{LSRT}_{\text{AICc}}$ and LSRT_{BIC}, the impact of two different information criteria is tested. To obtain meaningful results in accuracy and interpretability, the root mean squared error $\bar{E}$ and the model complexity $|\bar{T}|$ are averaged over 100 runs. The complexity $|\bar{T}|$ is measured by the number of nodes in the model and controlled by Breiman's *Minimal Cost Complexity Pruning* [1].

3.1   Synthetic Data Sets

To analyze the model behavior, a common test function from [5] is chosen and extended to

$$f(\mathbf{x}) = \sum_{i=1}^{I} \frac{10}{i^2}\sin(\pi x_{5i-4}x_{5i-3}) + \frac{20}{i^2}(x_{5i-2} - 0,5)^2 + \frac{10}{i^2}x_{5i-1} + \frac{5}{i^2}x_{5i} \quad (8)$$

with $\{\mathbf{x} \in \mathbb{R}^M \mid M = I \cdot 5\}$ to vary the dimensionality of $\mathbf{x}$ in discrete steps $I \in \mathbb{N}$ of five. For x_m the input space is limited to $0 \leq x_m \leq 1$. To construct N_{tr} samples for training, an optimized *Latin Hypercube Design* [3] is used to cover the whole input space. Furthermore, the impact of noise is analyzed either by adding a white Gaussian noise ε to $f(\mathbf{x})$ with a variance of $\sigma_\varepsilon^2 = 0.5$ and a mean of zero or by adding M_{n} ineffective noisy input variables. These ineffective noisy input variables are generated by white Gaussian noise and have no impact on $f(\mathbf{x})$. For testing, 1000 randomly generated samples are used.

Table 1 shows the experimental results for nine different synthetic data sets. It can be recognized that both $\text{LSRT}_{\text{AICc}}$ and LSRT_{BIC} are more accurate and less complex than CART. A comparison of $\text{LSRT}_{\text{AICc}}$ and CART shows that on average the error of $\text{LSRT}_{\text{AICc}}$ is 17% lower and its complexity is 44% lower. Especially if the number of samples is low or the dimensionality of the input space is high, which is caused by the interpolation capability of LS and the reduction to a single dimension. For $N_{\text{tr}} = 25$ the error of $\text{LSRT}_{\text{AICc}}$ is 20% lower and for $M = 15$ its complexity is 53% lower. Furthermore, it can be seen that the size of $\text{LSRT}_{\text{AICc}}$ is always smaller than the size of LSRT_{BIC}, which can be explained by the complexity penalty of BIC leading to less complex splits. For this reason, LSRT_{BIC} performs better under the impact of noise. In case of

Table 1. Experimental results for three different trees $LSRT_{AIC_C}$, $LSRT_{BIC}$ and CART using synthetic data sets. The elements in the brackets (left column) indicate the properties of the nine different synthetic data sets. The best results for model complexity $|\bar{T}|$ and test error $\bar{E}$ are in bold print.

Functionsettings	$LSRT_{AIC_C}$		$LSRT_{BIC}$		CART	
$\{N_{tr}, M, M_n, \sigma_\varepsilon^2\}$	$\|\bar{T}\|$	$\bar{E} \pm \sigma$	$\|\bar{T}\|$	$\bar{E} \pm \sigma$	$\|\bar{T}\|$	$\bar{E} \pm \sigma$
$\{25, 5, 0, 0\}$	**3.9**	**3.49 ± 0.21**	4.0	3.54 ± 0.21	6.0	4.32 ± 0.30
$\{50, 5, 0, 0\}$	**5.2**	**3.18 ± 0.24**	5.4	3.29 ± 0.23	8.9	3.88 ± 0.32
$\{100, 5, 0, 0\}$	**8.1**	**2.86 ± 0.18**	9.1	2.90 ± 0.21	17.4	3.36 ± 0.21
$\{200, 5, 0, 0\}$	**18.2**	2.58 ± 0.14	22.2	**2.56 ± 0.15**	28.6	3.08 ± 0.17
$\{300, 5, 0, 0\}$	**34.0**	2.39 ± 0.14	35.9	**2.36 ± 0.14**	44.3	2.89 ± 0.13
$\{300, 10, 0, 0\}$	**10.1**	**2.80 ± 0.11**	17.4	**2.80 ± 0.12**	20.1	3.39 ± 0.13
$\{300, 15, 0, 0\}$	**7.9**	**2.91 ± 0.09**	11.2	2.93 ± 0.10	17.0	3.48 ± 0.13
$\{300, 5, 5, 0\}$	**10.3**	2.67 ± 0.10	20.9	**2.53 ± 0.14**	22.0	3.04 ± 0.13
$\{300, 5, 0, 0.5\}$	**28.7**	2.47 ± 0.14	30.6	**2.42 ± 0.14**	38.8	2.96 ± 0.17

$M_n = 5$, $LSRT_{BIC}$ leads to a more generalized structure due to less complex splits. In addition, for a lower number of samples, $LSRT_{AIC_C}$ performs better than $LSRT_{BIC}$. This is caused by the bias-correction of AIC_C. After the method is analyzed using synthetic data sets, the experimental analysis is continued on real-world data sets.

3.2   Real-World Data Sets

In contrast to synthetic data sets, real-word data sets are usually more difficult to handle because their properties cannot be varied and the data sets may contain outliers and incomplete samples. For the following analysis, four different real-world data sets are used, which are common when comparing decision tree algorithms [6,12,14–16]. These data sets contain from 209 to 1599 samples and 6 to 24 input variables. To achieve comparable results to Sect. 3.1, the data sets are limited to numerical input variables. Furthermore, the input variables of the Tecator data set are reduced by the untransformed absorbance spectrum. Due to the small sample size of Tecator ($N = 240$) and CPU-Performance ($N = 209$), k-fold cross-validation was used. The results are determined by averaging the model complexity and predicting the entire data set using the k models. For Tecator and CPU-Performance a size of $k = 10$, for Boston Housing of $k = 5$ and for Redwine Quality of $k = 2$ was selected.

The experimental results using the four real-world data sets are shown in Table 2. Similar to the previous results, $LSRT_{AIC_C}$ and $LSRT_{BIC}$ are always much smaller and more accurate than CART. Over all four data sets, $LSRT_{AIC_C}$ is on average 35% less complex and 21% more accurate than CART. Especially for the small data sets Tecator and CPU-Performance, a significant improvement in prediction accuracy is achieved. For Tecator, the prediction accuracy of

$\text{LSRT}_{\text{AIC}_{\text{C}}}$ is 39% higher than the accuracy of CART. Due to Tecators dimension of $M = 24$ this indicates that the curse of dimensionality is weakened. Furthermore, because of the bias-correction of AIC_{C}, $\text{LSRT}_{\text{AIC}_{\text{C}}}$ achieves higher accuracy than LSRT_{BIC} for the small data sets. In summary, both the model behavior of LSRT and the impact of AIC_{C} and BIC are comparable to the observations in Sect. 3.1.

Table 2. Experimental results for three different trees $\text{LSRT}_{\text{AIC}_{\text{C}}}$, LSRT_{BIC} and CART using four real-world data sets.

Data sets	$\text{LSRT}_{\text{AIC}_{\text{C}}}$		LSRT_{BIC}		CART	
	$|\bar{T}|$	$\bar{E} \pm \sigma$	$|\bar{T}|$	$\bar{E} \pm \sigma$	$|\bar{T}|$	$\bar{E} \pm \sigma$
Tecator	38.5	**1.24** ± 0.07	**37.6**	1.30 ± 0.07	54.8	2.02 ± 0.11
Boston Housing	**10.5**	4.74 ± 0.20	11.8	**4.70** ± 0.21	15.4	4.97 ± 0.27
Redwine Quality	**5.1**	**0.68** ± 0.01	5.2	0.69 ± 0.01	9.8	0.71 ± 0.01
CPU-Performance	8.8	**68.96** ± 7.96	**8.3**	73.24 ± 9.15	11.9	90.24 ± 7.40

4   Conclusion

To improve the performance of *Regression Trees* by using multivariate splits, the direction of these splits needs to adapt to the process function gradient in the corresponding partition, even if the input space is high dimensional or noise effects occur. In this paper, a novel approach for determining the direction of a multivariate split was presented. The direction is determined by a first-order LS model estimated with a forward selection method. In this way, the direction adapts to the function gradient in a local area, high dimensional data can be handled and as much interpretability as possible is maintained. In addition, due to the forward selection method, a generalized split is achieved. The approach was implemented as an extension of CART, which results in the *Oblique Regression Tree* or *Mixed Regression Tree* LSRT. Finally, in an extensive experimental analysis, LSRT was tested on synthetic and real-world data sets and compared with univariate CART. The experimental analysis has shown that the approach improves the prediction accuracy of CART significantly and leads to much smaller trees, even if the dimension of the input space increases or noise effects occur. Furthermore, to obtain a generalized model in the case of small data sets or noise effects, different information criteria of the forward selection method can be used. In conclusion, the presented method is an efficient approach for multivariate split selection with respect to the analyzed data sets.

For further research, the LS approach should be implemented in a tree construction algorithm that contains more complex local models. Instead of the forward selection method, other methods could be used to improve generalization such as *Ridge Regression* or *Lasso Regression*. Moreover, the LS model that determines the split direction could be extended to a nonlinear LS model. This could lead to better results for processes with high nonlinearity. Furthermore, it should be investigated how categorical input variables can be integrated in an interpretable way into the multivariate split.

Acknowledgements. This work was supported by the EFRE-NRW funding programme"Forschungsinfrastrukturen" (grant no. 34.EFRE–0300180).

References

1. Breiman, L., Friedman, J., Stone, C.J., Olshen, R.: Classification and Regression Trees. Chapman and Hall/CRC, New York (1984)
2. Brodley, C.E., Utgoff, P.E.: Multivariate decision trees. Mach. Learn. **19**(1), 45–77 (1995)
3. Ebert, T., Fischer, T., Belz, J., Heinz, T.O., Kampmann, G., Nelles, O.: Extended deterministic local search algorithm for maximin latin hypercube designs. In: IEEE Symposium Series on Computational Intelligence, pp. 375–382 (2015)
4. Eriksson, L., Trygg, J., Wold, S.: PLS-trees® a top-down clustering approach. J. Chemometr. **23**, 569–580 (2009)
5. Friedman, J.H., Grosse, E., Stuetzle, W.: Multidimensional additive spline approximation. SIAM J. Sci. Stat. Comput. **4**(2), 291–301 (1983)
6. Gijsbers, P.: OpenML wine-quality-red. https://www.openml.org/d/40691. Accessed 21 May 2020
7. Evolutionary Decision Trees in Large-Scale Data Mining. SBD, vol. 59. Springer, Cham (2019). https://doi.org/10.1007/978-3-030-21851-5_8
8. Li, K.C., Lue, H.H., Chen, C.H.: Interactive tree-structured regression via principal hessian directions. J. Am. Stat. Assoc. **95**, 547–560 (2000)
9. Lindsey, C., Sheather, S.: Variable selection in linear regression. Stata J. **10**(4), 650–669 (2010)
10. Loh, W.Y.: Fifty years of classification and regression trees. Int. Stat. Rev. **82**, 329–348 (2014)
11. Nelles, O.: Nonlinear System Identification. Springer, Berlin Heidelberg (2001)
12. van Rijn, J.: OpenML machine_cpu. https://www.openml.org/d/230. Accessed 21 May 2020
13. Shang, C., You, F.: Data analytics and machine learning for smart process manufacturing: recent advances and perspectives in the big data era. Engineering **5**(6), 1010–1016 (2019)
14. Vanschoren, J.: OpenML boston. https://www.openml.org/d/531. Accessed 21 May 2020
15. Vanschoren, J.: OpenML tecator. https://www.openml.org/d/505. Accessed 21 May 2020
16. Vanschoren, J., van Rijn, J.N., Bischl, B., Torgo, L.: OpenML: networked science in machine learning. SIGKDD Explor. **15**(2), 49–60 (2013)

A Framework for the Multi-modal Analysis of Novel Behavior in Business Processes

Antonino Rullo[1]([✉]), Antonella Guzzo[1], Edoardo Serra[2], and Erika Tirrito[1]

[1] University of Calabria, Rende, Italy
n.rullo@dimes.unical.it, antonella.guzzo@unical.it,
erika.tirrito@hotmail.it
[2] Boise State University, Boise, ID, USA
edoardoserra@boisestate.edu

Abstract. Novelty detection refers to the task of finding observations that are new or unusual when compared to the 'known' behavior. Its practical and challenging nature has been proven in many application domains while in process mining field has very limited researched. In this paper we propose a framework for the multi-modal analysis of novel behavior in business processes. The framework exploits the potential of representation learning, and allows to look at the process from different perspectives besides that of the control flow. Experiments on a real-world dataset confirm the quality of our proposal.

Keywords: Process mining · Multi-modality · Novelty detection · Trace embedding

1 Introduction

Process mining aims at extracting useful knowledge from real-world event data recording process executions. Complex processes involve different subjects, resources and activities, while the rate and duration of their executions can vary according to several (possibly unknown) factors. Thus, heterogeneity is a key aspect of complex processes, and it characterizes both single process execution and the process as a whole itself. In fact, the data generated by a process execution (a.k.a. *trace*) are different in nature, and traces of the same process can be very different from each other. In the last decade, researchers have spent great effort in proposing process mining techniques that explain process behavior. To this end, the idea of *concept drift* [2] has been introduced into the process mining context to enable the analysis of changes in business processes *over time*, with the major focus on the *control-flow* perspective, i.e. which activities are executed in the process traces and in which order. However, the process behavior can also be analysed along different dimensions/perspectives, besides that of the control-flow. The main insights refer to: *organizational* perspective, focusing on the resources, i.e. which actors, systems, roles, departments, etc. are

© Springer Nature Switzerland AG 2020
C. Analide et al. (Eds.): IDEAL 2020, LNCS 12489, pp. 51–63, 2020.
https://doi.org/10.1007/978-3-030-62362-3_6

involved in the process and how they are related; *data* perspective, focusing on the data used and generated during the process execution, i.e. traces related attributes, and event related attributes; *time* perspective, focusing on the relation between processes traces/data and time, e.g. how traces are distributed in the time domain, how much activity occurrences are time dependent, etc. Furthermore, drift detection tasks restrict the analysis of changes along the time dimension only, leaving aside changes that happen in different feature domains. It follows that to accomplish a wider analysis of processes a more flexible approach is needed. To this end we move in the domain of the *novelty detection*, i.e. the task of finding observations that are new or unusual when compared with a *known behaviour*. In process mining, this translates into identifying how processes executions changes with respect to a reference data. In this regard, we propose a novelty detection framework that exploits the potential of multi-modal representation learning in detecting novel process behavior. We show how to extend known embedding techniques to work in a multi-perspective fashion, in particular we obtain the embeddings in an unsupervised way by automatically integrating the multi-modal aspect of the traces. The analysis gains quality by exploiting knowledge from all the perspectives, and, beside that of time, any of the process features can be adopted as reference axis along which looking for process changes. To the best of our knowledge, the multi-modal detection of novel processes behavior has never been applied to process mining tasks. As the first to deal with this topic, we emphasize the applicability of our approach that is supported by the results obtained from its application to a real word dataset.

2   Preliminaries

Event Log. Process-oriented commercial systems usually store information about process enactments by tracing the events related to their executions in the form of an *event log*. By abstracting from the specificity of the various systems, we can view a log $\mathcal{L}$ as a set of execution traces $\{T_1, \ldots, T_m\}$ of a specific process, with $m > 0$. An execution trace can be represented as a pair (A, S) where $A = \{ta_1, \ldots, ta_z\}$ is a set of trace attributes describing the global properties of the trace (e.g. identifier, beginning time), and $S = <e^1, \ldots, e^n>$ is a temporal ordered sequence of events mapping the execution of activities performed in the process with their attributes. Specifically, each event $e^i \in S$ is characterized by a timestamp t^i and a set of event attributes $\{ea_1^i, \ldots, ea_l^i\}$ describing the intrinsic properties of the event (e.g. activity's name, performing resource). Given two events e^i and e^j with $i > j$, we have that $t^i \geq t^j$.

Trace Embedding. Due to the heterogeneity and the high-dimensionality of the information stored in the process traces, process mining algorithms can be hardly effective if directly applied to the raw data. It comes therefore that a crucial task in these applications is to embed the input data into a suitable vectorial representation that can be more conveniently provided to the mining algorithms. We refer to this kind of approaches as *trace embedding* methods.

The learning task applied to vectorial representations is called *representation learning*. Actually, most of the representation learning methods proposed for business processes come in the form of feature selection algorithms which look at a single perspective at a time. These approaches are ineffective to deal with high-dimensional and multivariate data [7]. To overcome these drawbacks, a method called *Trace2Vec* [3] has been recently proposed. *Trace2Vec* does not focus on the extraction of specific features, but rather computes an embedding of the input data into a vectorial space. This method deals with the control-flow perspective only, however, the authors left open to develop embedding methods that simultaneously take into account the various process perspectives. Besides *Trace2Vec*, other techniques have been proposed for the vectorial represenation of process traces. The authors of [4] proposed the usage of *n-grams* for a trace classification task. In particular, traces are converted in 2-grams vectors, so as 2-grams are the features representing event orders in each trace, and used as input to a decision tree algorithm for trace classification.

Multi-modality. With the term *multi-modal analysis* we refer to an analysis performed in a multi-perspective fashion. Whereas initially the primary focus of the process mining research was on the control-flow perspective of a process (i.e. to the formal specification of the dependencies that hold over the events of a process), now most of the research has profitably enriched by looking at other perspectives, such as time (i.e., timing and frequency of events), resources (i.e., people, systems, and roles) and data involved in the execution (i.e. cases). These perspectives offer the analyst a deeper insight of the process as a whole. For instance, the set of process executions being analyzed may present the same sequence of events (control-flow perspective) but with different duration (data perspective). This understanding can be only acquired when working in a multi-modal fashion. To implement a representation learning task in a multi-modal fashion, besides the sequence of events, trace attributes, event attributes, and a set of derived attributes of interest must be encoded into the numerical vector as well. More formally, for each trace $T_j = (A, S) \in \mathcal{L}$, the features to consider are:

- the set of trace's attributes $A = \{ta_1, \ldots, ta_z\}$;
- for each event $e^i \in S$, the set of event's attributes $\{ea_1^i, \ldots, ea_l^i\}$;
- a set of derived attributes. These attributes may depend on the specific process domain and include, but are not limited to:
  - trace duration as the difference between the timestamps of the last and the first event;
  - number of events per trace;
  - elapsed time between two events: specific pairs of activities may be of particular interest for the analyst who may want to monitor how the temporal relation between them varies over time;
  - the relation between a specific activity attribute and the time the activity occurs (e.g. hour of the day, day of the week, month, etc.);
  - the time instant (e.g. hour of the day, day of the week, etc.) at which traces start and/or end;
  - the average elapsed time between any two consecutive activities.

3   The Framework

The proposed analysis framework is depicted in Fig. 1. It consists of six steps, each denoted with a red circle. In the first step two main data categories are selected from the event log, namely *reference data*, used as a reference to learn normal situations, and *analysis data*. The objective of the framework is to determine to what extent each process execution that belongs to the analysis data can be considered novel, i.e. different from the process executions of the reference data. The selection is performed according to a specific dimension, which depends on the objective of the analysis. For instance, if we were interested in finding the process changes between two time periods, then we would select the two data categories on the basis of the trace timestamps, so as the two dataset contain traces belonging to different time ranges; on the contrary, if we wanted to determine the differences between process executions bootstrapped by two different operators, then we would differentiate data based on the feature which identifies the operators. In the second step, the trace embeddings of both the reference and the analysis data are generated. In the third and fourth steps, the reference trace embeddings are used to train a novelty detection model and to score the traces of the analysis data on the basis of their novelty degree w.r.t. the reference data, respectively. Known novelty detection algorithms are Gaussian Mixture Model, One Class Support Vector Machine, Isolation Forest, and K-Nearest Neighbors, to cite a few. These approaches build a model representing normal behavior in a semi-supervised way, i.e. from a given labeled training data set (third step), and then compute the likelihood of a unlabeled data instance to be generated by the learnt model (fourth step). In the fifth step the analysis data scores are plotted in a specific feature space (which depends on the objective's analysis) in order to detect anomalous traces. Such traces may represent temporary process malfunctions or evolution anomalies that can be excluded from the analysis and/or analyzed separately. In the last step, analysis tools analyze the traces and find the properties that justify the different scores.

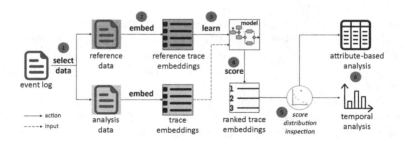

Fig. 1. The proposed framework.

3.1 Multi-modal Trace Embedding

In this Section we show how to implement a multi-modal trace embedding with the *n-grams* and *Trace2Vec* techniques. In both approaches a numerical vector is obtained as the concatenation of the vector emb^e inferred from the sequence of events' attributes, and the vector emb^t inferred from the set of trace's attributes. As the difference between the two techniques is in the computation of emb^e, we first show how to compute emb^t, and then we explain how to implement emb^e for each approach.

emb^t. Numerical trace's attributes values are appended to emb^t as they are. Since machine learning algorithms cannot work directly with nominal data, we do need to perform some preprocessing in order to encode nominal and boolean attributes values into numerical ones. To this end we apply the One-Hot-Encoding technique, that is, we transform the attribute into m binary features which can only contain a value of 1 or 0. Each observation in the categorical feature is thus converted into a vector of size m with only one of the values as 1. This procedure is extended to derived attributes too.

emb^e **with n-grams.** The procedure by which we compute the set of emb^e is the following. For each trace $T^k = (A, S) \in \mathcal{L}$ we partition $S = < e^1, \ldots, e^n >$ vertically, i.e. we extract a sequence EA_i^k of attribute values for each event's attribute ea_i, such that EA_i^k contains all the values of the attribute ea_i for each event $e^j \in S$. Then, for a given value of n we compute the n-grams on the set of all EA_i^k that contain nominal or boolean values, for each event's attribute ea_i. Then, for each trace T^k we count the occurrences o of each n-gram contained in the sequences EA_i^k, and append o to emb^e. Finally, for each numerical attribute, we aggregate EA_i^k values according to a specific aggregation function agg_i (e.g. $sum()$, $max()$, $min()$, $avg()$, etc.), which may vary according to the attribute and may depend on the attribute nature, then we append the result to emb^e. A graphical example of this procedure is sketched in Fig. 2, which depicts a trace expressed in a XES-like format on the left, and the EA_i computed over the set of event attributes, along with the associated 2-grams and the resulting embedding emb^e on the right.

emb^e **with Trace2Vec.** We extend the implementation provided by the authors of *Trace2Vec* [3]. This implementation is based on the Doc2Vec approach which extends the Continuous Bag of Words (CBOW) architecture with a paragraph vector [5]. CBOW is a neural network architecture based on the principle that a word can be predicted from its context (i.e. the words appearing before and after the focus word). In the *Trace2Vec* approach traces are considered as sentences and events as words. Given that *Trace2Vec* includes a representation of traces (based on the trace identifier), it allows for joint learning of representations of activities and traces. A document is represented as a list of traces, which in turn, are represented as lists of events' names, each encoded with the *unicode* encoding. Then, a model of the document is learned using the Doc2Vec implementation of the Gensim library, which generates a vectorial representation for each trace of the document. However, this implementation does not take into account any event's attribute except one, namely the event's activity. We extend *Trace2Vec* in

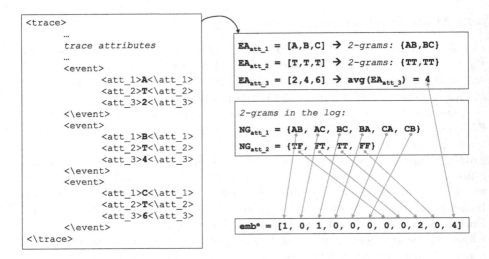

Fig. 2. Generation of emb^e with 2-grams: att_1, att_2, and att_3 are nominal, boolean, and numerical attributes, respectively.

such a way that it can work in a muti-modal fashion. In particular, we extends the list a of events' activities with the list of nominal and boolean attribute values. Such lists are extracted from the sequence of events S as for the *n-gram* based technique described above. Then, for each list of numerical values, we aggregate such values according to a specific aggregation function, and we append the result to a. Finally, a document with a list a for each trace is generated, and the vectorial representations emb^e are produced as the result of the learning task.

3.2 Score Distribution Inspection

This preliminary analysis allows to detect observations that diverge from the overall pattern on the analysis data. Such observations are *anomalous* (i.e. rare) process executions. The term anomalous denotes an exception with respect to the context, i.e. traces considerably different from neighboring ones, with the distance between traces computed on the basis of their novelty degree. Based on the objective's analysis and on the business scenario, one can handle anomalous traces in different ways. If the use of the framework is intended to study the progress of the process as a whole, then anomalous traces can be considered as *outliers* and filtered out in order to accomplish a more realistic analysis of the process evolution. Typically outliers are temporary malfunctions or evolution anomalies, i.e. process executions that did not contribute to the evolution of the process itself. Nevertheless, the understanding of anomalies may also help recognizing critical points in the process that could yield invalid or inappropriate behavior. In this case, anomalies can be considered as exceptional behaviour, can be kept together with in-cluster traces for subsequent analyses, and additionally, analyzed separately in order to understand the causes that led to a such behavior.

As results of the novelty detection task we have that each trace is associated with a score measuring its novelty degree with respect to the reference data. Given a certain feature of interest (e.g. time, see Fig. 3 for an example), we plot the obtained scores in the feature space; then, we detect the anomalies as the points/traces that are not included in any of the clusters formed by the plotted data points (if any).

3.3   Analyses

Attribute-based Analysis. In this analysis, we identify the causes of the changes in the process by looking at the attribute values of the most novel traces. Traces are ranked according to the obtained score, and the top k traces (i.e. the most novel ones) are compared with the traces used to train the model. In particular, a statistical analysis is conducted on the trace attributes, the event attributes, and the derived attributes values, by which we observe how frequent the attribute values are in the top k traces compared to the reference ones.

Temporal Analysis. This analysis shows how the process behaved over time. A first investigation an analyst may be interested in, is *how much* process executions varied over time, i.e. the novelty trend within a given time range. This can be achieved by plotting the trace scores over a time line, where each time instant denotes the starting or ending time of a trace. Considered that more than one trace may be associated to the same time instant, scores must be averaged so as to associate a single score to each time point. This way we are provided with a broad view of the novelty trend, and we can verify whether the process change was actually time dependent, or rather time and novelty are totally uncorrelated.

4   Experimental Results

In this Section we show an application of the proposed framework to a real case scenario. We first provide a broad view of the dataset used in our experimental study, then we show how to accomplish the six steps of the proposed framework.

4.1   Dataset

To show the capabilities of the proposed framework we adopted the *BPI Challenge 2018* dataset [11], a real-life event log over a period of three years (from year 2015 to year 2017) which covers the handling of applications for European Union direct payments for german farmers from the European Agricultural Guarantee Fund. In [9] a through analysis over the data and the process has been performed, thus we are aware of the kind of process changes across the years that we assume as ground truth. We performed a preprocessing to filter out attributes useless for the learning process. In particular we omitted all attributes with identifiers, since their values are unique for every activity, the year attribute, as the

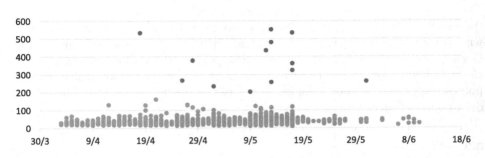

Fig. 3. Score distribution analysis: red points denote anomalous traces.

process changes every year and thus it would bias the model, and a set of one-value attributes, as their presence is pointless because they would not affect the model. In the resulting dataset, the set of attributes consists in 3 event nominal attributes and 7 trace attributes.

4.2 Framework Implementaion

1 - Select data. We reproduced the case of an analyst who wants to find out how the process evolved in the second year (2016) compared to the first year (2015) of its life cycle. To this end, we selected the subset of traces spanning from 2015 to 2016, omitting the data that belong to year 2017. Then we split the resulting dataset along the time dimension, i.e. we separated traces that start in year 2015 from traces that start in year 2016, and used these two dataset as the reference data and the analysis data, respectively. Both dataset contains 15000 traces circa. In the rest of this Section we denote these two dataset as to $BPIC\_15$ and $BPIC\_16$.

2 - Embed. First, we computed two derived attributes: `trace_duration`, as the difference between the timestamps of the last and the first event of a trace; and `events_per_trace`, as the length of the sequence of events S. Then, we used the n-grams-based technique described in Sect. 3.1 for generating the vectorial representations of both $BPIC\_15$ and $BPIC\_16$, with $n = 3$.

3 - Learn. We adopted the K-Nearest Neighbors as novelty detection algorithm. We applied a 10-fold cross-validation on the training data, where 90% of $BPIC\_15$ was used to train the model and the remaining 10% was merged with $BPIC\_16$ for comparison purposes (more details in step **6.b - Temporal Analysis** at the end of this Section).

4 - Score. The obtained model was used to score the $BPIC\_16$ traces plus the 10% of $BPIC\_15$ traces excluded from the training task. Traces were scored according to their novelty degree with respect to the reference traces, such that higher score were assigned to most novel traces.

5 - Score Distribution Inspection. We plotted the scores of $BPIC\_16$ traces as explained in Sect. 3.2, the result is shown in Fig. 3. We can notice that the

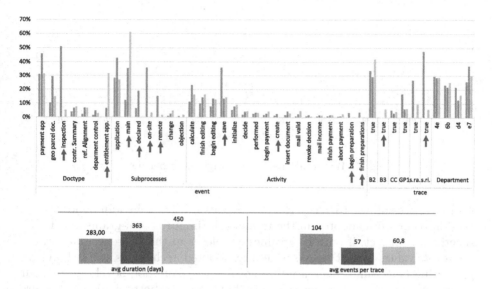

Fig. 4. Attribute-based analysis: event and trace attributes (top), derived attributes (bottom).

majority of traces forms a cluster of points between values 0 and 200. Traces with score equal or greater than 200 (13 traces in total) were considered as outliers and excluded from the analysis.

6.a - Attribute-based Analysis. We ranked the traces of $BPIC\_16$ on a score basis, and we compared the 1000 traces with higher score (i.e. the 1000 most novel traces) with those of $BPIC\_15$ (the reference data). Then, for each nominal and boolean attribute, we computed the percentage by which each of the trace and event attribute values appeared in both the $BPIC\_16$ top-1000 and the $BPIC\_15$ traces. The resulting plot is shown in Fig. 4 (top). We can notice that there are some attribute values for which the difference between the two dataset is marked more with respect to other values. Such attribute values are pointed out with red arrows. In particular, certain values are much more present in $BPIC\_16$ traces, as for the value `inspection` of the event attribute `Doctype`; while for certain other values it holds the contrary, as for the value `main` of the event attribute `Subprocess`. Some values, instead, are present *only* in one dataset, as for the trace attribute B3, which is `true` in $BPIC\_15$ traces, and `false` in $BPIC\_16$. Finally, there are attributes that do not affect process changes, as the trace attribute `Department` where the percentages are similar for all its values. For numerical attributes, instead, we plotted the attribute values, as for the two derived attributes `trace_duration` and `events_per_trace` shown in Fig. 4 (bottom). Besides the $BPIC\_16$ top-1000 and the $BPIC\_15$ traces, we also plotted the $BPIC\_16$ bottom-1000 traces, i.e. the 1000 traces more similar to the reference data. This gives an extent about the maximum gap between the traces belonging to the analysis data. Furthermore, looking at the

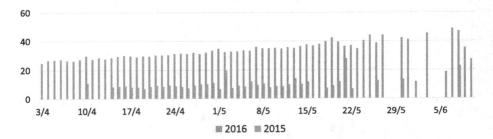

Fig. 5. Temporal analysis: average score per day.

bottom-k traces allows to verify the effectiveness of the embedding and novelty detection techniques adopted in the framework (in our case n-grams and KNN, respectively). In fact, if bottom-k values are closer to the reference data values than top-k values are, it means that the adopted techniques worked well in representing traces and differentiating novel traces from non-novel ones. In our experiments, in the vast majority of attributes values, bottom-1000 traces are closer to $BPIC\_15$ traces than $BPIC\_16$ traces are, which means that n-grams and KNN techniques are a good combination for this specific scenario.

6.b - Temporal Analysis. Scores of $BPIC\_16$ traces were plotted on the time axis in order to look at the novelty trend within year 2016. Figure 5 reveals that the novelty is time dependent, in particular that the novelty degree grows over time in a linear fashion, which means that the most novel traces occurred at the end of the time range taken into account. This first result, crossed with those obtained in the attribute-based analysis, suggests that the process has been made more efficient over time. In fact, Fig. 4 (bottom) shows that the average trace duration gradually decreases from year 2015 to the end of year 2016, and that the number of events per trace increases. Scores of the 10% $BPIC\_15$ traces excluded from the training task have been plotted too, in order to provide an extent of the gap between $BPIC\_15$ and $BPIC\_16$ data. We can notice that these traces have considerable lower score than $BPIC\_16$, confirming that the traces are actually different, and that the KNN algorithm was able to differentiate well between the two classes.

4.3   Model Performances

Here we propose a further analysis to show the performances of different novelty detection algorithms when applied to vectorial representation of traces generated with the *3-grams* and *Trace2Vec* embedding techniques. In particular, we evaluated the Gaussian Mixture Model, the One Class Support Vector Machine, the Isolation Forest, and the K-Nearest Neighbors algorithms. To measure the performances of the different models we used two metrics, namely the Area Under the Receiver Operating Characteristic Curve (AUROC) and the Average Precision (AP), the latter computed for both known data and unknown data. The analysis is to validate the choice of the novelty detection algorithm

Table 1. Comparison results.

	3-grams			Trace2Vec		
	AUROC	AP 15	AP 16	AUROC	AP 15	AP 16
GMM	0,856	0,910	0,926	0,769	0,842	0,818
OC_SVM	0,710	0,693	0,623	0,801	0,843	0,740
KNN	**0,977**	**0,986**	**0,943**	0,911	0,938	0,860
IF	0,763	0,943	0,882	0,679	0,868	0,741

and of the embedding technique used in our experiments. We applied a 10-folds cross-validation on $BPIC\_15$ traces, and we validated each model on $BPIC\_16$ traces. The Table 1 below shows the comparison results among the different novelty detection algorithms and the two multi-modal embedding techniques. All the results in the table are the average of the metrics measured for each fold. It is evident that KNN combined with 3-grams outperforms all the other cases. This further validates the choice of these techniques in our experiments.

5 Related Works

The practical importance and challenging nature of novelty detection has been evidenced in various application domains, such as faults and failure detection, health care, image processing and text mining. However, in the process mining field, novelty detection has not been investigated at all, while most of the research effort has been spent on two learning problems, namely anomaly behavior detection and concept drift. In [10] the authors analyze temporal behavior of individual activities to identify anomalies in single process instances. However, malicious users can split an attack on different process executions, thus an anomaly detection approach able to consider multiple process instances is required. To this end, Böhmer and Rinderle-Ma [1] proposed an unsupervised anomaly detection heuristic that exploits the temporal dependencies between multiple instances. This approach must rely to a model built from anomaly-free data, which is unlikely in real case scenarios. To overcome this issue, Nolle et al. [8] proposed a system relying on the autoencoder neural network technology, that is able to deal with the noise in the event log and learn a representation of the underlying model. Business process drift detection has been introduced in [2], where a method to detect process drifts based on statistical testing over feature vectors is proposed, and challenges and main learning problems in handling concept drifts have been identified. Maaradji et al. [6] presented a framework for detecting a rich set of process drifts based on tracking behavioral relations over time using statistical tests. However, in practice the existence of different types of drifts in a business process is not known beforehand. The authors of [12] addressed this issue by clustering declarative process constraints discovered from event logs based on their similarity, and then applying change point detection on the identified clusters to detect drifts. In drift detection literature, drifts are identified by looking for changes in the control flow that happen over time. Our analysis framework differs from these approaches in two main aspects: first, it is not able to explain the drift as a change in the relationships between activities, rather it identifies which activities occur more in novel traces w.r.t. the reference ones; second, it allows to analyze the process behavior from different perspectives (besides that of the control-flow), and the detection of changes are not limited to the time domain but can be extended to other feature spaces.

6   Conclusions

In the context of process changes analysis, the literature provides anomaly detection and drift detection techniques that mainly allow to find changes affecting the control flow over time. With the purpose of providing a wider analysis of processes, in this paper we proposed a framework for the multi-modal analysis of novel behavior. The framework presents two main strengths: first, it allows to look for changes in different feature spaces beside that of the process activities (control flow); second, beside that of time, any of the process features can be adopted as reference axis along which looking for process changes. As our framework exploits the potential of representation learning, in addition we provided the guidelines for extending two known embedding techniques, namely n-grams and Trace2Vec, to work in a multi-modal fashion. The overall approach was evaluated over a real world dataset from the BPI challenge 2018, revealing that our methods is effective in evidencing novel process behaviour from different perspectives w.r.t. that observed in a reference dataset.

References

1. Böhmer, K., Rinderle-Ma, S.: Multi instance anomaly detection in business process executions. In: Carmona, J., Engels, G., Kumar, A. (eds.) BPM 2017. LNCS, vol. 10445, pp. 77–93. Springer, Cham (2017). https://doi.org/10.1007/978-3-319-65000-5_5
2. Bose, R.P.J.C., van der Aalst, W.M.P., Žliobaitė, I., Pechenizkiy, M.: Handling concept drift in process mining. In: Mouratidis, H., Rolland, C. (eds.) CAiSE 2011. LNCS, vol. 6741, pp. 391–405. Springer, Heidelberg (2011). https://doi.org/10.1007/978-3-642-21640-4_30
3. De Koninck, P., vanden Broucke, S., De Weerdt, J.: act2vec, trace2vec, log2vec, and model2vec: representation learning for business processes. In: Weske, M., Montali, M., Weber, I., vom Brocke, J. (eds.) BPM 2018. LNCS, vol. 11080, pp. 305–321. Springer, Cham (2018). https://doi.org/10.1007/978-3-319-98648-7_18
4. Horita, H., Hirayama, H., Hayase, T., Tahara, Y., Ohsuga, A.: Business process verification and restructuring LTL formula based on machine learning approach. In: Lee, Roger (ed.) Computer and Information Science. SCI, vol. 656, pp. 89–102. Springer, Cham (2016). https://doi.org/10.1007/978-3-319-40171-3_7
5. Le, Q., Mikolov, T.: Distributed representations of sentences and documents. In: International conference on machine learning, pp. 1188–1196 (2014)

6. Maaradji, A., Dumas, M., La Rosa, M., Ostovar, A.: Detecting sudden and gradual drifts in business processes from execution traces. IEEE Trans. Knowl. Data Eng. **29**(10), 2140–2154 (2017)
7. Mhaskar, H., Poggio, T.A.: Deep vs. shallow networks : an approximation theory perspective. CoRR abs/1608.03287 (2016)
8. Nolle, T., Luettgen, S., Seeliger, A., Mühlhäuser, M.: Analyzing business process anomalies using autoencoders. Mach. Learn. **107**(11), 1875–1893 (2018). https:// doi.org/10.1007/s10994-018-5702-8
9. Pauwels, S., Calders, T.: Detecting and explaining drifts in yearly grant applications. arXiv preprint arXiv:1809.05650 (2018)
10. Rogge-Solti, A., Kasneci, G.: Temporal anomaly detection in business processes. In: Sadiq, S., Soffer, P., Völzer, H. (eds.) BPM 2014. LNCS, vol. 8659, pp. 234–249. Springer, Cham (2014). https://doi.org/10.1007/978-3-319-10172-9_15
11. Van Dongen, B.F. (Boudewijn), Borchert, F. (Florian): Bpi challenge (2018) https://data.4tu.nl/repository/uuid:3301445f-95e8-4ff0-98a4-901f1f204972
12. Yeshchenko, A., Di Ciccio, C., Mendling, J., Polyvyanyy, A.: Comprehensive process drift detection with visual analytics. In: Laender, A.H.F., Pernici, B., Lim, E.-P., de Oliveira, J.P.M. (eds.) ER 2019. LNCS, vol. 11788, pp. 119–135. Springer, Cham (2019). https://doi.org/10.1007/978-3-030-33223-5_11

A Novel Metaheuristic Approach for Loss Reduction and Voltage Profile Improvement in Power Distribution Networks Based on Simultaneous Placement and Sizing of Distributed Generators and Shunt Capacitor Banks

Mohammad Nasir[1], Ali Sadollah[2(✉)], Eneko Osaba[3], and Javier Del Ser[4]

[1] Materials and Energy Research Center, Dezful Branch, Islamic Azad University, Dezful, Iran
[2] Department of Mechanical Engineering, University of Science and Culture, Tehran, Iran
ali_sadollah@yahoo.com
[3] TECNALIA, Basque Research and Technology Alliance (BRTA), 48160 Derio, Spain
[4] University of the Basque Country (UPV/EHU), 48013 Bilbao, Spain

Abstract. In this paper, Neural Network Algorithm is employed for simultaneous placing and sizing Distributed Generators and Shunt Capacitors Banks in distribution network to minimize active power loss and improve the voltage profile. The NNA is a novel developed optimizer based on the concept of artificial neural networks which benefits from its unique structure and search operators for solving complex optimization problems. The difficulty of tuning the initial parameters and trapping in local optima is eliminated in the proposed optimizer. The capability and effectiveness of the proposed algorithm are evaluated on IEEE 69-bus distribution system with considering nine cases and the results are compared with previous published methods. Simulation outcomes of the recommended algorithm are assessed and compared with those attained by Genetic Algorithms, Grey Wolf Optimizer, and Water Cycle Algorithm. The analysis of these results is conclusive in regard to the superiority of the proposed algorithm.

Keywords: Distributed generations · Shunt capacitors banks · Power loss · Voltage profile · Neural network algorithm

1 Introduction

Distributed Generators (DGs) and Shunt Capacitor Banks (SCBs) are some of the essential components that play an important role in smart distribution systems. Smart grids require integrated solutions to achieve the goals of efficiency through loss minimization and high-quality power delivered to the users [1]. Optimal DGs and SCBs placements and sizing in distribution systems have many advantages including line losses reduction, voltage profile improvement, relieving the overloading of distribution lines, increased

© Springer Nature Switzerland AG 2020
C. Analide et al. (Eds.): IDEAL 2020, LNCS 12489, pp. 64–76, 2020.
https://doi.org/10.1007/978-3-030-62362-3_7

overall energy efficiency, etc. [2]. The optimal generation of active and reactive power in delivery networks from such tools eliminates power inputs from the substation and controls the flow of supply electricity. The placement of DGs is achieved by managing reactive power flow, while SCBs are similarly regulated by monitoring active energy flow.

In view of the advantages of using DGs and SCBs for distribution networks, a number of researchers recently proposed different techniques to simultaneously determine both sites and sizes for improving the voltage profile, the release of systems capacity, minimizing energy loss and increasing reliability. Many of these works use methods classified in the category of bioinspired computation [3]. In [4], for example, a two stage Grasshopper Optimization Algorithm based Fuzzy multi-objective approach is proposed for optimum DGs, SCBs and Electric Vehicle (EV) charging stations placement and sizing in distribution systems. In [5, 6] authors present a Genetic Algorithm (GA) based method to operate the distribution network with the consideration of DGs and SCBs. These studies aim to minimize grid feeder current and annual electricity losses as well as to reduce actual power loss and increase the voltage profile. In [7], authors use the Particle Swarm Optimization algorithm to find the optimal location and size of DG and SCB in IEEE 12, 30, 33 and 69-bus networks in order to minimize losses. Hybrid Artificial Bee Colony and Artificial Immune System algorithms [8] have been tested on IEEE 33-bus system for placing and sizing DGs and SCBs. The results of the simulation show that in comparison with different methods the proposed approach provides improved energy loss reductions and voltage profile improvement. In a research published by Kanwar and et al. [9], improved versions of GA, Particle Swarm Optimization and Cat Swarm Optimization have been used to solve the simultaneous allocation of SCBs and DGs problem in distribution systems with the consideration of variable load. Also, the authors in [10, 11] utilized a G-best-guided Artificial Bee Colony algorithm and Intersect Mutation Differential Evolution algorithm to tackle the mentioned problem. However, in aforementioned studies, restricted combinations of DGs and SCBs with low dimension for evaluating distribution system performance in different scenarios and cases have been considered.

To the best of our knowledge, no research work has been conducted focused on the application of the Neural Network Algorithm (NNA, [12]) to solve the simultaneous location and sizing of DGs and SCBs problem in distribution networks. In the designed experimentation, we consider nine different cases. Furthermore, simulation outcomes of the recommended algorithm are assessed and compared with those attained by a Genetic Algorithm, a Grey Wolf Optimizer (GWO, [13]), and a Water Cycle Algorithm (WCA, [14]). Obtained results demonstrate that the NNA is more effective in comparison with these methods. For this reason, we conclude that the proposed method is a promising technique for solving the faced problem.

The remainder of the article is organized as follows: Sect. 2 introduces the mathematical formulation of the problem to be solved. In Sect. 3, the NNA is described. After this, the implementation conducted for tackling the problem at hand is detailed in Sect. 4. The experimentation conducted is shown and discussed in Sect. 5. Lastly, Sect. 6 concludes this work paper by drawing conclusions and outlining some significant future research lines.

Symbols			
n	Number of buses	$X_{i,i+1}$	Reactance between bus ith and (i + 1)th
$P_{Loss}(i, i+1)$	Active power losses between $(i)th$ and $(i+1)th$ buses	P_{DG}	Real power of DG (kw)
$Q_{Loss}(i, i+1)$	Reactive power losses between $(i)th$ and $(i+1)th$ buses	Q_C	Capacitive compensation (kvar)
P_{TLoss}	Active power losses	V_i	voltage of ith bus (pu)
Q_{TLoss}	Reactive power losses	P_L	The active power loss of line
P_i	Active power flow from ith bus	Q_L	The reactive power loss of line
Q_i	Reactive power flow from ith bus	Q_{Li+1}	Reactive load of (i + 1)th bus
P_{Li+1}	Active load of (i + 1)th bus (kw)	$R_{i,i+1}$	Resistance between bus ith and (i + 1)th

2   Mathematical Formulation of Problem

This section is devoted to the mathematical formulation of the problem to solve. First, the objective function of the simultaneous optimal placement and sizing of DGs and SCBs problem is minimizing the total active power losses as follows:

$$\text{Minimize } P_{TLoss} = min \sum_{i=0}^{n-1} P_{Loss}(i, i+1) \tag{1}$$

Regarding the equality and inequality constraints handled in the optimal placement and sizing of DGs and SCBs problem there are a) the power flow across the lines, b) the buses voltage levels and c) the minimum and the maximum capacity available for installing DGs and SCBs. These considered constraints are formulated as follows:

Formulation of Power Flow: power flow problem, and consequently active and reactive power flow through the branches and bus voltage, are assessed by considering an equivalent single line diagram of distribution system as shown in Fig. 1.

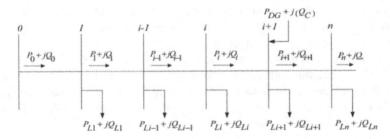

Fig. 1. Single line diagram of distribution system.

In this research, the backward and forward load flow method is employed for analyzing the real and reactive power flow through line ij using (2) and (3) respectively, whilst, the voltage level of $(j)th$ bus is calculated by according to (4) [15].

$$P_{i+1} = P_i + P_{DG} - P_{Li+1} - R_{i,i+1} \frac{(P_i^2 + Q_i^2)}{|V_i^2|} \tag{2}$$

$$Q_{i+1} = Q_i + Q_C - Q_{Li+1} - X_{i,i+1} \frac{(P_i^2 + Q_i^2)}{|V_i^2|} \tag{3}$$

$$|V_{i+1}|^2 = |V_i|^2 - 2(R_{i,i+1}.P_i + X_{i,i+1}.Q_i) + (R_{i,i+1}^2 + X_{i,i+1}^2).\frac{P_i^2 + Q_i^2}{|V_i|^2} \tag{4}$$

The active and reactive power loss in line is calculated as follows:

$$P_L = R_{i,i+1}.\frac{P_i^2 + Q_i^2}{|V_i^2|} \tag{5}$$

$$Q_L = X_{i,i+1}.\frac{P_i^2 + Q_i^2}{|V_i^2|} \tag{6}$$

As a result, the total active and reactive power loss of the network is obtained using (7) and (8), respectively:

$$P_{TLoss} = \sum_{i=0}^{n-1} P_L(i, i+1) \tag{7}$$

$$Q_{TLoss} = \sum_{i=0}^{n-1} Q_L(i, i+1) \tag{8}$$

Voltage Limitation: The voltage magnitude of all the buses in distribution network must lie between minimum and maximum range as follows:

$$V_i^{min} \leq V_i \leq V_i^{max} \tag{9}$$

Furthermore, line limitation for all the branches in distribution network should be less than or equal to the thermal limitation of the line:

$$I_{ij} \leq I_{ij}^{max} \tag{10}$$

DGs Capacity and SCBs Compensation Limitations: on the one hand, the DGs active power should lie between their minimum and maximum amount.

$$P_{DG}^{min} \leq P_{DG} \leq P_{DG}^{max} \tag{11}$$

On the other hand, the total compensation limitation provided by the SCBs must be lie between their minimum and maximum reactive power:

$$Q_C^{min} \leq Q_C \leq Q_C^{max} \tag{12}$$

3 Neural Network Algorithm

Neural Network Algorithm [12] is a metaheuristic inspired by the complicated structure of the Artificial Neural Networks (ANNs) and their operators. Similar to other meta-heuristic optimization algorithms, the NNA begins with an initial population of pattern solutions. Thebest solution obtained at each iteration (i.e., temporal optimal solution) is assumed as target data, and the aim is to reduce the error among the target data and other predicated pattern solutions (i.e., moving other predicted pattern solutions towards the target solution). Indeed, population of pattern solutions corresponds to input data in the ANNs.

The NNA resembles ANNs having N_{pop} input data, having D dimension(s) and only one target data (response). After setting the target solution (X^{Target}) among the other pattern solutions, the target weight (W^{Target}, the weight corresponding to the target solution) has to be selected from the population of weights. Initial weights in ANNs are random numbers, and as the iteration number increases, they are updated considering the calculated error of the network. Back to the NNA, initial weights are a $N_{pop} \times N_{pop}$ square matrix which generates uniform random samples drawn from [0,1] during iterations. However, there is an additionally imposed constraint: the summation of weights for a pattern solution should not exceed 1. Mathematically:

$$\sum_{j=1}^{Npop} w_{ij}(t) = 1, \quad i = 1, 2, 3, \ldots, N_{pop} \tag{13}$$

$$w_{ij} \in U(0, 1) \quad i, j = 1, 2, 3, \ldots, N_{pop} \tag{14}$$

After forming the weight matrix (W), new pattern solutions (X^{New}) are calculated by using the following equation inspired by the weight summation technique used in the ANNs:

$$\vec{X}_j^{New}(t + 1) = \sum_{i=1}^{Npop} w_{ij}(t) \times \vec{X}_i(t), \quad j = 1, 2, 3, \ldots, N_{pop} \tag{15}$$

$$\vec{X}_i(t + 1) = \vec{X}_i(t) + \vec{X}_i^{New}(t + 1), \quad i = 1, 2, 3, \ldots, N_{pop} \tag{16}$$

where t is an iteration index. After creating the new pattern solutions from the previous population of patterns, based on the best weight value so called "target weight", the weight matrix should be updated as well. The following equations suggest an updating equation for the weight matrix:

$$\vec{W}_i^{Updated}(t + 1) = \vec{W}_i(t) + 2 \times rand \times (\vec{W}^{Target}(t) - \vec{W}_i(t)), \quad i = 1, 2, 3, \ldots, N_{pop} \tag{17}$$

As global searching operator, the bias operator in the NNA is another way to explore the search space (exploration process) and it acts similarly to the mutation operator in the Genetic Algorithms. In general, the bias operator prevents the algorithm from premature convergence. The bias operator β is a modification factor, which determines the percentage of the pattern solutions that should be altered. Initially $\beta = 1$ (meaning a 1.0 of probability of modifying all individuals in population) and its value is adaptively reduced at each iteration using a reduction formulation as the following one:

$$\beta(t+1) = \beta(t) \times 0.99 \quad t = 1, 2, 3, \ldots, Max\_Iteration \tag{18}$$

Accordingly, as local searching operator, a transfer function operator transfers the new pattern solutions in the population to new positions, in order to update and generate better quality solutions toward the target solution. Therefore, the following equation is defined as a transfer function operator (TF) for the proposed method given as follows:

$$\vec{X}_i^*(t+1) = TF(\vec{X}_i(t+1)) = \vec{X}_i(t+1) + 2 \times rand \times (\vec{X}^{T \ \arg \ et}(t) - \vec{X}_i(t+1)), \quad i = 1, 2, 3, \ldots, N_{pop} \tag{19}$$

The whole procedures of the NNA are illustrated in Fig. 2, including all its compounding processes.

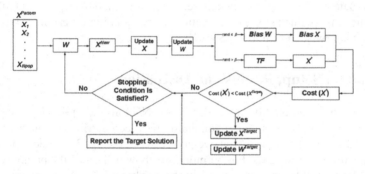

Fig. 2. Diagram showing the search workflow of the NNA algorithm.

4 Implementing NNA to Address the Considered Problem

In this section, a step-by-step process of NNA implementation for solving the optimal placement and sizing of DGs and SCBs problem is summarized as follows:

Step 1: Input information of feeder; the DGs number and the SCBs number.
Step 2: Choose the number of pattern solutions (i.e., population size) and maximum number of iterations (i.e., NFEs). In this study, population size and NFEs are 50 and 25000, respectively.
Step 3: Randomly generate an initial population of pattern solution between LB and UB for decision variables.
Step 4: Run load flow program in the presence of DGs and SCBs.
Step 5: Calculate active and reactive power, and voltage profile using Eqs. (1-6).
Step 6: Calculate total active and reactive power, and loss function using Eqs. (1) and (7–8).
Step 7: Calculate the fitness of initial pattern solutions.

Step 8: Randomly generate the weight matrix (initialization phase) between zero and one considering the imposed constraint (see Eqs. (13) and (14)).

Step 9: Set target solution (X^{Target}) (the minimum value for minimization problems) and its corresponding target weight (W^{Target}).

Step 10: Generate new pattern solutions (X^{New}) and update the pattern solutions using Eqs. (15) and (16).

Step 11: Update the weight matrix (W) using Eq. (17) considering the applied constraints (see Eqs. (13) and (14)).

Step 12: Check the bias condition. If $rand \leq \beta$, performs the bias operator for both new pattern solutions and updated weight matrix.

Step 13: Otherwise ($rand > \beta$), apply the transfer function operator (TF) for updating new position of pattern solutions (X_i^*) using Eq. (19).

Step 14: Calculate the objective function value for all updated pattern solutions.

Step 15: Update the target solution (i.e., temporal best solution) and its corresponding target weight.

Step 16: Update the value of β using any reduction formulation (e.g., Eq. (18))

Step 17: Check predefined stopping condition. If the stopping criterion is satisfied, the NNA stops. Otherwise, return to the Step 10.

5   Simulation Setup, Results and Discussion

In order to evaluate the quality of the developed NNA in the optimal placement and sizing of DGs and SCBs problem, a simulation has been carried out on IEEE 69-bus system with the aim of minimizing active power loss (see Fig. 3). To do that, we have deemed nine different test cases. Furthermore, we show in Table 1 the parameterization of the optimizers used for the comparison of the results.

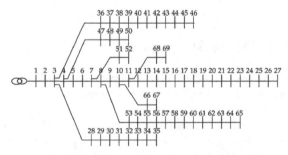

Fig. 3. IEEE 69-bus distribution system.

The 69-bus system has 3.8 MW active and 2.69 MVAr reactive load powers. The data of the system is derived from [10]. Also, following assumptions are considered: a) the test network is deemed as balance; b) first bus is chosen as a slack/substation bus and its voltage is 1 p.u; c) the test network is free from harmonics; d) shunt conductances and susceptances of all distribution lines are considered as negligible; and e) computer simulations on the problem are performed with normal load.

Table 1. Optimal values of user parameters used in the reported optimizers.

Methods	Parameters	Optimal Values
GA	N_{pop}	50
	P_c	0.8
	P_m	0.3
GWO	N_{pop}	50
WCA	N_{pop}	50
	N_{sr}	4
	$dmax$	$1e-16$

In this research, different and diverse combinations of DGs and SCBs are considered. In addition, it is assumed that DG power factor is set at unity for cases 1 to 7 and is set at less than unity for cases 8 and 9. These combinations are as follows: **Base:** there are no DGs and SCBs in distribution network; **Case 1:** Only capacitor is located in distribution network; **Case 2:** Only DG is located in distribution network. **Case 3:** Both DG and SCB are located simultaneously; **Case 4:** Two SCBs are located in distribution network; **Case 5:** Two DGs are located in distribution network; **Case 6:** Combination of two DGs and two SCBs are located in distribution network simultaneously; **Case 7:** Three DGs with unity power factor are located in distribution network; **Case 8:** Three DGs with power factor less than unity are located in distribution network; **Case 9:** Combination of three DGs and three SCBs with power factor less than unity are located in distribution network simultaneously.

The simulation results for 69-bus distribution system are indicated in Tables 2, 3, 4 and 5. Table 2 shows a) the network active power loss, b) the loss reduction percent compared to the base case, c) the weakest bus voltage and place, d) the DGs size and place, and e) the SCBs size and place. The results for base case in Table 2 indicate power losses (224.59 kW) and the weakest bus (0.9102) in system without any DGs and SCBs. As can be seen in this table, all cases show a decrease in the active power loss. Also, cases 8 and 9 illustrate the highest loss reduction and improvement in the weakest bus voltage amplitude among all cases. Tables 3, 4 and 5 present and compare the simulation results of the NNA with other algorithms proposed in the literature. Analyzing the obtained outcomes, we can conclude that NNA performs the best among the rest of alternatives.

Table 2. The simulation results of the NNA on IEEE 69-bus network for different cases

Particulars	Base	Case 1	Case 2	Case 3	Case 4	Case 5	Case 6	Case 7	Case 8	Case 9
Power Losses (KW)	224.59	134.535	79.315	22.349	129.743	68.897	7.139	66.976	4.241	3.305
Loss reduction (%)	–	40.09	64.68	90.04	42.23	69.32	96.82	70.17	98.11	98.52
Vworst in pu (Bus No)	0.9102 (65)	0.9353 (65)	0.9692 (27)	0.9731 (27)	0.9355 (65)	0.9794 (65)	0.9942 (69)	0.9794 (65)	0.9943 (50)	0.9970 (65)
CSB1 size in MVar	–	1.2966 (61)	–	1.2966 (61)	1.2372 (61)	–	0.3524 (17)	–	–	0.1629 (12)
CSB2 size in MVar	–	–	–	–	0.3524 (17)	–	1.2372 (61)	–	–	0.1352 (21)
CSB3 size in MVar	–	–	–	–	–	–	–	–	–	0.5121 (50)
DG1-P size in MW (Bus No)	–	–	1.8196 (61)	1.8196 (61)	–	0.5196 (17)	0.5196 (17)	0.4938 (11)	–	–
DG2-P size in MW (Bus No)	–	–	–	–	–	1.7319 (61)	1.7319 (61)	0.3783 (18)	–	–
DG3-P size in MW (Bus No)	–	–	–	–	–	–	–	1.6725 (61)	–	–
DG1-PQ size in MVA (Bus No)	–	–	–	–	–	–	–	–	0.5274 (11)	0.4937 (11)
DG2-PQ size in MVA (Bus No)	–	–	–	–	–	–	–	–	0.3447 (21)	0.3774 (18)
DG3-PQ size in MVA (Bus No)	–	–	–	–	–	–	–	–	1.6725 (61)	1.6727 (61)

Table 3. The active power loss (KW) in 69-bus network for different algorithms

Particulars	CTLBO [16]	QOTLBO [17]	TLBO [17]	ALOA [18]	BB-BC [19]	MOEA/D [20]	SKHA [21]	KHA [21]	ABC [22]
Case 1	–	–	–	–	–	–	–	–	–
Case 2	–	–	–	81.776	83.2246	–	81.6003	81.6003	83.31
Case 3	–	–	–	–	–	23.17	–	–	–
Case 4	–	–	–	–	–	–	–	–	–
Case 5	–	–	–	70.750	71.776	–	70.4092	77.0354	–
Case 6	–	–	–	–	–	7.20	–	–	–
Case 7	69.388	71.625	72.406	–	–	–	68.1523	69.1977	–
Case 8	–	–	–	–	–	–	–	–	–
Case 9	–	–	–	–	–	4.25	–	–	–

Table 4. The active power loss (KW) in 69-bus network for different algorithms (continues).

Particulars	PSO [6]	IMDE [10]	IWD [23]	SOS [24]	HPSO [25]	RPSO [26]	GA	GWO	WCA	Proposed (NNA)
Case 1	–	–	–	–	–	–	134.5353	134.5353	134.5353	134.5353
Case 2	83.17	–	–	–	87.13	83.22	83.2239	83.2239	83.2240	79.3152
Case 3	–	–	–	–	–	–	25.8474	23.1704	23.1703	22.3492
Case 4	–	145.5310	–	–	–	–	132.7517	130.9017	130.0196	129.7438
Case 5	71.96	70.926	–	–	86.68	–	72.5642	69.8564	69.0747	68.8973
Case 6	25.9	13.833	–	–	–	–	16.6769	7.9446	7.4157	7.1399
Case 7	69.89	–	73.55	69.4286	87.0	–	67.5810	68.8862	68.8823	66.9769
Case 8	–	–	–	–	–	–	6.0135	5.3370	7.0947	4.2410
Case 9	–	–	–	–	–	–	4.1208	5.8781	6.2688	3.3050

Table 5. The weakest bus voltage (pu) and place in 69-bus network for studied algorithms

Particulars	GA	GWO	WCA	Proposed (NNA)
Case 1	0.9353 (65)	0.9353 (65)	0.9353 (65)	0.9353 (65)
Case 2	0.9683 (27)	0.9683 (27)	0.9680 (27)	0.9692 (27)
Case 3	0.9723 (27)	0.9724 (27)	0.9724 (27)	0.9731 (27)
Case 4	0.9843 (65)	0.9356 (65)	0.9356 (65)	0.9355 (65)
Case 5	0.9777 (65)	0.9796 (65)	0.9794 (65)	0.9794 (65)
Case 6	0.9828 (65)	0.9942 (50)	0.9943 (50)	0.9942 (69)
Case 7	0.9786 (65)	0.9794 (65)	0.9794 (65)	0.9794 (65)
Case 8	0.9931 (65)	0.9943 (50)	0.9943 (69)	0.9943 (50)
Case 9	0.9951 (50)	0.9953 (50)	0.9904 (27)	0.9970 (65)

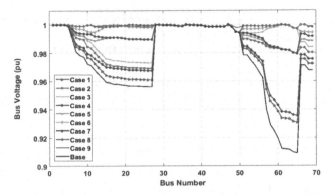

Fig. 4. The IEEE 69-bus network voltage profile for different cases.

Additionally, Fig. 4 illustrates the comparison of the bus voltage profile for base case and cases 1–9. It can be observed in this figure that by using the NNA to determine the size and location of the three DGs and three SCBs, the bus voltage profile is significantly improved.

6 Conclusions and Future Research Lines

In this paper, Neural Network Algorithm (NNA) is employed for solving the optimal placement and sizing of DGs and SCBs problem in distribution network for minimizing active power loss. In order to prove the efficiency and effectiveness of the presented method, the NNA is implemented on IEEE 69-bus system with the consideration of nine cases. Obtained outcomes are compared to competitive bio-inspired algorithms [27] such as Genetic Algorithm (GAs), Grey Wolf Optimizer (GWO) and Water Cycle Algorithm (WCA), which have demonstrated good performances in other fields [28–30]. A deeper comparison is also made with algorithms previously studied in the literature. The results of comparative analysis for different cases verify the superiority of the NNA to other approaches. Hence, the NNA can be considered as a desired optimization method for solving the reported problem against other existing optimizers in the literature.

We have planned a diversity of future research lines rooted on the conclusions disclosed in this study. First, we will analyze whether the superior performance of the developed NNA holds over additional test cases of higher complexity. Moreover, a critical next step is to assess if NNA performs competitively with respect to other bioinspired methods. In the long term, problems stemming from additional research fields will be also explored by using this meta-heuristic solver, with emphasis on combinatorial problems for which the extrapolation of the search operators within the NNA solver is not straightforward (e.g. orienteering, vehicle routing and other related formulations).

Acknowledgments. J. Del Ser and E. Osaba would like to thank the Spanish Centro para el Desarrollo Tecnologico Industrial (CDTI, Ministry of Science and Innovation) through the "Red Cervera" Programme (AI4ES project), as well as by the Basque Government through EMAITEK and ELKARTEK (ref. 3KIA) funding grants. J. Del Ser also acknowledges funding support from the Department of Education of the Basque Government (Consolidated Research Group MATHMODE, IT1294-19).

References

1. El-Hawary, M.E.: The smart grid-state-of-the-art and future trends. Electr. Power Compon. Syst. **42**(3–4), 239–250 (2014)
2. Srinivasa Rao, R., Ravindra, K., Satish, K., Narasimham, S.V.L.: Power loss minimization in distribution system using network reconfiguration in the presence of distributed generation. IEEE Trans. Power Syst. **28**(1), 317–325 (2012)
3. Del Ser, J., et al.: Bio-inspired computation: where we stand and what's next. Swarm and Evol. Comput. **48**, 220–250 (2019)
4. Gampa, S.R., Jasthi, K., Goli, P., Das, D., Bansal, R.C.: Grasshopper optimization algorithm based two stage fuzzy multiobjective approach for optimum sizing and placement of distributed generations, shunt capacitors and electric vehicle charging stations. J. Energy Storage **27**, 101117 (2020)
5. Das, S., Das, D., Patra, A.: Operation of distribution network with optimal placement and sizing of dispatchable DGs and shunt capacitors. Renew. Sustain. Energy Rev. **113**, 109219 (2019)
6. Reza, E.H., Darijany, O., Mohammadian, M.: Optimal placement and sizing of DG units and capacitors simultaneously in radial distribution networks based on the voltage stability security margin. Turkish Journal of Electrical Engineering & Computer Science, pp. 1–14 (2014)
7. Aman, M.M., Jasmon, G.B., Solangi, K.H., Bakar, A.H.A., Mokhlis, H.: Optimum simultaneous DG and capacitor placement on the basis of minimization of power losses. Int. J. Comput. Electr. Eng. **5**(5), 516 (2013)
8. Muhtazaruddin, M.N.B., Tuyen, N.D., Fujita, G., Jamian, J.J.B.: Optimal distributed generation and capacitor coordination for power loss minimization. IEEE PES T&D Conference and Exposition, pp. 1–5 (2014)
9. Kanwar, N., Gupta, N., Niazi, K.R., Swarnkar, A.: Improved meta-heuristic techniques for simultaneous capacitor and DG allocation in radial distribution networks. Int. J. Electr. Power Energy Syst. **73**, 653–664 (2015)
10. Dixit, M., Kundu, P., Jariwala, H.R.: Incorporation of distributed generation and shunt capacitor in radial distribution system for techno-economic benefits. Eng. Sci. Technol. Int. J. **20**(2), 482–493 (2017)
11. Khodabakhshian, A., Andishgar, M.H.: Simultaneous placement and sizing of DGs and shunt capacitors in distribution systems by using IMDE algorithm. Int. J. Electr. Power Energy Syst. **82**, 599–607 (2016)
12. Sadollah, A., Sayyaadi, H., Yadav, A.: A dynamic metaheuristic optimization model inspired by biological nervous systems: neural network algorithm. Appl. Soft Comput. **71**, 747–782 (2018)
13. Sulaiman, M.H., Mustaffa, Z., Mohamed, M.R., Aliman, O.: Using the gray wolf optimizer for solving optimal reactive power dispatch problem. Appl. Soft Comput. **32**, 286–292 (2015)
14. Eskandar, H., Sadollah, A., Bahreininejad, A., Hamdi, M.: Water cycle algorithm–A novel metaheuristic optimization method for solving constrained engineering optimization problems. Comput. Struct. **110**, 151–166 (2012)
15. Haque, M.: Efficient load flow method for distribution systems with radial or mesh configuration. IEEE Proc. Gener. Transm. Distrib. **143**, 33–38 (2012)

16. Quadri, I.A., Bhowmick, S., Joshi, D.: A comprehensive technique for optimal allocation of distributed energy resources in radial distribution systems. Appl. Energy **211**, 1245–1260 (2018)
17. Sultana, S., Roy, P.K.: Multi-objective quasi-oppositional teaching learning-based optimization for optimal location of distributed generator in radial distribution systems. Int. J. Electr. Power Energy Syst. **63**, 534–545 (2014)
18. Ali, E.S., Elazim, S.M.A., Abdelaziz, A.Y.: Ant lion optimization algorithm for optimal location and sizing of renewable distributed generations. Renew. Energy **101**, 1311–1324 (2017)
19. Abdelaziz, A., Hegazy, Y., El-Khattam, W., Othman, M.: A multi-objective optimization for sizing and placement of voltage-controlled distributed generation using supervised Big Bang-Big Crunch method. Electr. Power Compon. Syst. **43**(1), 105–117 (2015)
20. Biswas, P., Mallipeddi, R., Suganthan, P.N., Amaratunga, G.A.J.: A multi-objective approach for optimal placement and sizing of distributed generators and capacitors in distribution network. Appl. Soft Comput. **60**, 268–280 (2017)
21. Chithra Devi, S.A., Lakshminarasimman, L., Balamurugan, R.: Stud Krill herd Algorithm for multiple DG placement and sizing in a radial distribution system. Eng. Sci. Technol. Int. J. **20**(2), 748–759 (2017)
22. Abu-Mouti, F.S., El-Hawary, M.E.: Optimal distributed generation allocation and sizing in distribution systems via artificial Bee colony algorithm. IEEE Trans. Power Deliv. **26**(4), 2090–2101 (2011)
23. Rama Prabha, D., Jayabarathi, T., Umamageswari, R., Saranya, S.: Optimal location and sizing of distributed generation unit using intelligent water drop algorithm. Sustain. Energy Technol. Assess. **11**, 106–113 (2015)
24. Das, B., Mukherjee, V., Das, D.: DG placement in radial distribution network by symbiotic organisms search algorithm for real power loss minimization. Appl. Soft Comput. **49**, 920–936 (2016)
25. Aman, M.M., Jasmon, G.B., Bakar, A.H.A., Mokhlis, H.: A new approach for optimum simultaneous multi-DG distributed generation units placement and sizing based on maximization of system load ability using HPSO (hybrid particle swarm optimization) algorithm. Energy **66**, 202–215 (2014)
26. Jamian, J.J., Mustafa, M.W., Mokhlis, H.: Optimal multiple distributed generation output through rank evolutionary particle swarm optimization. Neurocomput. **152**, 190–198 (2015)
27. Osaba, E., Del Ser, J., Camacho, D., Bilbao, M.N., Yang, X.S.: Community detection in networks using bio-inspired optimization: Latest developments, new results and perspectives with a selection of recent meta-heuristics. Appl. Soft Comput. **87**, 106010 (2020)
28. Osaba, E., Del Ser, J., Sadollah, A., Bilbao, M.N., Camacho, D.: A discrete water cycle algorithm for solving the symmetric and asymmetric traveling salesman problem. Appl. Soft Comput. **71**, 277–290 (2018)
29. Precup, R.-E., David, R.-C., Petriu, E.M., Szedlak-Stinean, A.-I., Bojan-Dragos, C.-A.: Grey wolf optimizer-based approach to the tuning of PI-fuzzy controllers with a reduced process parametric sensitivity. IFAC-PapersOnLine **49**(5), 55–60 (2016)
30. Precup, R.-E., David, R.-C.: Nature-Inspired Optimization Algorithms for Fuzzy Controlled Servo Systems. Butterworth-Heinemann, Elsevier, Oxford, UK (2019)

Simple Effective Methods for Decision-Level Fusion in Two-Stream Convolutional Neural Networks for Video Classification

Rukiye Savran Kızıltepe[✉] and John Q. Gan

School of Computer Science and Electronic Engineering,
University of Essex, Colchester, UK
{rs16419,jqgan}@essex.ac.uk

Abstract. Convolutional Neural Networks (CNNs) have recently been applied for video classification applications where various methods for combining the appearance (spatial) and motion (temporal) information from video clips are considered. The most common method for combining the spatial and temporal information for video classification is averaging prediction scores at softmax layer. Inspired by the Mycin uncertainty system for combining production rules in expert systems, this paper proposes using the Mycin formula for decision fusion in two-stream convolutional neural networks. Based on the intuition that spatial information is more useful than temporal information for video classification, this paper also proposes multiplication and asymmetrical multiplication for decision fusion, aiming to better combine the spatial and temporal information for video classification using two-stream convolutional neural networks. The experimental results show that (i) both spatial and temporal information are important, but the decision from the spatial stream should be dominating with the decision from temporal stream as complementary and (ii) the proposed asymmetrical multiplication method for decision fusion significantly outperforms the Mycin method and average method as well.

Keywords: Deep learning · Video classification · Action recognition · Convolutional neural networks · Decision fusion

1 Introduction

Over the past few years, research interest in the detection and recognition of human actions from video data streams has increased substantially. While humans have adapted to understanding the behaviour of other humans (at least within their cultural framework), the identification and classification of human activities have remained a signification challenge for Computer Vision research. This problem is exacerbated when these actions are mirrored within different environmental settings. Furthermore, understanding multilevel human actions

© Springer Nature Switzerland AG 2020
C. Analide et al. (Eds.): IDEAL 2020, LNCS 12489, pp. 77–87, 2020.
https://doi.org/10.1007/978-3-030-62362-3_8

within complex video data structures is crucial to meaningfully apply these techniques to a wide range of real-world domains.

There have been several approaches to classify human actions throughout a video. One of the most popular approaches is an extension of single image classification tasks to multiple image frames, followed by combining the predictions from each frame. Despite the fact that there has been recent success in using deep learning for image classification, more research should be focused on deep neural network architecture design and representation learning for video classification.

Real-world video data involves a complex data structure, which generally consists of two basic components, spatial and temporal features. Research has shown that only using spatial features in video classification, by extending the approaches from image-based architectures, has not achieved the same level of interpretative accuracy in multi-image domains compared to single image domains.

Convolutional neural networks (CNNs) have been widely utilised in video classification due to its powerful structure and robust feature representation capability. However, one of the main issues is the methodology behind combining appearance and motion information in CNN architectures, which is by information fusion in general. The fusion of information can take place at different levels: signal, image, feature, and symbol (decision) level. Signal-level fusion is the combination of multiple signals to produce a new signal, having the same format as the input signals. Fusion at image-level, or pixel-level, combines a set of pixels from each input image. Feature-level fusion refers to combining different features extracted from multiple images. Decision-level fusion provides the fused representations from several inputs at the highest level of abstraction. The input image(s) are normally analysed individually for feature extraction and classification. The results are multiple symbolic representations which are then combined based on decision rules. The highest level of fusion is obviously symbolic fusion as the existence of more complexity at this level.

The main contribution of this paper is to have proposed simple but effective methods for decision-level fusion in two-stream convolutional neural networks for video classification, which have taken into account the asymmetry of the spatial and temporal information in videos about human actions. This paper is organized as follows. Section 2 presents a brief overview of the existing fusion approaches in two-stream neural networks for video classification. The proposed method is described in Sect. 3, the results are presented in Sect. 4, and conclusions are drawn.

2   Related Work

Action recognition is attracting widespread interest due to the need to capture context information from an entire video instead of just extracting information from each frame (a process that is an extension of image-based classification). CNN has received increased attention in video classification research over the last decade because of its powerful architecture in image-based tasks [5]. A common extension of image-based approaches is to extend 2D CNNs into time dimension to capture spatiotemporal information with the first layer of CNNs [4, 16]. The progress of video classification architectures has been slower than image classification architectures [6, 14]. The possible challenges in video classification are huge computational cost, difficulty in complex architectures, overfitting, and capturing spatiotemporal information [1].

Karpathy et al. explored several fusion approaches to sample temporal information from consecutive frames using pre-trained 2D CNNs: single frame, late, early, and slow fusion [5]. The single frame model utilises single stream network which combines all frames at the final step whereas the late fusion approach utilises two networks with shared parameters a distance of 15 frames apart and then fuses predictions at the last stage. The early fusion involves pixel-level fusion by combining information from the entire video time window. The slow fusion model is a form of a balanced mixture of the late and early fusion models by fusing temporal information throughout the network. The authors highlighted that the spatiotemporal feature representations could not capture motion features properly as the single frame method outperformed the spatiotemporal network architectures [5].

Simmoyan and Zisserman implemented a deep two stream architecture to learn motion features by modelling these features in stacked optical flow vectors [11]. Spatial and temporal streams tackle with spatial and temporal information in separate networks and the fusion occurs in the last classification layer of the network. Although the temporal information is absent, video-level predictions were gathered after averaging predictions over video clips. Their achievement on the UCF-101 and HMDB-51 over the models consisting either spatial or temporal stream has led to researchers to investigating multiple deep neural networks for video classification.

As opposed to stream-based design, Yue-Hei et al. employed Recurrent Neural Network (RNN) on trained feature maps to explore the use of Long Short-Term Memory (LSTM) cell to capture temporal information over video clips [18]. They also explored the necessity of motion information and highlighted the use of optical flow in action recognition problems, although using optical flow is not always beneficial, especially for the videos taken from wild. Donahue et al. built an end-to-end training architecture by encoding convolution blocks and a form of LSTM as a decoder [2]. They also showed the weighted average between spatial and temporal networks returned average scores. A temporal attention mechanism within 3D CNN and RNN encoder-decoder architecture was proposed by Yao et al. to tackle local temporal modelling and learn how to select the most relevant temporal segments given the RNN [17]. Sun et al. [13] proposed factorized spatio-temporal CNNs that factorize the original 3D convolution kernel learning, as a sequential process, which learns 2D spatial kernels in the lower layers by using separate spatial and temporal kernels rather than different networks. They also proposed a score fusion scheme based on the sparsity concentration index which puts more weights on the score vectors of class probability that have a higher degree of sparsity [13].

Feichtenhofer et al. explored how the fusion of spatial and temporal streams affects to the performance from spatio-temporal information gathered throughout the networks [3]. They proposed a novel spatio-temporal architecture involving convolutional fusion layer between spatial and temporal networks with a temporal fusion layer after extensively examining feature-level fusion on feature maps using sum, max, concatenation, convolutional, bilinear, and 3D CNN and 3D Pooling fusion [3]. Moreover, spatial-temporal information fusion with frame attention has been proposed by Ou and Sun [9].

Combining both spatial and temporal features has been a common problem for video classification. Researchers use pixel-level, feature-level and decision-level fusion approaches to take the best advantage of spatiotemporal information [7,8,19]. In this paper, we investigate how decision-level fusion affects the final prediction performance without using complex architectures. Based on the fact that both spatial and temporal information are useful for video classification and they play an asymmetrical role in recognising human actions from videos and inspired by the Mycin uncertainty system [10] for combining production rules in expert systems, we aim to propose simple but effective decision fusion methods in two-stream convolutional neural networks for video classification.

3   Methods

In this section, we present the architecture of the two-stream neural networks, the decision fusion functions, the dataset, the details of preprocessing and feature extraction and the performance evaluation approach. We implement the two-stream architecture proposed in [11] with some modifications. The architecture is one of the most commonly used networks in computer vision applications as it is the first implementation of decision fusion on the top of two-stream neural networks by averaging the scores from both streams. Nonetheless, it is limited in the temporal domain as the input of spatial stream is a single frame and that of the temporal stream is a stack of fixed frames. Due to its popularity, researchers commonly use averaging for the final decision by taking both spatial and temporal information equally.

3.1   Two-Stream Convolutional Neural Networks

Two stream neural networks have been implemented by using Tensorflow-gpu v.1.13 on NVidia GTX1080Ti GPU using CUDA v9.0 toolkit. For the spatial stream network, pre-trained Inception v3 on ImageNet is employed to extract high level feature representations from video clips. With all convolutional layers of Inception pre-trained network frozen, its fully connected layers are trained by initializing weights randomly. After that, the top 2 and 3 inception blocks are trained by freezing the rest of the network separately. The temporal stream network is created by mitigating the same structure defined by Simonyan and

Zisserman [11] for the two-stream CNN as shown in Fig. 1. The model is evaluated on a holdout validation dataset after every epoch and the performance of the model is monitored based on the loss during the training by using early stopping technique.

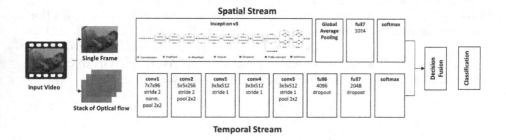

Fig. 1. The architecture of convolutional neural networks used in the experiment

3.2 Decision Fusion

In this study, we focus on decision fusion in two-stream neural networks to investigate the effect of different operations on the final decision at the highest level of abstraction. Many decision fusion rules have been proposed to combine information gathered from different sources based on the same input in video classification applications. The common technique is averaging or weighted averaging which is generally used to get better representation for the combination of spatial and temporal information extracted by two different input representations.

The softmax activation function is widely employed in the output layer of a deep neural network to present a categorical distribution of a list of labels and to calculate the probabilities of each input element belonging to a category. The softmax function $S : \mathbb{R}^K \to \mathbb{R}^K$ is defined by the formula

$$S_i(x) = \frac{exp(x_i)}{\Sigma_j^K exp(x_j)} \tag{1}$$

where $i = 1, ..., K$ and $x = x_1, ..., x_K$ is the output of the final dense layer of the neural network.

The softmax scores from the networks can be fused to give a better representation than could be obtained from any individual stream. Spatial features are mainly used along with temporal features in many video classification applications. These spatial features have taken the same importance as temporal information in the same scope, by using average function to make the final decision. In this paper, it is argued that the temporal information is also important for video classification, but it may not be as important as spatial information for video classification, that is, they are asymmetrical. Inspired by the Mycin uncertainty system for combining production rules in expert systems, this paper proposes using the Mycin formula for decision fusion in two-stream convolutional neural networks. Based on the intuition that spatial information is more useful than temporal information for video classification, this paper also proposes multiplication and asymmetrical multiplication for decision fusion, aiming to better combine the spatial and temporal information for video classification using two-stream convolutional neural networks. To test the above ideas, this study proposes to investigate the following four decision fusion functions for the two-stream neural networks for video classification: Average (A), Mycin (B), Multiplication (C), and Asymmetrical Multiplication (D).

$$A_{X,Y} = \frac{X+Y}{2} \tag{2}$$

$$B_{X,Y} = X + Y - XY \tag{3}$$

$$C_{X,Y} = XY \tag{4}$$

$$D_{X,Y} = X(1 - min(X,Y)) \tag{5}$$

where X and Y represent appearance and motion softmax scores, respectively. When the softmax scores X and Y are asymmetrical, using min(X, Y) in the decision fusion function would bring about extra information in combining them for decision fusion. This will be investigated further by experiments in Sect. 4.

3.3   Dataset

The dataset used in this study is the UCF-101 dataset which was provided by the Center for Research in Computer Vision in 2012 [12]. The UCF-101 contains 101 categories of human actions divided into five types: 1) Human-Object Interaction, 2) Body-Motion Only, 3) Human-Human Interaction, 4) Playing Musical Instruments, and 5) Sports. This dataset is one of the most common action recognition datasets with 101 categories and over 13000 clips. The number of clips per action category and the distribution of clip duration are demonstrated in Fig. 2. The temporal resolution of all clips is of 25 frames per second and the frame size of 320×240 pixels. Three training and test splits officially released are adopted to test the neural networks used to classify human actions from video clips.

3.4   Preprocessing and Feature Extraction

The first step of preprocessing was extracting and saving video frames from video samples. Then, we resized the frames by cropping the centre of frames from the original rectangular image size 320×240 to a 240×240 square. To feed the temporal stream with the temporal information between the frame sequences, we used the preprocessed tvl1 optical flow representations [3]. The Inception v3 is a Convolutional Neural Network which is pretrained on ImageNet dataset containing 1,000 image categories. The pretrained Inception v3 was employed as a spatial feature extractor in the spatial stream network [15].

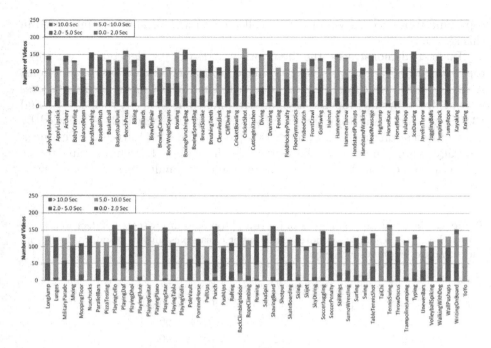

Fig. 2. The number of clips and the distribution of clip duration in the UCF-101 [12]

3.5   Performance Evaluation

In this experiment, classification accuracy is used as performance metric, which is the percentage of correctly classified video instances. In the experiment, four different decision fusion functions were compared based on their accuracy scores through statistical significance test. 48 training, validation, and testing accuracy scores were collected under the same conditions but with either different data splits for training, validation and testing data or different configuration. A significance level of 0.05 was considered during the analysis. The distribution of performance data was evaluated using the Kolmogorov-Smirnov Test to check whether the accuracy scores are normally distributed. As for the test of homogeneity of variances, the Levene statistic was applied to the dependent variables. Furthermore, ANOVA test has been conducted to compare the variance differences to figure out whether the results are significant. Afterwards, the Tukey's Honest Significant Difference (HSD) test was run to determine whether the specific groups' means are different. The results are presented in Sect. 4.

4   Results

We compared four different decision-level fusion functions for the two-stream neural network for video classification. 12 different runs were conducted for each function over the 3 training, validation and testing splits. In total, 48 sets of top-1 and top-5 accuracy scores were collected, and the results are presented in this section.

Table 1 demonstrates the descriptive statistics of dependent variables, top-1 and top-5 accuracy scores, obtained from the decision fusion functions, A, B, C, and D which are defined in Sect. 3. The highest mean accuracy scores were achieved by the proposed asymmetrical multiplication, D, with 59.14% and 83.94% for top-1 and top-5 accuracy scores, respectively, which is followed by the multiplication function, C, at 58.94% (top-1) and 83.75% (top-5). The mean accuracy score achieved by the average is the lowest. It is interesting to see the performance gap between the average and the multiplication functions is over 3%.

Table 1. The descriptive statistics of top-1 and top-5 accuracy scores of decision fusion approaches: N, $\bar{x}$, σ, and $\sigma_{\bar{x}}$ denotes the number of samples, mean, standard deviation, and standard error, respectively.

	Group	N	$\bar{x}$	σ	$\sigma_{\bar{x}}$
top-1	A	12	55.56%	.01492	.00431
	B	12	55.54%	.01259	.00363
	C	12	58.94%	.01252	.00361
	D	12	59.14%	.01352	.00390
top-5	A	12	80.87%	.01285	.00371
	B	12	80.67%	.01127	.00325
	C	12	83.75%	.01309	.00378
	D	12	83.94%	.01642	.00474

As shown in Table 2, there was a statistically significant difference at the $p < 0.05$ level in top-1 accuracy scores achieved by the four decision-level fusion functions: $F~(3,44) = 27.137$, $p = 0.00$. The mean top-5 accuracy scores varied significantly as well: $F~(3,44) = 20.755$, $p = 0.00$.

Additional comparisons using the Tukey's HSD test, as shown in Table 3, indicated that there were significant differences in prediction performance between the proposed asymmetrical multiplication and the average function (mean difference = 0.03607, p = 0.00), and between the average and multiplication functions (mean difference = 0.03403, p = 0.00). Although there was a significant difference between the proposed asymmetrical multiplication and Mycin functions (mean difference = 0.03582, p = 0.00), results did not support the significant difference between the average and Mycin functions (mean difference = 0.00025, p = 1.00). The experimental results have strengthened our argument that the importance of spatial features is not the same as that of temporal features and the decision fusion making use of this asymmetry works well.

Table 2. One-way between-groups ANOVA of performance by fusion functions where df, SS, MS and F refer to degrees of freedom, sum of squares, mean sum of squares, and F score, respectively.

		SS	df	MS	F	Sig.
top-1	Between groups	.015	3	.005	27.137	.000
	Within groups	.008	44	.000		
	Total	.023	47			
top-5	Between groups	.011	3	.004	20.755	.000
	Within groups	.008	44	.000		
	Total	.019	47			

Table 3. Post hoc comparisons using Tukey's HSD. The values are significant at the 0.05 level

Method		Mean Difference (I-J)	Std. Error	Significance	95% Confidence Interval	
(I)	(J)				Lower bound	Upper bound
A	B	.00025	.00548	1.000	−.0144	.0149
	C	−.03378*	.00548	.000	−.0484	−.0192
	D	−.03582*	.00548	.000	−.0505	−.0212
B	A	−.00025	.00548	1.000	−.0149	.0144
	C	−.03403*	.00548	.000	−.0487	−.0194
	D	−.03607*	.00548	.000	−.0507	−.0214
C	A	.03378*	.00548	.000	.0192	.0484
	B	.03403*	.00548	.000	.0194	.0487
	D	−.00204	.00548	.982	−.0167	.0126
D	A	.03582*	.00548	.000	.0212	.0505
	B	.03607*	.00548	.000	.0214	.0507
	C	.00204	.00548	.982	−.0126	.0167

* The mean difference is significant at the 0.05 level.

5   Conclusion

This paper has investigated decision fusion methods for combining spatial and temporal information in two-stream neural networks for video classification. The proposed asymmetrical multiplication function for decision fusion outperformed the other three methods compared in this paper, significantly better than the famous Mycin method and commonly used averaging method. The findings of this study support the idea that spatial information should be treated as dominant while the temporal information must be complementary. Our work has some limitations. The most important limitation is that the two-stream neural network used is not the state-of-the-art due to the limited computing facility used in our experiment and cannot produce the best results. Future work could be focused on more powerful feature learning in deep neural networks and combining decision fusion with feature fusion for further improving video classification performance.

Acknowledgments. This research was partially supported by the Republic of Turkey Ministry of National Education. We gratefully acknowledge the help provided by Fırat Kızıltepe and Edward Longford.

References

1. Castro, D., Hickson, S., Sangkloy, P., Mittal, B., Dai, S., Hays, J., Essa, I.: Convnet architecture search for spatiotemporal feature learning. arXiv preprint arXiv:1708.05038 (2017)
2. Donahue, J., et al.: Long-term recurrent convolutional networks for visual recognition and description. In: Proceedings of the IEEE Conference on Computer Vision and Pattern Recognition, pp. 2625–2634 (2015)
3. Feichtenhofer, C., Pinz, A., Zisserman, A.: Convolutional two-stream network fusion for video action recognition. In: Proceedings of the IEEE Conference on Computer Vision and Pattern Recognition, pp. 1933–1941 (2016)
4. Ji, S., Xu, W., Yang, M., Yu, K.: 3D convolutional neural networks for human action recognition. IEEE Trans. Pattern Analy. Mach. Intell. **35**(1), 221–231 (2012)
5. Karpathy, A., Toderici, G., Shetty, S., Leung, T., Sukthankar, R., Fei-Fei, L.: Large-scale video classification with convolutional neural networks. In: Proceedings of the IEEE Conference on Computer Vision and Pattern Recognition, pp. 1725–1732 (2014)
6. Krizhevsky, A., Sutskever, I., Hinton, G.E.: Imagenet classification with deep convolutional neural networks. In: Advances in Neural Information Processing Systems, pp. 1097–1105 (2012)
7. Li, M., Leung, H., Shum, H.P.: Human action recognition via skeletal and depth based feature fusion. In: Proceedings of the 9th International Conference on Motion in Games, pp. 123–132 (2016)
8. Liu, L., Shao, L.: Learning discriminative representations from RGB-D video data. In: Proceedings of the Twenty-Third International Joint Conference on Artificial Intelligence, pp. 1493–1500 (2013)
9. Ou, H., Sun, J.: Spatiotemporal information deep fusion network with frame attention mechanism for video action recognition. J. Electron. Imaging **28**(2), 023009 (2019)
10. Shortliffe, E.H., Buchanan, B.G.: Rule-Based Expert Systems: The MYCIN Experiments of the Stanford Heuristic Programming Project. Addison-Wesley Publishing Company, Boston (1985)
11. Simonyan, K., Zisserman, A.: Two-stream convolutional networks for action recognition in videos. In: Advances in Neural Information Processing Systems, pp. 568–576 (2014)
12. Soomro, K., Zamir, A.R., Shah, M.: UCF101: A dataset of 101 human actions classes from videos in the wild. Center for Research in Computer Vision 2 (2012)
13. Sun, L., Jia, K., Yeung, D.Y., Shi, B.E.: Human action recognition using factorized spatio-temporal convolutional networks. In: Proceedings of the IEEE International Conference on Computer Vision, pp. 4597–4605 (2015)
14. Szegedy, C., et al.: Going deeper with convolutions. In: Proceedings of the IEEE Conference on Computer Vision and Pattern Recognition, pp. 1–9 (2015)
15. Szegedy, C., Vanhoucke, V., Ioffe, S., Shlens, J., Wojna, Z.: Rethinking the inception architecture for computer vision. In: Proceedings of the IEEE Conference on Computer Vision and Pattern Recognition, pp. 2818–2826 (2016)

16. Tran, D., Bourdev, L., Fergus, R., Torresani, L., Paluri, M.: Learning spatiotemporal features with 3D convolutional networks. In: Proceedings of the IEEE International Conference on Computer Vision, pp. 4489–4497 (2015)
17. Yao, L., et al.: Describing videos by exploiting temporal structure. In: Proceedings of the IEEE International Conference on Computer Vision, pp. 4507–4515 (2015)
18. Yue-Hei Ng, J., Hausknecht, M., Vijayanarasimhan, S., Vinyals, O., Monga, R., Toderici, G.: Beyond short snippets: Deep networks for video classification. In: Proceedings of the IEEE Conference on Computer Vision and Pattern Recognition, pp. 4694–4702 (2015)
19. Zhang, J., Li, W., Ogunbona, P.O., Wang, P., Tang, C.: RGB-D-based action recognition datasets: a survey. Pattern Recogn. **60**, 86–105 (2016)

LSI Based Mechanism for Educational Videos Retrieval by Transcripts Processing

Diana Iulia Bleoancă[1], Stella Heras[2], Javier Palanca[2], Vicente Julian[2], and Marian Cristian Mihăescu[1(✉)]

[1] University of Craiova, Craiova, Romania
bdianaiulia@yahoo.com, mihaescu@software.ucv.ro
[2] Universitat Politecnica de Valencia, Valencia, Spain
{sheras,jpalanca,vinglada}@dsic.upv.es

Abstract. Retrieval of relevant educational videos by NLP analysis of their transcripts represents a particular information retrieval problem that is found in many systems. Since various indexing techniques are available, finding the suitable ingredients that build an efficient data analysis pipeline represents a critical task. The paper tackles the problem of retrieving top-N videos that are relevant for a query provided in the Spanish language at Universitat Politècnica de València (UPV). The main elements that are used in the processing pipeline are clustering, LSI modelling and Wikipedia contextualizing along with basic NLP processing techniques such as bag-of-words, lemmatization, singularization and TF-IDF computing. Experimental results on a real-world dataset of 15.386 transcripts show good results, especially compared with currently existing search mechanism which takes into consideration only the title and keywords of the transcripts. Although live application deployment may be further necessary for further relevance evaluation, we conclude that current progress represents a milestone in further building a system that retrieves appropriate videos for the provided query.

Keywords: LSI · Clustering · Educational video transcripts

1 Introduction

In recent years, the context of higher education has changed enormously both in methodological and instrumental terms. Increasingly, universities around the world are offering online educational resources in the form of Massive Online Open Courses (MOOCs, e.g. Coursera, edX, Udacity, MiriadaX, Udemy) [16], or flipped classroom courses [22], where students have video lectures available to view on demand through different media platforms. In this way, students can visualise the theory of the course outside the classroom and dedicate the classes to solving exercises, practical cases and doubts that may arise. Although online

© Springer Nature Switzerland AG 2020
C. Analide et al. (Eds.): IDEAL 2020, LNCS 12489, pp. 88–100, 2020.
https://doi.org/10.1007/978-3-030-62362-3_9

education is not new, the actual panorama has undoubtedly led to what could be considered the most crucial innovation in education in the last 200 years[1].

However, a key point to ensure the success of any online educational platform is the ability to *understand how students learn and what they need to learn*. To facilitate this ability, any online educational platform must include a powerful search engine or recommendation tool that allows students to find the relevant videos to achieve their learning goals and improve their learning outcomes. This is a considerable challenge for the scientific community, especially in the field of educational data mining [20], educational recommender systems [8], and natural language processing (NLP) for information retrieval [13]. Most search engines are simple browsers that match a list of keywords typed by the students with specific attributes that characterize the available learning resources (e.g. the title or keywords of video lectures). Others use more advanced artificial intelligence techniques to cluster and classify these resources and retrieve those that are relevant for a specific query. However, suppose the ultimate goal of the system is to help students achieve their learning objectives. What is really important is that the system can interpret the context and infer the student's intentions when searching.

In this article, we present our current advances in this domain, in the context of a research project developed in the UPV, where 357 flipped classroom courses are being taught by 421 lecturers to more than 9000 students[2]. In addition, it has extensive experience in the production of multimedia classes and MOOC courses through its video lectures platform *UPV media*[3] and its MOOCs platform *UPVx*[4]. The latter is part of *edx.org*[5], a non-profit institution that is a reference in the world of MOOCs. Through these platforms, students are provided with videos and recorded classes that makeup teaching units. Overall, we have available a (UPV-Media dataset) which contains more than 55.000 short educational videos (on average 5–10 min each). The videos (most of them in Spanish) cover different topics that are taught in various courses in the university. Most videos have a set of keywords associated, which are entered by the teacher as free text and therefore, are not always good indicators of the video topic and lack from any standardisation. In addition, more than 15.000 include a (Spanish) transcript. Therefore, a differentiating feature of the University-Media dataset from other videos dataset available is the fact that they are short videos, in Spanish, that partly overlap in content, and that can be related to many different courses. Furthermore, title and keywords may have different semantic interpretations and are not enough to accurately determine the video topic.

This work presents a new approach to analyzing and contextualizing student searches on our dataset search engine. The ultimate goal is to accurately interpret

---

[1] https://www.technologyreview.com/s/506351/the-most-important-education-techn ology-in-200-years.

[2] https://www.upv.es/noticias-upv/noticia-10134-flipped-classr-en.html.

[3] https://media.upv.es/.

[4] https://www.upvx.es.

[5] https://www.edx.org/school/upvalenciax.

the student's intentions and provide them with the learning objects that will help improve their learning outcomes. The proposed approach continues previous works from [23] which designed and implemented a retrieval system based on using the video transcripts to build a bag-of-words, compute TF-IDF (term frequency-inverse document frequency) scores [18], create clusters of transcripts and use Latent Dirichlet Allocation (LDA) [5] for ranking and retrieval. The shortcomings of this method regard poor contextualization on the query analysis, the indexing, and the retrieval process. In the former, the problem consisted of the inability to effectively cluster the input query, given the fact that it usually consists of few words. The proposed solution is to build an *augmented query* by inferring the context of the search by using Wikipedia to automatically collect related articles [24], which are further used for determining the context of the query and for clustering. Regarding the clustering and retrieval method, our initial LDA-based algorithm did not take into consideration the importance of the words within the transcript, which gave rise to contextualization problems where the semantics of the students' query was not correctly understood.

2   Related Work

Our work proposes a novel combination of up-to-date NLP techniques to contextualize the students' queries and provide them with videos that better match their learning goals.

To the best of our knowledge, there are few attempts to build Spanish lemmatizers, among which is [2]. Also, a specific characteristic of the Spanish language is that it presents more lemma ambiguity than the majority of languages (14% [1]). In this work, we have built up a suitable lemmatizer for the Spanish transcription of the videos in our dataset.

Once lemmas have been identified, the next step in the NLP pipeline is to compute word embeddings to translate them to real number vectors that can be better processed by NLP algorithms. [10] provides a comprehensive literature review on word embedding algorithms. In the specific domain of video lessons analysis, the work presented in [9] uses word embeddings instead of traditional bag-of-words approaches to deal with the lecture video fragmentation problem.

Also, following an approach similar to ours, in [11], authors propose a video classification framework to organize content from MOOCs that uses Word2Vec [15] to generate word embeddings. A sub-product of this research has also been the creation of word embedding-topic distribution vectors for their MOOC English video lectures dataset [12]. They use an LDA algorithm to generate topic vectors. The same approach was used in *Videopedia* [4], a video lectures recommendation system for educational blogs. However, the system is designed to automatically recommend videos based on the content of these websites and not for individuals.

In our system, we instead use Latent Semantic Indexing (LSI) [7] combined with TF-IDF to take into consideration the importance of each word within the transcript and hence, better contextualize them [3,6].

Finally, an important feature that differentiates our work from other systems that deal with topic modelling from video transcripts and the following video lesson recommendation functionalities is the ability to understand the actual intentions of the student when he/she performs the query. With this aim, we augment the query with related data extracted from Wikipedia to better determine its context [24].

3   Proposed Approach

The current approach for video retrieval from the UPV's multimedia platforms takes into consideration only the *title* and *keyword* fields of the videos. Due to a large number of available videos, this simplistic approach makes searching a difficult task for a student. Therefore, the current search mechanism does not take into account any information regarding the semantics or the context for the input query or title and keywords for the available videos. Since titles and keywords of videos gather only minimal information about the video, which may not be accurate, the current retrieval mechanism commonly retrieves videos that have low relevance for students.

The main contributions in the proposed approach as compared with our previous works are:

1. Reprocessing of transcripts transforms nouns to their simplest form and removes stop-words.
2. The input query provided by the student is contextualized by means of Wikipedia articles.
3. The business logic uses LSI algorithms instead of LDA [3, 6].

We further present the main processing blocks from the data analysis pipeline (Fig. 1).

1. Reprocessing - Build the Bag of Words (BoW): The available transcripts of educational videos represent the raw input in the data analysis process. The first step is to build a BoW (bag-of-words) by tokenizing the transcripts. Because transcripts pertain to many lexical and semantic domains, we observe that there is a large number of distinct words that describe them. One possible solution to this issue is to narrow down the number of words by lemmatization that is grouping together the inflected forms of a word so they can be analyzed as a single item. Taking into consideration that we could not find a Spanish lemmatizer with satisfactory performance of handling Spanish, we decided to retrieve only the nouns from sentences. Thus, we use the *POSTaggerAnnotator* from StanfordCoreNLP [14] that also assigns parts of speech to each word.

We chose to keep only the nouns from transcripts for two main reasons. First, nouns have the most essential reference in context. At the same time, verbs or other parts of a sentence are not so specific and can target many other subjects. Another reason would be the fact that verbs can be found in many different forms, without a method to handle it.

We further create a list of stop-words using *nltk.corpus* package from NLTK [17] and then translate English words in Spanish using *gensim* library [19]. Finally, we manually add common words that did not exist in either one of the libraries (e.g., "example", "case") by using the Google Translate service.

For each word, we retrieve only the singular forms by using the *inflector* library [21]. Moreover, there is still an issue regarding a large number of words from the BoW and a medium-to-large number of transcripts, ending up in having a sufficiently large input dataset for clustering.

From the perspective of the application domain, there are few sizeable particular learning domains. Transcripts are grouped in clusters such that each cluster represents a learning area. We have all discipline regarding: Engineering & Architecture, Sciences (Biological Sciences), and Social & Legal & Humanities.

2. Embedding of the transcripts: For analyzing the transcripts, we must use an embedding that matches every word to score value. This step is compulsory because algorithms may not take as input words or sentences. Therefore, an embedding method is required to obtain a numerical representation of the transcripts. One option is to count the number of appearances of each distinct word and return a sparse vector filled with integers that are related to each word. This approach uses *doc2bow()* function from *gensim* library [19]. Still, it is not enough in capturing the importance of a word within a document. The solution is to apply a transformation on this simple representation such that newly obtained values may be successfully used for computing document similarities.

The *TF-IDF* weighting scheme takes into account the frequency of a word in all documents and its usage in a particular document, assigning them a score. Therefore, the words are ranked higher since they are considered keywords of the document even if they are not the most numerous as the more general-used ones. This score is represented to be a value between 0 and 1. This approach has been successfully used in [18] to determine word relevance in document queries.

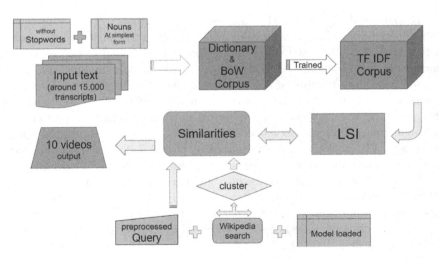

Fig. 1. Overview of the data analysis pipeline

3. Model each cluster and find its topics, scores and similarities:
The *transformed corpus* represents the text embedding with the TF-IDF values
computed previously, and that is ready for clustering and LSI modelling.

We use the *LSI Model* to determine the topics and associated scores for each
determined cluster. From the LSI perspective, the input is represented by the
transformed corpus as a sparse vector obtained in the second step of our pipeline.
The *id2word* method from *corpora* is used to get a mapping between words and
their IDs for the cluster in which a specific number of topics has been chosen.
As a result, LSI learns latent topics by performing a matrix decomposition on
the values of the terms matrix using SVD (Singular value decomposition).

For every cluster, the LSI model creates a list of topics parametrized by
scores, whose values add up to one. The topics generated are represented by
linear combinations of terms that create a 'low-rank approximation' (i.e., data
compression) of the 'term-document matrix' (i.e., the matrix with TF-IDF values). Lastly, the LSI model (i.e., the topic's scores and their list of coefficients
and words) is serialized and registered for later querying.

The next task is to determine the similitude between transcripts and queries
to return the first ten most relevant educational videos. To prepare for similarity
query between LSI vectors and the input query, we build an index matrix represented by cosine similarity against each transformed corpus. The final result
is represented by a NxD LSI matrix, where N represents the number of topics
determined by LSI and D represents the dimension of the BoW transformed
corpus after the second step of our pipeline.

4. Determine the cluster for the query provided by learner: At this
point, we have a language model for every cluster of transcripts, where each
cluster represents a domain. The next critical step is to assign the right cluster
to the input query so that the search can be performed in the correct domain. The
difficulty of this approach comes from the fact that we cannot reliably compute
a text embedding (i.e., TF-IDF values) for the query because it usually consists
of only a few words. Therefore, a context for the query needs to be determined,
and for that context the text embedding will be computed and further be used
to determine the right cluster.

We determine the context of the query by using Wikipedia. We extract the
first sequence from a Wikipedia web page related to the query performed as
returned by the Wikipedia API. By employing this contextualization mechanism,
we successfully increased the probability of determining the proper cluster for
the input query. From the business logic perspective, contextualization with
Wikipedia is used only for determining the cluster to which the query belongs.

5. Getting the list of results: top-N most significant videos: Given
the fact that the entire knowledge base is converted into the most simple format
(i.e., BoW of singular nouns), it will be necessary to pre-process the submitted
query similarly. Therefore, the input query is transformed into a BoW corpus
and further to an LSI space. The query is processed by removing the stop-words
and converting the words to their singular. At this step, we do not use Wikipedia
for contextualizing the query.

Finally, we compute the similarity scores between the input query and the corpus of the corresponding clusters and return the top N videos.

4   Experimental Results

The input dataset represents a *json* file that contains 15.386 video transcripts which are preprocessed for obtaining a corpus with TF-IDF values. Table 1 presents a sample corpus with computed TF-IDF values in descending order, showing the significance of the words not only in their transcript, but also in the cluster that the words belong to.

Table 1. Sample corpus computed by TF-IDF with their scores (English translation)

Word	Score	Word	Score	Word	Score
doc in Cluster BS		doc in Cluster E		doc in Cluster HA	
grupo (group)	0.551	canvas	0.465	cursiva (italics)	0.634
carboxilo (carboxyl)	0.423	script	0.366	letra (letter)	0.409
carbonilo (carbonyl)	0.323	awake	0.261	palabra (word)	0.179
ácido (acid)	0.164	botón (button)	0.245	anglicismo (anglicism)	0.153
propanoico (propanoic)	0.115	método (method)	0.161	publicación (publication)	0.041
oxo	0.115	interface	0.089	continuación (continued)	0.039
butanoato (botanoate)	0.115	acción (action)	0.068	bibliografía (bibliography)	0.038
sustituyente (substitute)	0.114	overlay	0.058	plató (set)	0.038
etilo (ethyl)	0.097	game	0.058	terminología (terminology)	0.035
molécula (molecule)	0.047	editarlo (edit)	0.052	documental	0.032
toxi	0.038	screen	0.052	diccionario (dictionary)	0.032
acetato (acetate)	0.038	objeto (object)	0.044	sintagma (syntagm)	0.032
propilo (propylene)	0.038	cabo (cape)	0.036	resumen (summary)	0.031
proxi (proxy)	0.038	animación (animation)	0.029	vehículo (vehicle)	0.026
oxígeno (oxygen)	0.032	evento (event)	0.021	símbolo (symbol)	0.023
derivado (derive)	0.024	mecanismo (mechanism)	0.021	televisión (television)	0.022
alcohol	0.024	selección (selection)	0.019	caso (case)	0.017
esquema (scheme)	0.018	práctica (practice)	0.018	especie (species)	0.016
piel (skin)	0.015	ciclo (cycle)	0.017	calidad (quality)	0.014
relación (relation)	0.011	control	0.017	formación (formation)	0.014
tratamiento (treatment)	0.010	presentación (presentation)	0.010	universidad (university)	0.011
hora (hour)	0.008	objetivo (objective)	0.006	concepto (concept)	0.008
caso (case)	0.008	elemento (element)	0.006	tiempo (time)	0.005

Every transcript is preprocessed (keeping the nouns to their simplest form and remove the common words) and transformed in a Bow Corpus. A value representing the number of appearances in the script is assigned to every word.

The total number of words per transcript (with an average of 1.653) was reduced to 340. All the values were saved in a matrix with the following structure: there are N rows (i.e., the number of transcripts), each of them containing M number of filtered "word, number of appearances" pairs. One matrix of this type is created for each cluster.

From the 15.386 available transcripts, after keeping only valid transcripts for processing, we remain with 7.935 transcripts, organized as 1.185 from cluster BS (i.e., Biology and Sciences)), 3.475 from cluster E (i.e., Engineering) and 3.275 from cluster HA (i.e., Humanities and Arts).

Firstly, every word is represented by its number of appearances. Still, we have to compute more specific values, where each transcripts' context must influence their weight. For this purpose, we determine the *TF-IDF* values on the lists and thus create a similar matrix with the same dimension.

Table 2. Sample LSI models with their topics (*English translation; n/t when there is no translation since it is only a Spanish sufix of a word*)

Cluster ID and LSI results: topics
Cluster BS: Biology and Science
Topic 0: -0.193*"úlcera" (*ulcer*) + -0.174*"presión" (*pressure*) + -0.136*"fibra" (*fiber*)+ -0.122*"cultivo" (*crop*) + -0.116*"plaga" (*plague*) + -0.114*"pec" (*pec*) + -0.112*"planta" (*plant*) + -0.107*"molécula" (*molecule*) + -0.104*"lesión" (*injury*) + -0.103*"piel" (*skin*)
Topic 1: -0.491*"úlcera" (*ulcer*) + -0.360*"presión" (*presure*) + -0.223*"piel" (*skin*) + -0.186*"desbridamiento" (*debridment*) + -0.179*"paciente" (*patient*) + -0.165*"prevención" (*prevention*) + 0.163*"fibra" (*fiber*) + -0.161*"apósito" (*dressing*) + -0.144*"lesión" (*injury*) + 0.119*"matriz" (*matrix*)
Topic 2: 0.289*"fibra" (*fiber*) + -0.236*"plaga" (*plague*) + -0.210*"cultivo" (*crop*) + 0.194*"átomo" (*atom*) + 0.193*"matriz" (*matrix*) + -0.179*"planta" (*plant*) + 0.173*"molécula" (*molecule*) + -0.173*"insecto" (*insect*) + -0.145*"especi" (*n/t*) + 0.135*"hidrógeno" (*hydrogen*)
Topic 3: 0.698*"fibra" (*fiber*) + -0.264*"átomo" (*atom*) + -0.241*"molécula" (*molecule*) + 0.237*"matriz" (*matrix*) + -0.180*"enlace" (*link*) + -0.177*"hidrógeno" (*hydrogen*) + -0.148*"electrón" (*electron*) + -0.125*"carbono" (*carbon*) + -0.120*"enlac" (*n/t*) + -0.097*"reacción" (*reaction*)
Topic 4: -0.475*"fibra" (*fiber*) + 0.387*"matriz" (*matrix*) + 0.158*"vector" (*vector*) + 0.149*"equi" (*equi*) + 0.144*"cero" (*zero*) + 0.138*"cuadrado" (*square*) + 0.136*"función" (*function*) + 0.134*"matric" (*n/t*) + -0.127*"molécula" (*molecule*) + -0.123*"átomo" (*atom*)

Finally, a *Transformed Corpus* is built and integrated with LSI making possible retrieval of the conceptual content of transcripts by establishing associations between terms in similar contexts (i.e., the queries). Table 2 presents the results of the LSI method with reduced number of dimensions to the number of desired topics (i.e, K dimensions). The LSI method uses singular-value decomposition (SVD) and decomposes the input matrix into three other matrices: M × K Word Assignment to Topics matrix, K × K Topic Importance matrix and K × N Topic Distribution Across Documents matrix. For this study we have used 5 topics.

Table 3. Example sequences from Wikipedia with their clusters (*English translation*)

Query (*Eng.*), [Cluster *without* wiki search, Cluster **with** wiki search]   Sequence extracted from Wikipedia (*English translation*)
celula madre (*stem cell*), [Humanities and Arts, **Biology and Science**]   'Las células madre son células que se encuentran en todos los organismos pluricelulares y que tienen la capacidad de dividirse (a través de la mitosis) y diferenciarse en diversos tipos de células...'   *'Stem cells are cells that are found in all multicellular organisms and have the ability to divide (through mitosis) and differentiate into various specialized cell types...'*
memoria principal (*main memory*), [*Biology and Science*, **Engineering**]   'Memoria primaria (MP), memoria principal, memoria central o memoria interna es la memoria de la computadora donde se almacenan temporalmente tanto los datos como los programas que la unidad central de procesamiento (CPU) ...'   *'Primary memory (PM), main memory, central memory or internal memory is the memory of the computer where both data and programs that the central processing unit (CPU) ...'*
leyes del estado republicano (*'republican state law'*), [*Biology and Science*, **Humanities and Arts**]   'El Partido Republicano (GOP) (Republican Party; también conocido como GOP, de Grand Old Party, Gran Partido Viejo) es un partido político de los Estados Unidos.'   *'The Republican Party (also known as the GOP, from Grand Old Party) is a political party in the United States.'*

Table 4. Validation scores for example of preprocessed query labeling

No of cluster	Accuracy (cross-validation)	Precision	Recall	F1-score	Support
Cluster 0	0.83	1.00	0.83	0.91	331
Cluster 1	0.90	1.00	0.90	0.95	352
Cluster 2	0.86	1.00	0.86	0.92	346

Once that we have determined the LSI models for every cluster, we have to match the query with the proper cluster such that searching will be performed within the correct set of transcripts. As many queries are ambiguous and can not be assigned directly to a specific domain (i.e., cluster), we use instead the first sentence from a relevant Wikipedia web page. Table 3 shows the impact on clustering results after using Wikipedia contextualization. Cross-validation results of clustering with Wikipedia contextualization are being presented in Table 4 for 350 video titles.

Once the cluster is obtained, the initial query is preprocessed (brought to the singular form) to be transformed into embedding results. The query is further converted to LSI space, and according to the cosine similarity metric, the most similar videos are being retrieved.

Table 5. List of results with their scores for *Biology and Science* cluster for querying "fibra" (en. *"fibre"*)

Id	Score	Transcript based search	Simple search after title
79	0.9975	Clasificación de las fibras textiles (*Classification of textile fibres*)	Fibra textil viscosa (*Viscous textile fiber*)
114	0.9974	Fibra textil seda (*Silk textile fiber*)	Fibra textil lino (*Linen textile fiber*)
213	0.9970	Fibra textil viscosa (*Viscous textile fiber*)	Fibra textil seda (*Silk textile fiber*)
900	0.9968	Fibra textil elastano (*Elastane textile fibre*)	Fibra textil acetato (*Acetate textile fibre*)
921	0.9961	Fibra textil lino (*Linen textile fibre*)	Fibra textil elastano (*Elastane textile fibre*)
544	0.9958	Fibra textil acetato (*Acetate textile fibre*)	Fibra textil poliolefina (*Polyolefin textile fibre*)
607	0.9954	Fibra textil acrílica (*Acrylic Textile Fiber*)	Fibra textil acrílica (*Acrylic Textile Fiber*)
67	0.9948	Fibra textil poliolefina (*Polyolefin textile fibre*)	Funcionamiento de una fibra óptica (*Operation of an optical fiber*)
725	0.9943	Influencia del substrato. Sección transversal (*Influence of the substrate. Cross section*)	Interacción entre la fibra y la matriz (*Fiber-Matrix Interaction*)
549	0.9934	Fibras textiles de algodón (*Cotton textile fibres*)	Conectorización de fibra óptica IV (*Fiber Optic Connectorization IV*)

By sorting the results, we can return the 10 most relevant videos. In Table 5 and 6 we perform 2 sample queries and present a comparative result of transcript-based search against simple search based on title of the video.

The results presented in Table 5 show that both methods produce similar results sue to the fact that the query is also found in the title of the videos. Still, the transcript-based search also finds appropriate videos from the same domain (i.e., cluster) and whose query may not be in the title. The results presented in Table 6 show a more precise indication that transcript-based search returns more appropriate videos than simple search mechanism.

Table 6. List of results with their scores for *Engineering* cluster for querying "Como leer de un archivo" (en. *"How to read from a file"*)

Id	Score	Transcript based search	Simple Search after title
558	0.9867	Preferencias de usuario. Menú FILE (II) en Blender (*User preferences. FILE (II) menu in Blender*)	Crear un archivo de Excel nuevo como duplicado de otro ya existente (*Create a new Excel file as a duplicate of an existing one*)
3334	0.9830	Guardar, exportar e imprimir presentaciones (*Save, export and print presentations*)	Un servicio como mecanismo de comunicación entre aplicaciones (*A service as a communication mechanism between applications*)
1016	0.9812	Uso de AutoCAD Design Center para operaciones con bloques (*Using AutoCAD Design Center for Block Operations*)	Un receptor de Anuncios como mecanismo de comunicación entre aplicaciones (*An ad receiver as a communication mechanism between applications*)
1326	0.9780	Uso de referencias externas (*Use of external references*)	Cerrar un acuerdo de compra (*Closing a Purchase Agreement*)
2230	0.9760	Preferencias de usuario. Menú FILE (I) en Blender (*User preferences. FILE (I) menu in Blender*)	Sistema tributario Español: elementos de un tributo (*Spanish tax system: elements of a tax*)
3255	0.9692	Lista de bases de datos: Guardar (*List of databases: Save*)	Archivo de la Experiencia (*Experience Archive*)
617	0.9584	Diseño General de la Presentación I (*General Design of the Presentation I*)	Modificar un archivo de plantilla Excel (*Modify an Excel template file*)
4	0.9583	Introducción al manejo de ficheros (*Introduction to file handling*)	Cómo citar adecuadamente bibliografía en un trabajo de investigación (*How to properly cite literature in a research paper*)
788	0.9581	Introducción a Powerpoint (*Introduction to Powerpoint*)	Cómo solicitar un aula de Videoapuntes (*How to request a video classroom*)
1407	0.9580	Uso de archivos de AutoCAD como bloques (*Using AutoCAD files as blocks*)	Criterios para la consideración de un suelo como contaminado (*Criteria for the consideration of a soil as contaminated*)

5 Conclusions

In this work, we present an NLP-based approach for retrieval appropriate educational videos based on transcripts processing. As principal ingredients in the data analysis process, we have used TF-IDF embedding, Wikipedia contextualization of the query and LSI for topic detection. We have found that our proposed mechanism retrieves more relevant videos as compared with the existing simple search mechanism. Although evaluation of proposed method has been performed only on two sample queries, the results are encouraging to continue improving the data analysis pipeline such that exhaustive validation mechanism may be implemented on queries performed by students at UPV. As future work we plan to test other embedding methods (i.e., BERT, USE, NNLM) and run online evaluation with actual students.

Acknowledgement. This work was partially supported by RTI2018-095390-B-C31-AR project of the Spanish government, and by the Generalitat Valenciana (PROME-TEO/2018/002) project.

References

1. State-of-the-art multilingual lemmatization. https://towardsdatascience.com/state-of-the-art-multilingual-lemmatization-f303e8ff1a8. Accessed 25 Feb 2020
2. Aker, A., Petrak, J., Sabbah, F.: An extensible multilingual open source lemmatizer. In: Proceedings of the International Conference Recent Advances in Natural Language Processing, RANLP 2017, pp. 40–45. ACL (2017)
3. Anaya, L.H.: Comparing Latent Dirichlet Allocation and Latent Semantic Analysis as Classifiers. ERIC (2011)
4. Basu, S., Yu, Y., Singh, V.K., Zimmermann, R.: Videopedia: lecture video recommendation for educational blogs using topic modeling. In: Tian, Q., Sebe, N., Qi, G.-J., Huet, B., Hong, R., Liu, X. (eds.) MMM 2016. LNCS, vol. 9516, pp. 238–250. Springer, Cham (2016). https://doi.org/10.1007/978-3-319-27671-7_20
5. Blei, D.M., Ng, A.Y., Jordan, M.I.: Latent dirichlet allocation. J. Mach. Learn. Res. **3**(Jan), 993–1022 (2003)
6. Cvitanic, T., Lee, B., Song, H.I., Fu, K., Rosen, D.: Lda vs lsa: a comparison of two computational text analysis tools for the functional categorization of patents. In: International Conference on Case-Based Reasoning (2016)
7. Deerwester, S., Dumais, S.T., Landauer, T.K., Furnas, G., Beck, F.d.L., Leighton-Beck, L.: Improvinginformation-retrieval with latent semantic indexing (1988)
8. Drachsler, H., Verbert, K., Santos, O.C., Manouselis, N.: Panorama of recommender systems to support learning. In: Ricci, F., Rokach, L., Shapira, B. (eds.) Recommender Systems Handbook, pp. 421–451. Springer, Boston, MA (2015). https://doi.org/10.1007/978-1-4899-7637-6_12
9. Galanopoulos, D., Mezaris, V.: Temporal lecture video fragmentation using word embeddings. In: Kompatsiaris, I., Huet, B., Mezaris, V., Gurrin, C., Cheng, W.-H., Vrochidis, S. (eds.) MMM 2019. LNCS, vol. 11296, pp. 254–265. Springer, Cham (2019). https://doi.org/10.1007/978-3-030-05716-9_21

10. Gutiérrez, L., Keith, B.: A systematic literature review on word embeddings. In: Mejia, J., Muñoz, M., Rocha, Á., Peña, A., Pérez-Cisneros, M. (eds.) CIMPS 2018. AISC, vol. 865, pp. 132–141. Springer, Cham (2019). https://doi.org/10.1007/978-3-030-01171-0_12

11. Kastrati, Z., Imran, A.S., Kurti, A.: Integrating word embeddings and document topics with deep learning in a video classification framework. Pattern Recogn. Lett. **128**, 85–92 (2019)

12. Kastrati, Z., Kurti, A., Imran, A.S.: Wet: word embedding-topic distribution vectors for MOOC video lectures dataset. Data Brief **28**, 105090 (2020)

13. Lewis, D.D., Jones, K.S.: Natural language processing for information retrieval. Commun. ACM **39**(1), 92–101 (1996)

14. Manning, C.D., Surdeanu, M., Bauer, J., Finkel, J.R., Bethard, S., McClosky, D.: The stanford corenlp natural language processing toolkit. In: Proceedings of 52nd Annual Meeting of the Association for Computational Linguistics: System Demonstrations, pp. 55–60 (2014)

15. Mikolov, T., Sutskever, I., Chen, K., Corrado, G.S., Dean, J.: Distributed representations of words and phrases and their compositionality. In: Advances in Neural Information Processing Systems, pp. 3111–3119 (2013)

16. Pappano, L.: The year of the MOOC. New York Times **2**(12), 2012 (2012)

17. Perkins, J.: Python 3 Text Processing with NLTK 3 Cookbook. Packt Publishing Ltd. (2014)

18. Ramos, J., et al.: Using TF-IDF to determine word relevance in document queries. In: Proceedings of the First Instructional Conference on Machine Learning, Piscataway, vol. 242, pp. 133–142 (2003)

19. Řehůřek, R., Sojka, P.: Software framework for topic modelling with large corpora. In: Proceedings of the LREC 2010 Workshop on New Challenges for NLP Frameworks, pp. 45–50. ELRA, Valletta, Malta, May 2010. http://is.muni.cz/publication/884893/en

20. Romero, C., Ventura, S.: Educational data mining: a review of the state of the art. IEEE Trans. Syst. Man Cybern. Part C (Appl. Rev.) **40**(6), 601–618 (2010)

21. Springmeyer, P.: Inflector for python (2019). https://pypi.org/project/Inflector/

22. Tucker, B.: The flipped classroom. Educ. Next **12**(1), 82–83 (2012)

23. Turcu, G., Mihaescu, M.C., Heras, S., Palanca, J., Julián, V.: Video transcript indexing and retrieval procedure. In: SoftCOM 2019, pp. 1–6. IEEE (2019)

24. Zhu, H., Dong, L., Wei, F., Qin, B., Liu, T.: Transforming wikipedia into augmented data for query-focused summarization. arXiv:1911.03324 (2019)

Visualization of Numerical Association Rules by Hill Slopes

Iztok Fister Jr.[1(✉)], Dušan Fister[2], Andres Iglesias[3,4], Akemi Galvez[3,4], Eneko Osaba[6], Javier Del Ser[5,6], and Iztok Fister[1,3]

[1] Faculty of Electrical Engineering and Computer Science, University of Maribor, Smetanova 17, 2000 Maribor, Slovenia
iztok.fister1@um.si
[2] Faculty of Economics and Business, University of Maribor, Razlagova 14, 2000 Maribor, Slovenia
[3] University of Cantabria, Avenida de los Castros, s/n, 39005 Santander, Spain
[4] Toho University, 2-2-1 Miyama, Funabashi 274-8510, Japan
[5] University of the Basque Country (UPV/EHU), Bilbao, Spain
[6] TECNALIA, Basque Research and Technology Alliance (BRTA), Derio, Spain

Abstract. Association Rule Mining belongs to one of the more prominent methods in Data Mining, where relations are looked for among features in a transaction database. Normally, algorithms for Association Rule Mining mine a lot of association rules, from which it is hard to extract knowledge. This paper proposes a new visualization method capable of extracting information hidden in a collection of association rules using numerical attributes, and presenting them in the form inspired by prominent cycling races (i.e., the Tour de France). Similar as in the Tour de France cycling race, where the hill climbers have more chances to win the race when the race contains more hills to overcome, the virtual hill slopes, reflecting a probability of one attribute to be more interesting than the other, help a user to understand the relationships among attributes in a selected association rule. The visualization method was tested on data obtained during the sports training sessions of a professional athlete that were processed by the algorithms for Association Rule Mining using numerical attributes.

Keywords: Association rule mining · Optimization · Sports training · Tour de France · Visualization

1 Introduction

Association Rule Mining (ARM) [1] is an important part of Machine Learning that searches for relations among features in a transaction database. The majority of the algorithms for ARM work on categorical features, like Apriori proposed by Agrawal [2], Eclat, introduced by Zaki et al. [20], and FP-Growth, developed by Han et al. [8]. Algorithms for dealing with the numerical features have also been developed recently [3,6].

© Springer Nature Switzerland AG 2020
C. Analide et al. (Eds.): IDEAL 2020, LNCS 12489, pp. 101–111, 2020.
https://doi.org/10.1007/978-3-030-62362-3_10

Typically, these algorithms generate a huge number of association rules, from which it is hard to discover the most important relations. In order to extract a knowledge from the collection of mined association rules, a lot of tools have emerged [7] that are able to cope with complex data structures, e.g., instance-relationship data [5], user generated context [9,12], and scanner data [14]. On the other hand, there are a number of papers proposing data visualization methods as a means for extracting meaningful results from highly complex settings [4].

This paper focuses on the visualization of ARM using numerical features. Thus, the relationships among these features are visualized using an inspiration taken from one of the most prominent cycling races in the world, i.e., the Tour de France (TDF). Similar as in the Tour de France cycling race, where the hill climbers have more chances to win the race, when the race contains more hills to overcome, the virtual hill slopes, reflecting a probability of one attribute to be more interesting than the other, help a user to understand the relationships among attributes in a selected association rule.

The proposed visualization method was applied on a transaction database consisting of data obtained by measuring data during the training sessions with a mobile device worn by a professional cyclist in the past three seasons. The results of visualization reveal that using the method in the real-world can improve the interpretation of the mining of the association rules, and direct the user to the more important ones.

The structure of the paper is as follows. Section 2 highlights the basic information needed for understanding the subject of the paper. In Sect. 3, an algorithm for visualizing the mined association rules is illustrated in detail. The results of the proposed visualization method are presented in Sect. 4. The paper is concluded by Sect. 5, which summarizes the work performed and outlines directions for the future.

2   Basic Information

This section is focused on the background information necessary for understanding the subject that follows. At first, the principles of TDF serving as an inspiration for visualization are highlighted, followed by describing the problem of discovering association rules.

2.1   Tour de France

Numerous research articles published in the past support the idea that good hill climbing abilities are a nuisance for winning the Tour De France (TDF). Climber specialists (fr. grimpeur), all-rounders (fr. rouleur) and time-trial specialists (fr. chronoman) usually fight for overall podium positions at the Champs-Élysées, contrary to the breakaway specialists (fr. baroudeur) and sprinters, who strive for glory at individual stages. Good hill climbing abilities come naturally come with suitable anthropometric stature: According to [16], climbers usually weigh 60–66 kg, with their BMI (Body Mass Index) reaching 19–20 kg/m^2. Compared to other specialists, climbers perform exceptionally well at maintaining high relative power output to their weight W/kg. From the characteristics of climbers, we deduce that steep climbs, where intensity is near maximum, determine (or are crucial for) the overall winner of TDF, and we further deduce that climbers are in a favorable role there.

Lucia et al. [10] introduce the TDF historical overview and climbing facts. Good climbing performance in the Alps and Pyrenees is highly correlated to good time-trial performance [18], which provides a good chance that a strong climber will perform solidly at the time-trial, too. Indeed, both are necessary to win the TDF [17]. However, Rogge et al. [13] agree that good performance at mountain stages is crucial for a favorable TDF result overall. Torgler [18] further exposes the difficulties of mountain stages, and emphasizes the high efforts needed to provide good General Classification (GC); the steep mountainous stages are supposed to be the most difficult ones among them all and, thus, are decisive for a successful GC; subsequent climbs with descents are found to be most exhaustive. It follows that, the more exhaustive the stage, the larger the time differences at the finish. Sanders and Heijboer [15] supported this idea, by finding out that mountain stages are of the highest intensity and exercise load among mass-start stage types. To a higher degree, this is because of the total elevation gain and highly alternating pace that occurs at hilltops. In our opinion, van Erp et al. [19] present the most comprehensive empirical study of individual load, intensity and performance characteristics of a single GC contender. Among many testing hypotheses, authors state that the most necessary to compete for victory in a Grand Tour is to achieve the highest power output possible (app. 5.7–6.0 W/kg) at key mountain stages.

Climbing to "Hors Catégorie" (HC) climbs is extremely specific. Such climbs are usually of extraordinary distance and elevation, and, thus, require extreme efforts. At high altitudes, moderate hypoxia can come into play, which tightens the cyclist's margins and increases physical fatigue even more. An example of an HC finish climb is shown in Fig. 1. These facts contribute easily to early exhaustion, or overreaching [10] and thus are critical for good overall GC classification. The lost time at mountainous stages usually cannot be recovered anymore (cyclists can barely limit only the losses).

On the other hand, competing to win the TDF is not only about climbing. The TDF is extremely psychologically and physically demanding, especially for GC contenders: (1) These need to be cautious of opponents at all times, (2) No relaxation days are allowed to them and (3) a single bad day or opponent's explosive burst may be devastating. Without mentioning high temperatures, team spirit, injuries, crashes and technical glitches, the psychological and physical tension to GC contenders are the highest at the high-intensity phases, such as steep hills [11].

2.2   Association Rule Mining

ARM can be defined formally as follows: Let us assume a set of objects $O = \{o_1, \ldots, o_M\}$ and transaction dataset $D = \{T\}$ are given, where each transaction

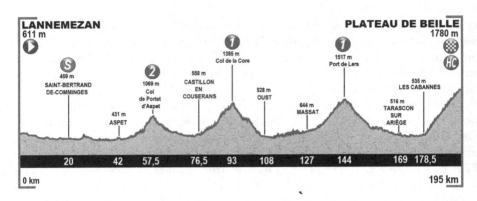

Fig. 1. Example of TDF Stage 12, TDF 2015. Image origin: https://commons. wikimedia.org/wiki/File:Profile_stage_12_Tour_de_France_2015.png, distributed under: Creative Commons Attribution-Share Alike 4.0 International License

T is a subset of objects $T \subseteq O$. Then, an association rule is defined as an implication:

$$X \Rightarrow Y, \tag{1}$$

where $X \subset O$, $Y \subset O$, and $X \cap Y = \emptyset$. In order to estimate the quality of a mined association rule, two measures are defined: A support and a confidence. The support is defined as:

$$supp(X) = \frac{|t \in T; X \subset t|}{|T|}, \tag{2}$$

while the confidence as:

$$conf(X \Rightarrow Y) = \frac{n(X \cup Y)}{n(X)}, \tag{3}$$

where function $n(.)$ calculates the number of repetitions of a particular rule within D, and $N = |T|$ is the total number of transactions in D. Let us emphasize that two additional variables are defined, i.e., the minimum confidence C_{min} and the minimum support S_{min}. These variables denote a threshold value limiting the particular association rule with lower confidence and support from being taken into consideration.

3   Description of Constructing the New Visualization Method for ARM

The problem of predicting the winner of the TDF can be defined informally as follows: Let us assume that cyclist X is a specialist for hill climbing. This cyclist started in n from the total N races and overcame a set of elite cyclists $Y_1, \ldots, Y_{m-1}$ $M_1, \ldots, M_{m-1}$-times, respectively, where m denotes the number of observed cyclists, i.e., not only hill climbers. Such framework could be applied to all cyclists in general. Based on these data, the question is, what is the probability for X to win the TDF this year?

The problem could be solved visually in the sense of the ARM as follows: The number of races in which the cyclist X started can be expressed as $supp(X)$. The same is also true for cyclist Y_i, where the number of his starts is expressed as $supp(Y_i)$. The numbers of races, where X overcame Y_i can be expressed as $conf(X \Rightarrow Y_i)$. Then, the equivalent relation

$$supp(X) \equiv conf(X \Rightarrow Y) \tag{4}$$

means that cyclist X overcame cyclist Y_i in all races in which they both started. This relation can be visualized as a rectangular triangle with two sides of equal length, $supp(Y_i)$ and $conf(X \Rightarrow Y_i)$, and the length of diagonal L is expressed by Pythagoras rule, as follows:

$$L = \sqrt{supp^2(Y_i) + conf^2(X \Rightarrow Y_i)}. \tag{5}$$

As a result, a sequence of triangles is obtained, where each triangle highlights the relationship between the two cyclists. If triangles are ordered according to their supports and put onto a line with spacing proportional to a $conf(X \Rightarrow Y_i)$, and the model triangle with two sides equal to $supp(X)$ is added, the new visualization method emerges, as presented in Fig. 2.

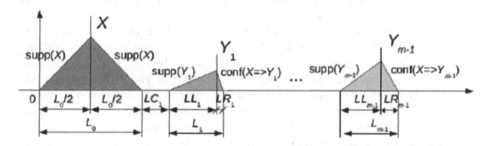

Fig. 2. Mathematical model of virtual hills, on which the new visualization method is founded.

As can be seen from Fig. 2, the model triangle representing the triangle with the largest area, is isosceles, and it is placed on the position $L_1/2$ distant from the origin of 0. However, the positions L_i for $i = 0, \ldots, m - 1$ are calculated according to Eq. 5. A corresponding length of the triangle is calculated for each of the observed cyclists Y_i. The position of the triangle on the line is determined as follows:

$$pos_i = L_0 + \sum_{j=1}^{i-1} (LC_j + LL_j + LR_j) + (LC_i + LL_i), \tag{6}$$

where L_0 denotes the diagonal length of the model triangle, $LC_j \propto conf(X \Rightarrow Y_j)$ is the distance between two subsequent triangles, LL_j, expressed as follows:

$$\cos\alpha = \frac{supp(Y_j)}{L_j}, \quad \text{for } j = 1,\ldots,m-1,$$

$$LL_j = supp(Y_j) \cdot \cos\alpha = \frac{supp^2(Y_j)}{L_j}, \tag{7}$$

while LR_j as:

$$\cos\beta = \frac{conf(X \Rightarrow Y_j)}{L_j}, \quad \text{for } j = 1,\ldots,m-1,$$

$$LR_j = conf(X \Rightarrow Y_j) \cdot \cos\beta = \frac{conf^2(X \Rightarrow Y_j)}{L_j}. \tag{8}$$

The interpretation of these triangles representing hills in the TDF is as follows: At first, the larger the area of the triangle for cyclist Y_i, the more chances for cyclist X to overcome this in the race. The same relationship is highlighted in the distance between these triangles. This means that the more the triangle is away from the model triangle, the higher is the probability that X will be overcome by Y_i. Let us emphasize that the discussion is valid for classical ARM using numerical attributes, where the mined association rules serve only as a basis for determining the best features. Thus, the implication relation in the rule is ignored and redefined by introducing the $m-1$ individual implication relations between pairs of features.

Similar as steep slopes have a crucial role in the TDF cycle race for determining the final winner, the virtual hill slopes help the user to understand the relations among features in the transaction database. Indeed, this visualization method can also be applied for visualizing features in the ARM transaction databases. Here, the features are taken into consideration instead of cyclists. On the other hand, interpretation of visualization is also slightly different from the TDF. Here, we are interested in those relations between features that most highlights the mined association rules. The larger the area of the triangle Y_i, the closer the relationship between the feature X.

4   Experiments and Results

The goal of our experimental work was to show that the mined association rules can be visualized using the proposed visualization method based on inspiration taken from the TDF cycling competition. In line with this, two selected association rules mined from a corresponding transaction database are visualized.

The association rules were mined using contemporary approaches using stochastic nature-inspired population-based algorithms, like Differential Evolution [6]. Then, the best mined association rules according to support and confidence are taken into consideration. In each specific association rule, the feature with the the best support is searched for. This feature serves as a model, with which all the other features in the observed rule are compared according to a confidence.

The transaction database consists of seven numerical features, whose domain of feasible values are illustrated in Table 1. The transaction database consists of

Table 1. Observed numerical features with their domain of values.

Nr.	Feature	Domain
F-1	Duration	[43.15, 80.683]
F-2	Distance	[0.00, 56.857]
F-3	Average HR	[72, 151]
F-4	Average altitude	[0.2278, 1857.256]
F-5	Maximum altitude	[0.0, 0.0]
F-6	Calories	[20.0, 1209]
F-7	Ascent	[0.00, 1541
F-8	Descent	[0.00, 1597]

700 transactions representing the results measured by a mobile device worn by a professional cyclist during the sports training session. These data were obtained in the last three cycling seasons. In a temporary sense, this means that we can start to set a get valuable visualization with full predictive power for the current season after lapse of three seasons. However, this does not mean that the method cannot be applied before elapsing three seasons, but the obtained results could be less accurate.

The best two association rules according to support are presented in Table 2, where they are denoted as visualization scenarios 1–2. Let us emphasize that the association rules are treated without implication relation. Here, the intervals of numerical attributes are important for the proposed visualization method only. On the other hand, the proposed approach is not limited by a huge number

Table 2. Mined numerical features in association rules.

Feature	Scenario 1	Scenario 2
Duration	[76.67, 78.07]	[46.95, 65.87]
Distance	[14.28, 26.32]	[26.24, 53.30]
Average HR	[78.79, 114.92]	[104.12, 141.40]
Average altitude	[631.70, 1809.21]	[17.59, 547.05]
Calories	[774.92, 1161.43]	[1096.82, 1209.00]
Ascent	[0.00, 10.00]	[0.00, 74.19]
Descent	[0.00, 54.19]	[0.00, 623.88]

of triangle diagrams, because here we are focused on visualization of the best association rules according to some criteria that normally have a limited number of numerical attributes.

Experimental figures were drawn using the Matlab software framework, using the colored 3-D ribbon plot. All the Figures start at *Location* = 0 and spread on the x axis. The height of the triangles is symbolized at the z axis, and the shades of color are represented by a vertical color-bar. On the other hand, the y axis does not include any meaning.

The visualization of the two mentioned scenarios are presented in the remainder of the paper.

4.1   Scenario 1

The best association rule is presented in Table 3, where the feature F-2 (i.e., "Distance") is compared with the closest features according to a confidence, i.e., "Average HR", "Average altitude", "Descent", "Duration", "Ascent", "Calories".

Table 3. Scenario 1 in numbers.

Scenario	$supp(X)$	$conf(X \Rightarrow Y_i)$					
1	F-2	F-3	F-4	F-8	F-1	F-7	F-6
	0.40	0.19	0.16	0.06	0.05	0.02	0.01

The corresponding visualization of these data are illustrated in Fig. 3, from which it can be seen that the feature "Distance" has higher interdependence with features "Average HR" and "Average altitude" only, but the relationships

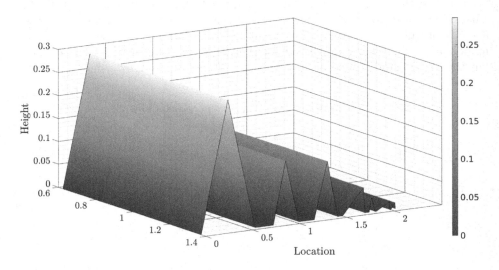

Fig. 3. Visualization of Scenario 1.

among the other features do not have a higher effect on the performance of the athlete.

4.2 Scenario 2

In this scenario, only six features are incorporated, because the feature R-6 (i.e., "Calories") does not affect the performance of the cyclist in training (Table 4) (Fig. 4).

Table 4. Scenario 2 in numbers.

Scenario	$supp(X)$	$conf(X \Rightarrow Y_i)$					
2	F-8	F-3	F-4	F-1	F-7	F-2	F-6
	0.93	0.85	0.75	0.57	0.26	0.22	0.00

As can be seen from Fig. 4, there are six hills, where the former three hills are comparable with the first one according to the area, while the last two expose the lower interdependence. Indeed, the higher interdependence is also confirmed by the larger distances between hills.

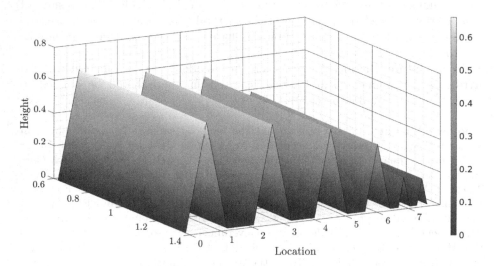

Fig. 4. Second visualization.

5   Conclusion

ARM using numerical attributes of features was rarely applied in practice. The task of the algorithm for ARM with numerical attributes is to find the proper boundary values of numerical features. Consequently, these values specify mined association rules using different values of support and confidence. Thus, the association rules with the best values of support and their closeness to the other features are the more interesting for the user. The users suffer from the lack of information that is hidden in association rules. Obviously, the solution of the problem presents various visualization tolls for extracting the knowledge hidden in data.

This paper proposes a new visualization method inspired by the TDF. Similar as in the TDF cycle race, where the hill climbers have more chances to win the race when the race contains more hills to overcome, the virtual hill slopes, reflecting a probability of one attribute to be more interesting than the other, help a user to understand the relationships among attributes in a selected association rule.

Thus, the relationships between features in the transaction database are illustrated using triangles, representing hills, that need to be overcome by the cyclists. The first triangle in a sequence is a model, because it contains the largest area. The other triangles represent the opponents in the following sense: The larger the area of a definite triangle, the easier it is for the opponent to overcome. In the ARM sense, this means the following: The larger the triangle, the closer the feature in the transaction database.

The visualization method was employed on a transaction database consisting of features characterizing the realized sports training sessions. Two scenarios were visualized, based on two selected mined association rules. The results of visualization showed the potential of the method, that is able to illustrate the hidden relationships in a transaction database in an easy and understandable way to the user.

In the future, the method could also be broadened for dealing with mixed attributes, i.e., numerical and categorical. The method should be applied to another transaction databases.

Acknowledgments. Iztok Fister thanks the financial support from the Slovenian Research Agency (Research Core Funding No. P2-0042 - Digital twin). Iztok Fister Jr. thanks the financial support from the Slovenian Research Agency (Research Core Funding No. P2-0057). Dušan Fister thanks the financial support from the Slovenian Research Agency (Research Core Funding No. P5-0027). J. Del Ser and E. Osaba would like to thank the Basque Government through EMAITEK and ELKARTEK (ref. 3KIA) funding grants. J. Del Ser also acknowledges funding support from the Department of Education of the Basque Government (Consolidated Research Group MATHMODE, IT1294-19). Andres Iglesias and Akemi Galvez acknowledge financial support from the project PDE-GIR of the European Union's Horizon 2020 research and innovation programme under the Marie Sklodowska-Curie grant agreement No 778035, and the Spanish Ministry of Science, Innovation, and Universities (Computer Science National Program) under grant #TIN2017-89275-R of the Agencia Estatal de Investigación and European Funds EFRD (AEI/FEDER, UE).

References

1. Agrawal, R., Imieliński, T., Swami, A.: Mining association rules between sets of items in large databases. In: Proceedings of the 1993 ACM SIGMOD International Conference on Management of Data, SIGMOD 1993, pp. 207–216. ACM, New York (1993). http://doi.acm.org/10.1145/170035.170072
2. Agrawal, R., Srikant, R.: Fast algorithms for mining association rules. In: Proceedings of 20th International Conference on VLDB, pp. 487–499 (1994)
3. Altay, E.V., Alatas, B.: Performance analysis of multi-objective artificial intelligence optimization algorithms in numerical association rule mining. J. Ambient Intell. Hum. Comput. **11**, 1–21 (2019). https://doi.org/10.1007/s12652-019-01540-7
4. Arrieta, A.B., et al.: Explainable artificial intelligence (xai): Concepts, taxonomies, opportunities and challenges toward responsible ai. Inf. Fusion **58**, 82–115 (2020)
5. Fader, P.S., Hardie, B.G.S., Shang, J.: Customer-base analysis in a discrete-time noncontractual setting. Market. Sci. **29**(6), 1086–1108 (2010)
6. Fister Jr., I., Iglesias, A., Galvez, A., Del Ser, J., Osaba, E., Fister, I.: Differential evolution for association rule mining using categorical and numerical attributes. In: International Conference on Intelligent Data Engineering and Automated Learning, pp. 79–88 (2018)
7. Hahsler, M., Karpienko, R.: Visualizing association rules in hierachical groups. J. Bus. Econ. **87**, 317–335 (2017)
8. Han, J., Pei, J., Yin, Y.: Mining frequent patterns without candidate generation. In: Proceedings of the 2000 ACM SIGMOD International Conference on Management of Data, SIGMOD 2000, pp. 1–12. Association for Computing Machinery, New York (2000). https://doi.org/10.1145/342009.335372
9. Lee, T.Y., Bradlow, E.T.: Automated marketing research using online customer reviews. J. Market. Res. **48**(5), 881–894 (2011)
10. Lucía, A., Earnest, C., Arribas, C.: The Tour de France: a physiological review (2003)
11. Lucía, A., Hoyos, J., Santalla, A., Earnest, C., Chicharro, J.L.: Tour de France versus Vuelta a España: which is harder? Med. Sci. Sports Exerc. **35**(5), 872–878 (2003)
12. Netzer, O., Feldman, R., Goldenberg, J., Fresko, M.: Mine your own business: market-structure surveillance through text mining. Market. Sci. **31**(3), 521–543 (2012)
13. Rogge, N., Reeth, D.V., Puyenbroeck, T.V.: Performance evaluation of tour de france cycling teams using data envelopment analysis. Int. J. Sport Finance **8**(3), 236–257 (2013)
14. Rooderkerk, R.P., Van Heerde, H.J., Bijmolt, T.H.: Optimizing retail assortments. Market. Sci. **32**(5), 699–715 (2013)
15. Sanders, D., Heijboer, M.: Physical demands and power profile of different stage types within a cycling grand tour. Eur. J. Sport Sci. **19**(6), 736–744 (2019)
16. Santalla, A., Earnest, C.P., Marroyo, J.A., Lucía, A.: The Tour de France: an updated physiological review (2012)
17. Sundhagen, T.A.: Lance Armstrong: an American Legend? (2011)
18. Torgler, B.: "La Grande Boucle" : determinants of success at the Tour de France. J. Sports Econ. **8**(3), 317–331 (2007)
19. Van Erp, T., Hoozemans, M., Foster, C., De Koning, J.J.: Case report: load, intensity, and performance characteristics in multiple grand tours. Med. Sci. Sports Exerc. **52**(4), 868–875 (2020)
20. Zaki, M.J., Parthasarathy, S., Ogihara, M., Li, W.: New Algorithms for Fast Discovery of Association Rules. Technical report, USA (1997)

Adaptation and Anxiety Assessment in Undergraduate Nursing Students

Ana Costa[1], Analisa Candeias[2] ⓘ, Célia Ribeiro[3], Herlander Rodrigues[2],
Jorge Mesquita[2], Luís Caldas[4], Beatriz Araújo[5], Isabel Araújo[6] ⓘ,
Henrique Vicente[7,8] ⓘ, Jorge Ribeiro[9] ⓘ, and José Neves[6,8(✉)] ⓘ

[1] Hospital Senhora da Oliveira, Guimarães, Portugal
a45330@gmail.com
[2] Escola Superior de Enfermagem, Universidade do Minho, Braga, Portugal
acandeias@ese.uminho.pt, twinscorpion@gmail.com
[3] Hospital da Misericórdia, Vila Verde, Portugal
celia.ribeiro1984@gmail.com
[4] Centro Hospitalar de Setúbal, Setúbal, Portugal
luiscaldas@gmail.com
[5] Centro de Investigação Interdisciplinar em Saúde, Universidade Católica Portuguesa,
Lisbon, Portugal
bea9araujo@gmail.com
[6] CESPU, Instituto Universitário de Ciências da Saúde, Famalicão, Portugal
isabel.araujo@ipsn.cespu.pt
[7] Departamento de Química, Escola de Ciências e Tecnologia, REQUIMTE/LAQV,
Universidade de Évora, Évora, Portugal
hvicente@uevora.pt
[8] Centro Algoritmi, Universidade do Minho, Braga, Portugal
jneves@di.uminho.pt
[9] Instituto Politécnico de Viana do Castelo, Rua da Escola Industrial e Comercial de
Nun'Álvares, 4900-347 Viana do Castelo, Portugal
jribeiro@estg.ipvc.pt

Abstract. The experiences and feelings in a first phase of transition from undergraduate to graduate courses may lead to some kind of anxiety, depression, malaise or loneliness that are not easily overwhelmed, no doubt the educational character of each one comes into play, since the involvement of each student in academic practice depends on his/her openness to the world. In this study it will be analyzed and evaluated the relationships between academic experiences and the correspondent anxiety levels. Indeed, it is important not only a diagnose and evaluation of the students' needs for pedagogical and educational reorientation, but also an identification of what knowledge and attitudes subsist at different stages of their academic experience. The system envisaged stands for a Hybrid Artificial Intelligence Agency that integrates the phases of data gathering, processing and results' analysis. It intends to uncover the students' states of Adaptation, Anxiety and Anxiety Trait in terms of an evaluation of their entropic states, according to the 2nd Law of Thermodynamics, i.e., that energy cannot be created or destroyed; the total quantity of energy in the universe stays the same. The logic procedures are based on a Logic Programming approach to Knowledge Representation and Reasoning complemented with an Artificial Neural Network approach to computing.

© Springer Nature Switzerland AG 2020
C. Analide et al. (Eds.): IDEAL 2020, LNCS 12489, pp. 112–123, 2020.
https://doi.org/10.1007/978-3-030-62362-3_11

Keywords: Adaptation · Anxiety · Anxiety trait · Artificial Intelligence · Entropy · Logic Programming · Artificial neural networks

1 Introduction

On the one hand, the quality of academic life and the experience of students in the transition from basic courses to higher education prove to be an area of interest for researchers [1]. On the other hand, the diversity of student profiles that reach our universities, be it cultural, social or educational, indicates a review of the adaptation and learning processes. Indeed, promoting the quality of academic experience during the training course is an incentive for a responsibility that the institution tends to share through its organization and services; of employees and teachers; the family; the environment; the students or their whereabouts. It is therefore essential to diagnose the needs of students with a view to reorienting them at the pedagogical, educational level or determining what knowledge and attitudes are required in the various phases of academic experience [2]. In order to achieve an optimal adaptation to higher education and the highest quality of all academic experiences, strategies to control feelings, attitudes and signs that show fear must also be developed. This can pose a threat to student performance because it is an emotion that symbolizes concern and fear. This threat can serve as a model for changing the student's emotional state at a particular time and adding inherent factors to the person, their environment and their ability to be in harmony with themselves [3]. The aim of our study is undeniably to promote the best possible adaptation of the student to the bachelor nursing courses, to the institution and its training and to reduce the fear in the academic environment. The document is divided into 4 (four) chapters. The former one is dedicated to the theoretical revision of the literature. Chapter 2 introduces the theoretical questions, namely how the questions of qualitative thinking, uncertainty and vagueness are dealt with and how they relate to Logic Programming [4]. Chapter 3 contains a case study and its development into a complete computer system [5, 6]. Chapter 4 contains the presentation and discussion of the results as well as perspectives for further research.

2 Fundamentals

Knowledge Representation and Reasoning (KRR) practices may be understood as a process of energy devaluation [7]. A data item is to be understood as being in a given moment at a particular entropic state as untainted energy, whose values range in the interval $0...1$. According to the First Law of Thermodynamics it stands for an amount of energy that cannot be consumed in the sense of destruction, but may be consumed in the sense of devaluation. It may be presented in the form, viz.

- *exergy*, sometimes called available energy or more precisely available work, is the part of the energy which can be arbitrarily used or, in other words, the entropy generated by it. In Fig. 1 it is given by the gray colored areas;

- *vagueness*, i.e., the corresponding *energy values* that *may or may not have been consumed*. In Fig. 1 it is given by the gray colored areas with spheres; and
- *anergy*, that stands for an *energetic potential* that was not yet consumed, being therefore available, i.e., *all of energy that is not exergy*. In Fig. 1 it is given by the white colored areas [7].

As a data collection instrument, it will be used the *QVA-r Questionnaire – Six Dimensions (QVAQ – 6)*, a self-report questionnaire that seeks to assess how young people adapt to some of the demands of academic life. However, it will be extended with another dimension in order to assess the student's integration process. *QVAQ – 6* is now presented in the form, viz.

Q1 – Are you aware of aspects such as your emotional balance, affective stability, optimism, decision-making and self-confidence that may affect your Personal Dimension?
Q2 – Are you aware of aspects such as your emotional balance, affective stability, optimism, decision-making and self-confidence that may affect your Inter-personal Dimension?
Q3 – Are you aware of your feelings related to the course attended and career prospects that includes satisfaction with the course and perception of skills for the course that stand for the Career Dimension?
Q4 – Are you aware of aspects that refers to study habits and time management, that includes study routines, time planning, use of learning resources, and test preparation, among others, that may affect your Study Dimension?
Q5 – Are you aware of aspects that refers to the appraisal of the educational institution attended, that includes feelings related to the institution, the desire to stay or change institutions, knowledge and appreciation of the infrastructure? and
Q6 – The IT organization of your educational institution has a problem-solving strategy or methodology that ensures the security of our IT systems and data?

In order to answer if the organization's goals are well defined and agree with the students' ones, on the assumption that the organization and the students have a very good match, i.e., that high scores are to be found in the answers to the questionnaire, leading to positive results and benefits at the organizational level. Therefore, it was decided to use the scale, viz.

strongly agree (4), *agree* (3), *disagree* (2), *strongly disagree* (1), *disagree* (2), *agree* (3), *strongly agree* (4)

in order to assess the student's work (i.e. the attained entropic state), measured in terms of the energy consumed to answer the questions. Moreover, it is included a neutral term, *neither agree nor disagree*, which stands for *uncertain* or *vague*. The reason for the individual's answers is in relation to the query, viz.

As a member of the academic organization, how much would you agree with each one of QVAQ – 6 referred to above?

In order to transpose the qualitative information into a quantitative one all calculation details for $QVAQ - 6$ are presented, in order to illustrate the process. The answers of the student #1 to the $QVAQ - 6$ group are present in Table 1, where the input for $Q1$ means that he/she *strongly agrees* but does not rule out the possibility that he/she will *agree* in certain situations. The inputs are be read from left to right, i.e., from *strongly agree* to *strongly disagree* (with increasing entropy), or from *strongly disagree* to *strongly agree* (with decreasing entropy), i.e., in terms of Fig. 1 the markers on the axis correspond to any of the possible scale options, which may be used from *bottom* → *top* (from *strongly*

Table 1. $QVAQ - 6$ Single Student Answers.

Questions	Scale							
	(4)	(3)	(2)	(1)	(2)	(3)	(4)	vagueness
Q1					×		×	
Q2	×	×						
Q3					×	×		
Q4						×		
Q5								×
Q6						×		

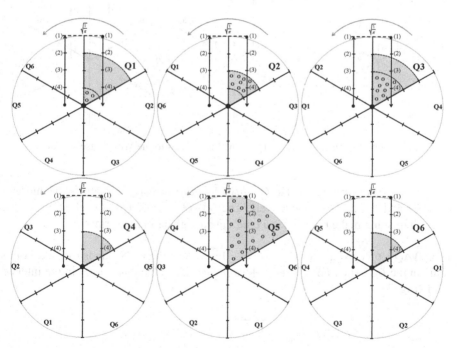

Fig. 1. A graphic representation of students' answers to the Best-case scenario.

agree to *strongly disagree*), indicating that the performance of the system (set below as *Program 1* with the acronym *LOP*) as far as a student's evaluation of the degree of *Adaptation, Anxiety* and *Anxiety Trait* decreases as entropy increases, or used from *top* → *bottom* (from *strongly disagree* to *strongly agree*), indicating that the performance of the system increases as entropy decreases). Figure 1 and Fig. 2 show the conversion of the information contained in Table 1 into the diverse types of energy described above (*exergy, vagueness* and *anergy*) for the *Best* and *Worst* case scenarios. The contribution of each individual to the system entropic state as untainted energy is evaluated as shown in Table 2 for both scenarios.

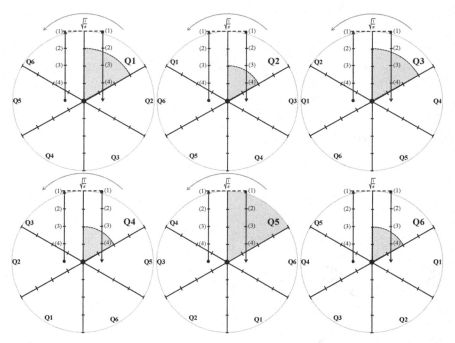

Fig. 2. A graphic representation of students' responses to the Worst-case scenario.

The *qvaq − 6 performance* and its *Quality-of-Information (QoI)* range in the intervals *0.88...0.92* and *0.53...0.61*, respectively. The *QVAQ − 6* and *QoI's* evaluation for the different items that make the *qvaq − 6 predicate* are given in the form, viz.

- The *QVAQ − 6* is given as $QVAQ - 6 = \sqrt{1 - ES^2}$, where *ES* stands for an exergy's value in the *BCS* (i.e., *ES = exergy + vagueness*), a value that ranges in the interval 0...1 (Fig. 3).

$$QVAQ - 6 = \sqrt{1 - (0.14 + 0.25)^2} = 0.92$$

- *QoI* is evaluated in the form, viz.

Table 2. Best and Worst-case scenarios (BCS and WCS).

Questions	Best Case Scenario	Worst Case Scenario
Q1	$exergy_{Q1} = \frac{1}{6}\pi r^2 \big]_0^{\frac{1}{4}\sqrt{\frac{1}{\pi}}} = 0.01$	$exergy_{Q1} = \frac{1}{6}\pi r^2 \big]_0^{\frac{3}{4}\sqrt{\frac{1}{\pi}}} = 0.09$
	$vagueness_{Q1} = \frac{1}{6}\pi r^2 \big]_0^{\frac{1}{4}\sqrt{\frac{1}{\pi}}} = 0.01$	$vagueness_{Q1} = \frac{1}{6}\pi r^2 \big]_{\frac{1}{4}\sqrt{\frac{1}{\pi}}}^{\frac{1}{4}\sqrt{\frac{1}{\pi}}} = 0$
	$anergy_{Q1} = \frac{1}{6}\pi r^2 \big]_{\frac{1}{4}\sqrt{\frac{1}{\pi}}}^{\sqrt{\frac{1}{\pi}}} = 0.16$	$anergy_{Q1} = \frac{1}{6}\pi r^2 \big]_{\frac{3}{4}\sqrt{\frac{1}{\pi}}}^{\sqrt{\frac{1}{\pi}}} = 0.08$
Q2	$exergy_{Q2} = \frac{1}{6}\pi r^2 \big]_0^{\frac{1}{4}\sqrt{\frac{1}{\pi}}} = 0.01$	$exergy_{Q2} = \frac{1}{6}\pi r^2 \big]_0^{\frac{2}{4}\sqrt{\frac{1}{\pi}}} = 0.04$
	$vagueness_{Q2} = \frac{1}{6}\pi r^2 \big]_{\frac{1}{4}\sqrt{\frac{1}{\pi}}}^{\frac{2}{4}\sqrt{\frac{1}{\pi}}} = 0.03$	$vagueness_{Q2} = \frac{1}{6}\pi r^2 \big]_{\frac{2}{4}\sqrt{\frac{1}{\pi}}}^{\frac{2}{4}\sqrt{\frac{1}{\pi}}} = 0$
	$anergy_{Q2} = \frac{1}{6}\pi r^2 \big]_{\frac{1}{4}\sqrt{\frac{1}{\pi}}}^{\sqrt{\frac{1}{\pi}}} = 0.16$	$anergy_{Q2} = \frac{1}{6}\pi r^2 \big]_{\frac{2}{4}\sqrt{\frac{1}{\pi}}}^{\sqrt{\frac{1}{\pi}}} = 0.13$
The evaluation of Q3 is similar to Q1		
Q4	$exergy_{Q4} = \frac{1}{6}\pi r^2 \big]_0^{\frac{2}{4}\sqrt{\frac{1}{\pi}}} = 0.04$	$exergy_{Q4} = \frac{1}{6}\pi r^2 \big]_0^{\frac{2}{4}\sqrt{\frac{1}{\pi}}} = 0.04$
	$vagueness_{Q4} = \frac{1}{6}\pi r^2 \big]_{\frac{2}{4}\sqrt{\frac{1}{\pi}}}^{\frac{2}{4}\sqrt{\frac{1}{\pi}}} = 0$	$vagueness_{Q4} = \frac{1}{6}\pi r^2 \big]_{\frac{2}{4}\sqrt{\frac{1}{\pi}}}^{\frac{2}{4}\sqrt{\frac{1}{\pi}}} = 0$
	$anergy_{Q4} = \frac{1}{6}\pi r^2 \big]_{\frac{2}{4}\sqrt{\frac{1}{\pi}}}^{\sqrt{\frac{1}{\pi}}} = 0.13$	$anergy_{Q4} = \frac{1}{6}\pi r^2 \big]_{\frac{2}{4}\sqrt{\frac{1}{\pi}}}^{\sqrt{\frac{1}{\pi}}} = 0.13$
Q5	$exergy_{Q5} = \frac{1}{6}\pi r^2 \big]_0^{0} = 0$	$exergy_{Q5} = \frac{1}{6}\pi r^2 \big]_0^{\sqrt{\frac{1}{\pi}}} = 0.17$
	$vagueness_{Q5} = \frac{1}{6}\pi r^2 \big]_0^{\sqrt{\frac{1}{\pi}}} = 0.17$	$vagueness_{Q5} = \frac{1}{6}\pi r^2 \big]_{\sqrt{\frac{1}{\pi}}}^{\sqrt{\frac{1}{\pi}}} = 0$
	$anergy_{Q5} = \frac{1}{6}\pi r^2 \big]_{\sqrt{\frac{1}{\pi}}}^{\sqrt{\frac{1}{\pi}}} = 0$	$anergy_{Q5} = \frac{1}{6}\pi r^2 \big]_{\sqrt{\frac{1}{\pi}}}^{\sqrt{\frac{1}{\pi}}} = 0$
The evaluation of Q6 is similar to Q4		

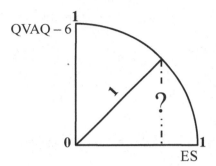

Fig. 3. *QVAQ – 6* evaluation.

$$QoI = 1 - (exergy + vagueness) = 1 - (0.14 + 0.25) = 0.61$$

On the other hand, the data collected above can be structured in terms of the extent of predicate *qvaq – 6*, viz.

qvaq − 6 : *EXergy, VAgueness, ANergy, QVAQ* − *6's Performance,*

Quality − *of* − *Information* → *{True, False}*

and depicted as *Logic Program 1*, below.

3   Case Study

3.1   State and Anxiety Trait

In order to achieve this goal, it will be used the Spielberger – Form Y STAI, a personal reporting instrument [8]. Due to its psychometric qualities, STAI is an invaluable gain in psychological evaluation. It consists of two questionnaires, designed for the self-reported assessment of the intensity of feelings. As such, the STAI adopts a psychological reasoning approach. Thus, it is a dimension that needs to be evaluated in different contexts (e.g., psychology, health, forensic, educational). The fact that STAI provides two measures of assessment, the *State of Anxiety* and the *Anxiety Trace*, multiplies its potential as a tool not only for evaluation but also for research.

State of Anxiety. *State of Anxiety* is defined as a transitory, emotional condition characterized by subjective feelings of tension and apprehension. It will be assessed in terms of the *State of Anxiety Questionnaire Four–Dimensions (SAQ – 4)*, viz.

Q1 – Do you feel strained?
Q2 – Do you feel satisfied?
Q3 – Do you feel frightened? and
Q4 – Do you I feel indecisive?

Table 3. Individual Student Answers to *SAQ – 4* and *ATQ – 6* questionnaires.

Questionnaire	Questions	*Scale*							
		(4)	(3)	(2)	(1)	(2)	(3)	(4)	*vagueness*
SAQ – 4	Q1				×		×		
	Q2		×						
	Q3								×
	Q4						×		
ATQ – 5	Q1						×	×	
	Q2					×			
	Q3		×	×					
	Q4						×		
	Q5								×
	Q6		×	×					

The state scale is designed to assess anxiety as it relates to a specifically experienced situation in terms of apprehension, worry, tension and nervousness. Here, the scale is scored on an eight-point Likert-type (Table 3).

Anxiety Trait. *Anxiety Trait* is defined as anxiety-proneness, i.e., an individual's tendency to respond to stressful situations with raised *State Anxiety*. It is designed to assess the person's general level of anxiety in terms of the general tendency to respond fearfully to a number of aversive stimuli. It will be assessed in terms of the *Anxiety Trait Questionnaire Six–Dimensions (ATQ – 6)* (Table 3), whose scale is scored on an eight-point Likert-type, viz.

Q1 – Do you feel nervous and restless?
Q2 – Do you feel satisfied with yourself?
Q3 – Do you feel like a failure?
Q4 – Do you have disturbing thoughts?
Q5 – Do you make decisions easily? and
Q6 – Do you feel inadequate?

In addition to Tables 1, 3 and 4 show student # 1's answers to the *SAQ – 4* and *ATQ – 6*.
In Table 4 *PP* stands for *predicate's performance* and therefore endorsing the *qvaq – 6*, *saq – 4* and *atq – 6 predicates' performance*.

3.2   Computational Make-up

One's work that went step-by-step to understand the problem and come up with a solution was possible due to the power of *Computational Logic* or *Computational Thinking*, i.e.,

Table 4. The extent of the *qvaq – 6, saq – 4* and *atq – 6* predicates obtained according to an individual student's answer to the *QVAQ – 6, SAQ – 4 and ATQ – 6* questionnaires.

Questionnaire	EX BCS	VA BCS	AN BCS	PP BCS	QoI BCS	EX WCS	VA WCS	AN WCS	PP WCS	QoI WCS
QVAQ – 6	0.14	0.25	0.86	0.92	0.61	0.47	0	0.53	0.88	0.53
SAQ – 4	0.28	0.24	0.72	0.71	0.48	0.75	0	0.25	0.66	0.25
ATQ – 6	0.34	0.12	0.66	0.93	0.54	0.78	0	0.22	0.62	0.22

a term that describes the decision-making progress used in programming and to turn up with algorithms. Here it is used deduction, i.e., starting from a conjecture and, according to a fixed set of relations (axioms and inference rules), try to construct a proof of the conjecture. It is a creative process. *Computational Logic* works with a proof search for a defined strategy. If one knows what this strategy is, one may implement certain algorithms in logic and do the algorithms with proof-finding [9–11].

{
/* The sentence below states that the extent of predicate ¬ qvaq − 6 is made on the clauses that are explicitly stated plus the ones that cannot be discarded */
  ¬ qvaq − 6 (EX, VA, AN, QVAQ − 6, QoI)
      ← not qvaq − 6 (EX, VA, AN, QVAQ − 6, QoI),
          not abducible_{qvaq−6} (EX, VA, AN, QVAQ − 6, QoI)
/* the sentence below denotes a qvaq − 6 axiom*/
  qvaq − 6 (0.14, 0.25, 0.86, 0.92, 0.61).

/* The sentence below states that the extent of predicate ¬ saq − 4 is made on the clauses that are explicitly stated plus the ones that cannot be discarded */
  ¬ saq − 4 (EX, VA, AN, SAQ − 4, QoI)
      ← not saq − 4 (EX, VA, AN, SAQ − 4, QoI),
          not abducible_{saq−4} (EX, VA, AN, SAQ − 4, QoI)
/* the sentence below denotes a saq − 4 axiom*/
  saq − 4 (0.28, 0.24, 0.72, 0.71, 0.48).

/* The sentence below states that the extent of predicate ¬ atq − 6 is made on the clauses that are explicitly stated plus the ones that cannot be discarded */
  ¬ atq − 6 (EX, VA, AN, ATQ − 6, QoI)
      ← not atq − 6 (EX, VA, AN, ATQ − 6, QoI),
          not abducible_{atq−6} (EX, VA, AN, ATQ − 6, QoI)
/* the sentence below denotes a atq − 6 axiom*/
  atq − 6 (0.34, 0.12, 0.66, 0.93, 0.54).
}

It is now possible to generate the data sets that will allow one to train an *ANN* [5, 6] (Fig. 4), where the nursing students' degrees of *Adaptation, Anxiety* and *Anxiety Trait* are used as input to the *ANN*, that presents as output an assessment of the **Lo**gic **P**rogram *1* (**LoP**) performance referred to above in measuring the student's degree of *Adaptation, Anxiety* and *Anxiety Trait* plus a measure of its **Sus**T*ainability* (**SusT**). Indeed, **SusT** may be seen as a logic value that ranges in the interval 0…1. On the other hand, considering that one has a group of 20 (twenty) students, the *ANN* training sets may be obtained by making the sentence depicted below obvious (i.e., object of formal proof), viz.

$$\forall(EX_1, VA_1, AN_1, LoP_1, SusT_1, \cdots, EX_3, VA_3, AN_3, LoP_3, SusT_3),$$

$$(qvaq - 6(EX_1, VA_1, AN_1, LoP_1, SusT_1), \cdots, atq - 6(EX_3, VA_3, AN_3, LoP_3, SusT_3))$$

in every possible way, i.e., generating all the different possible sequences that combine the extent of predicates $qvaq - 6$, $saq - 4$, and $atq - 6$, therefore leading to a number of 1140 sets that are presented in the form, viz.

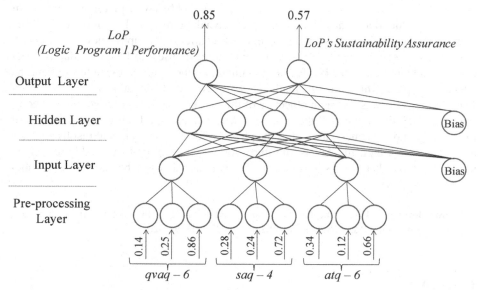

Fig. 4. A creative (abstract) view of the *ANN* topology to assess **Lo**gic **P**rogram *1 Performance* (**LoP**) and of its *Sustainability* (**SusT**)

$$\{\{qvaq - 6(EX_1, VA_1, AN_1, LoP_1, SusT_1), saq - 4(EX_2, VA_2, AN_2, LoP_2, SusT_2),$$
$$atq - 6(EX_3, VA_3, AN_3, LoP_3, SusT_3)\}, \cdots\}$$

$$\approx \{\{(qvaq - 6(0.14, 0.25, 0.86, 0.92, 0.61), saq - 4(0.28, 0.24, 0.72, 0.71, 0.48),$$
$$atq - 6(0.34, 0.12, 0.66, 0.93, 0.54)\}, \cdots\}$$

where *{}* is the expression for sets, and $\approx$ stands for itself. It serves as input (75% for training, 25% for testing) for the *ANN* (Fig. 4). On the other hand, and still on the issue of the *ANN* training process, the assessment of *Logic Program 1* performance (*LoP*) and of its sustainability (*SusT*) are considered, and weighed in the form, viz.

$$\{\{(LoP_{qvaq-6} + LoP_{saq-4} + LoP_{atq-6})/3\}, \cdots\}$$
$$\approx \{\{(0.92 + 0.71 + 0.93)/3 = 0.85\}, \cdots\}$$

and, viz.

$$\{\{(SusT_{qvaq-6} + SusT_{saq-4} + SusT_{atq-6})/3\}, \cdots\}$$
$$\approx \{\{(0.61 + 0.48 + 0.54)/3 = 0.57\}, \cdots\}$$

4 Conclusions and Future Work

In the transition to higher education, students experience many changes and ups and downs in their lives, including those in development and behavior that reflect their transition to adulthood and the emergence of new social, cultural and training needs. They have to change their way of living, accept new challenges, adapt to new experiences, develop new behavioral, cognitive, and emotional responses. In fact, the adaptive success of students becomes a key factor in maintaining their perseverance throughout the educational path, which must be mathematically evaluated and subject to formal proof in real time. This was the main objective of this work, which was carried out as a computing agency that integrated the phases of data collection, a logical representation of uncertainty and vagueness, as well as the phases of data processing and results analysis and their evaluation using ANNs. In future work, a vision of multi-valued logic will be pursued, a topic that has just been mentioned here and that will be carried out without leaving first-order logic.

Acknowledgments. This work has been supported by FCT – Fundação para a Ciência e Tecnologia within the R&D Units Project Scope: UIDB/00319/2020.

References

1. Admi, H.: Nursing student's stress during the initial clinical experience. J. Nurs. Educ. **36**, 323–327 (1997)
2. Barros, M.: The relevance and quality of academic experience: An active training path. In: Pouzada, A.S., Almeida, L.S., Vasconcelos, R.M. (eds.) Contexts and Dynamics of Academic Life, pp. 99–106. University of Minho, Guimarães (2002)
3. Evans, N.J., Forney, D.S., Guido, F.M., Patton, L.D., Renn, K.A.: Student Development in College: Theory, Research and Practice. Jossey–Bass, San Francisco (2010)
4. Neves, J., et al.: Entropy and organizational performance. In: Pérez García, H., Sánchez González, L., Castejón Limas, M., Quintián Pardo, H., Corchado Rodríguez, E. (eds.) HAIS 2019. LNCS (LNAI), vol. 11734, pp. 206–217. Springer, Cham (2019). https://doi.org/10.1007/978-3-030-29859-3_18

5. Fernandes, B., Vicente, H., Ribeiro J., Capita, A., Analide, C., Neves, J.: Fully informed vulnerable road users – simpler, maybe better. In: Proceedings of the 21st International Conference on Information Integration and Web-based Applications & Services (iiWAS2019), pp. 600–604. Association for Computing Machinery, New York (2020)

6. Fernández-Delgado, M., Cernadas, E., Barro, S., Ribeiro, J., Neves, J.: Direct Kernel Perceptron (DKP): ultra-fast kernel ELM-based classification with non-iterative closed-form weight calculation. J. Neural Netw. **50**, 60–71 (2014)

7. Wenterodt, T., Herwig, H.: The entropic potential concept: a new way to look at energy transfer operations. Entropy **16**, 2071–2084 (2014)

8. Spielberger, C.D., Sarason, I.G. (eds.): Stress and Emotion: Anxiety, Anger, and Curiosity, vol. 16. Taylor & Francis, New York (1996)

9. Neves, J.: A logic interpreter to handle time and negation in logic databases. In: Muller, R., Pottmyer, J. (eds.) Proceedings of the 1984 Annual Conference of the ACM on the 5th Generation Challenge, pp. 50–54. ACM, New York (1984)

10. Figueiredo, M., Fernandes, A., Ribeiro, J., Neves, J., Dias, A., Vicente, H.: An assessment of students' satisfaction in higher education. In: Vittorini, P., Di Mascio, T., Tarantino, L., Temperini, M., Gennari, R., De la Prieta, F. (eds.) MIS4TEL 2020. AISC, vol. 1241, pp. 147–161. Springer, Cham (2020). https://doi.org/10.1007/978-3-030-52538-5_16

11. Fernandes, A., Figueiredo, M., Ribeiro, J., Vicente, D., Neves, J., Vicente, H.: Psychosocial risks management. Proc. Comput. Sci. **176**, 743–752 (2020). https://doi.org/10.1016/j.procs.2020.09.069

Text Similarity Between Concepts Extracted from Source Code and Documentation

Zaki Pauzi[1]([✉]) and Andrea Capiluppi[2]([✉])

[1] Software and Platform Engineering Chapter, BP, London, UK
zaki.pauzi@bp.com
[2] Department of Computer Science, University of Groningen,
Groningen, The Netherlands
a.capiluppi@rug.nl

Abstract. *Context*: Constant evolution in software systems often results in its documentation losing sync with the content of the source code. The traceability research field has often helped in the past with the aim to recover links between code and documentation, when the two fell out of sync.

Objective: The aim of this paper is to compare the concepts contained within the source code of a system with those extracted from its documentation, in order to detect how similar these two sets are. If vastly different, the difference between the two sets might indicate a considerable ageing of the documentation, and a need to update it.

Methods: In this paper we reduce the source code of 50 software systems to a set of key terms, each containing the concepts of one of the systems sampled. At the same time, we reduce the documentation of each system to another set of key terms. We then use four different approaches for set comparison to detect how the sets are similar.

Results: Using the well known Jaccard index as the benchmark for the comparisons, we have discovered that the cosine distance has excellent comparative powers, and depending on the pre-training of the machine learning model. In particular, the SpaCy and the FastText embeddings offer up to 80% and 90% similarity scores.

Conclusion: For most of the sampled systems, the source code and the documentation tend to contain very similar concepts. Given the accuracy for one pre-trained model (e.g., FastText), it becomes also evident that a few systems show a measurable drift between the concepts contained in the documentation and in the source code.

Keywords: Information Retrieval · Text similarity · Natural language processing

1 Introduction

To understand the semantics of a software, we need to comprehend its concepts. Concepts extracted through its keywords from a single software derived from its

© Springer Nature Switzerland AG 2020
C. Analide et al. (Eds.): IDEAL 2020, LNCS 12489, pp. 124–135, 2020.
https://doi.org/10.1007/978-3-030-62362-3_12

source code and documentation have to be consistent to each other and they must be aligned as they define the software's identity, allowing us to classify in accordance to its domain, among others. The importance of similarity stems from the fact that software artefacts should represent a single system where the different cogs are meshed together in synchrony, enabling the moving parts to work together collectively. This also plays a pivotal role in software management and maintenance as software development is usually not a solo effort.

Past research that has attempted this link involved automatic generation of documentation from code such as with SAS [15] and Lambda expressions in Java [1], automatic generation of code from documentation such as with CON-CODE [8] and visualising traceability links between the artefacts [7].

Using a sample of 50 open source projects, we extracted the concepts of each project from two sources: the source code and documentation. On one hand, we extracted the concepts emerging from the keywords used in the source code (class and variable names, method signatures, etc) and on the other hand, we extracted the concepts from the plain text description of a system's functionalities, as described in the main documentation file (e.g., the README file). With these two sets of concepts, we run multiple similarity measurements to determine the similitude of these building blocks of what makes a software, a software.

We articulate this paper as follows: in Sect. 3, we describe the methodology that we used to extract the concepts from the source code and the documentation of the sampled systems. In Sect. 4, we illustrate the methods that were implemented to determine the similarity of the two sets of concepts, per system. In Sect. 5, we summarise the results and Sect. 6 discusses the findings of the study. Section 7 finally concludes.

2   Background Research

The study of the traceability links between code and documentation has been conducted for program comprehension and maintenance. As one of the first attempts to discover links between free-text documentation and source code, the work performed in [2] used two Information Retrieval (IR) models on the documentation and programs of one C++ and one Java system. The objective was to uniquely connect one source code item (class, package etc) to a specific documentation page. Our approach is slightly different, and it explores whether there is an overlap between the concepts expressed in the source code and those contained in the documentation.

The field of traceability has become an active research topic: in [12,13] for example, the authors show how traceability links can be obtained between source code and documentation by using the LSI technique (i.e., latent semantic indexing). In our work we show how latent semantic techniques have relatively poor performance as compared to more sophisticated techniques.

In the context of locating topics in source code, the work presented in [10] demonstrated that the Latent Dirichlet Allocation (LDA) technique has a strong applicability in the extraction of topics from classes, packages and overall systems. The LDA technique was used in a later paper [4] to compare how the topics from source code, and the reading of the main core of documentation, can be used by experts to assign a software system to an application domain. In this paper, however, we combine and enhance past research by using various techniques to extract concepts and keywords from source code, and we compare this set with the concepts and keywords extracted from the documentation.

3   Empirical Approach

In this section, we discuss how we sampled the systems to study and how the two data sources were extracted from the software projects. In summary, we extracted the plain description of software systems, alongside the keywords of their source codes. We then ran a variety of similarity measurements to determine the degree of similarity between the concepts extracted. Figure 1 represents the toolchain visualisation of the approach.

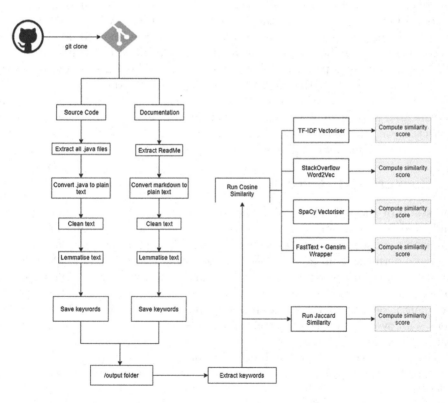

Fig. 1. Toolchain implemented throughout this paper

3.1 Definitions

This section defines the terminology as used throughout the paper. Each of the items described below will be operationalised in one of the subsections of the methodology.

– **Corpus keyword** (term) – given the source code contained in a class, a term is any item that is contained in the source code. We do not consider as a term any of the Java-specific keywords (e.g., `if`, `then`, `switch`, etc.)[1]. Additionally, the camelCase or PascalCase notations are first decoupled in their components (e.g., the class constructor *InvalidRequestTest* produces the terms *invalid, request* and *test*).
– **Class corpus** – with the term 'class corpus' we consider the set of all the terms contained within a class. We consider two types of class corpus, per class: the *complete* corpus, with all the terms contained; and the *unique* set of terms, that is, purged of the duplicates.

3.2 Sampling Software Systems and ReadMe Files

Leveraging the GitHub repository, we collected the project IDs of the 50 most successful[2] Java projects hosted on GitHub as case studies. As such, our data set does not represent a random sample, but a complete sub-population based on one attribute (i.e., success) that is related to usage by end users. As a result of the sampling, our selection contains projects that are larger in size than average.

The repository of each project was cloned and stored, and all the Java files (in the latest master branch) identified for further parsing. From each project's folder we extracted the main *ReadMe* file, that is typically assumed to be the first port of information for new users (or developers) of a project.

3.3 Concept Extraction

The extraction was executed in Python, which was then analysed using different text similarity measurements in a Jupyter notebook[3]. This extraction was carried out to build the class corpus for each project. Results were then compared and analysed. For the source codes, we extracted all the class names and identifiers that were used for methods and attributes. These terms are generally written with camel casing in Java and this was handled with the decamelize[4] plugin. Additionally, inline comments were extracted as well. This results in an

---

[1] The complete list of Java reserved words that we considered is available at https://en.wikipedia.org/wiki/List_of_Java_keywords. The `String` keyword was also considered as a reserved word, and excluded from the text parsing.

[2] As a measure of success, we used the number of *stars* that a project received from other users: that implies appreciation for the quality of the project itself.

[3] Available online at https://github.com/zakipauzi/text-similarity-code-documentation/blob/master/text_similarity.ipynb.

[4] https://github.com/abranhe/decamelize.

extraction that comprehensively represents the semantic overview of concepts whilst minimising noise from code syntax. The final part of the extraction is the lemmatisation of the terms using SpaCy's `token.lemma_`. Lemmatising is simply deriving the root word from the terms, thus enabling more matches when we compare from the different sources.

An excerpt of the complete corpus from the source code of `http-request` is shown at Fig. 2.

```
http request charset content type form content type json encoding
gzip gzip header accept accept header accept charset charset
header ...
```

Fig. 2. Complete class corpus from `http-request` source code (excerpt)

For the documentation, the extraction of concepts for the class corpus from the *ReadMe* markdown file was done by conventional methods such as removing stop words, punctuation and code blocks. All non-English characters were disregarded during the exercise by checking if the character falls within the ASCII Latin space. Figure 3 shows an excerpt of the complete corpus from the documentation of `http-request`.

```
http request simple convenience library use httpurlconnection
make request access response library available mit license usage
httprequest library available maven ...
```

Fig. 3. Complete class corpus from `http-request` documentation (excerpt)

4   Text Similarity

In this section, we address the question of similarity measurements in text. This is crucial to the goal, which is to understand the software's *definition* and be able to apply to it.

Similarity in text can be analysed through lexical, syntactical and semantic similarity methods[6]. Lexical and syntactical similarity depends on the word morphology whereas semantic similarity depends on the underlying meaning of the corpus keywords. The difference between these can be simply understood by the usage of **polysemous** words in different sentences, such as:

1. *"The boy went to the **bank** to <u>deposit</u> his money."*
2. *"The boy fell by the river **bank** and dropped his rent <u>deposit</u> money."*

By merely reading these sentences, we understand that they do not mean the same thing or even remotely similar scenarios. This is because we derive meaning by context, such as identifying how the other keywords in the sentence provide meaning to the **polysemous** words. We could also see how the position of <u>deposit</u> in the sentences plays a part in differentiating whether it is a noun or a verb. However, a lexical similarity measurement on these two sentences will yield a high degree of similarity given that multiple identical words are being used in both sentences.

These similarity types are a factor in determining the kind of similarity that we are trying to achieve. This is also evident in past research that involved different languages such as cross-lingual textual entailment[16]. In the following sections we describe four different similarity measures that were used to evaluate the two sets of concepts extracted from the software systems.

4.1   Jaccard Similarity

As a measure of similarity, we have used the *Jaccard Similarity* as our baseline measurement. This is one of the oldest similarity indexes [9], that purely focuses on the intersection over union based on the lexical structure of the concepts. This method converts the keywords to sets (thus removing duplicates) and measuring the similarity by the intersection. Figure 4 shows the Venn Diagram acting as a visual representation of the number of keywords in each unique corpus for one of the projects sampled (e.g., the `android-gpuimage` project). Although we could detect 98 *identical* terms in the intersection of the two sets, there are a further 265 terms that are only found in the source code, and some 80 terms that only appear in the documentation set.

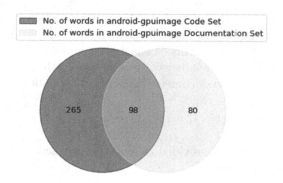

Fig. 4. Intersection of similar concepts by lexical structure

Given how the Jaccard similarity index is being computed, it is important to note that concepts extracted in both sources may not necessarily be written in an identical manner, but they may be meaning the same. This is reflected by the results, and it becomes apparent when we explore other similarity measurements and compare the results in Sect. 5.

4.2 Cosine Similarity

To handle cases where the keywords are not identical to each other character by character, we have to measure them differently and this is where cosine similarity coupled with word vectors come in. Cosine similarity is a distance metric irrespective of orientation and magnitude. This is particularly important because in a multi-dimensional space, magnitude will not necessarily reflect the degree of similarity, which may be measured using Euclidean distance.

The lower the angle, the higher the similarity. This is measured by:

$$sim(A, B) = cos(\theta) = \frac{A.B}{||A|| ||B||} = \frac{\sum_1^n A_i B_i}{\sqrt{\sum_1^n A_i^2} \sqrt{\sum_1^n B_i^2}}$$

Another measure of similarity is by the inverse of Word's Mover's Distance (WMD) as first introduced in [11]. This distance measures the *dissimilarity* between two text documents as the minimum amount of distance that the embedded words of one document need to "travel" to reach the embedded words of another document. This method will not be covered in this paper but may be explored in future work.

Term Frequency Inverse Document Frequency (TF-IDF) Vectorizer. Prior to measuring the cosine similarity, we need to embed the terms in a vector space. TF-IDF, short for Term Frequency Inverse Document Frequency, is a statistical measurement to determine how relevant a word is to the document. In this instance, TF-IDF was used to convert the concepts extracted into a Vector Space Model (VSM). Extracting important text features is crucial to the representation of the vector model and this is why we used TF-IDF Vectorizer as term frequency[5] alone will not factor in the inverse importance of commonly used concepts across the documents. As such, the complete corpus was used for this method. TF-IDF for corpus keyword ck is simply the product of the term frequency tf with its inverse document frequency df as follows:

$$ck(i, j) = tf(i, j) \times log(\frac{N}{df_i})$$

The `TFIDFvectorizer` is provided as a class in the scikit-learn[14] module[6]. The results are shown in Sect. 5.

---

[5] This can be achieved by the scikit-learn module `CountVectorizer`.
[6] The feature is available in the module named https://scikit-learn.org/stable/modules/generated/sklearn.feature_extraction.text.TF-IDFVectorizer.

Word Embeddings and Other Vector Space Models. There are numerous word Vector Space Models (VSM) that have been pre-trained such as word embeddings offered by Gensim[7], FastText[8] and SpaCy[9]. We looked into combining the word vectorisation from these modules and then running the cosine similarity measurement to determine the degree of similarity. We have chosen to use these three state-of-the-art word embeddings as they offer a more accurate word representation to compare similarity as opposed to the traditional vectoriser. Terms that were extracted mirror closely to that of the English dictionary of the pre-trained models' vocabulary (StackOverflow, OntoNotes, Wikipedia etc.), making them suitable and effective as word vectorisers.

For Gensim, we used the word2vec word embedding technique based on a pre-trained model on StackOverflow[10] [5] data. As we aim to bring the word vectorisation technique *closest* to home as possible, we decided to use a model that is trained within the software engineering space. On the other hand, as this is a word2vec model, we had to average the results from all the terms extracted as the model is specific for words and not documents. For SpaCy, we used the en_core_web_md model, which was trained on OntoNotes[11] and Glove Common Crawl[12].

Finally for FastText, we implemented the vectorisation through the pre-trained wiki.en.bin model [3] and loaded it through the Gensim wrapper. This model is trained on Wikipedia data. These vectors in a dimension space of 300 were obtained using the skip-gram model.

5   Results

In this section, we look at how these similarity measurements determine the degree of similarity between terms extracted from source code and documentation. Our baseline measurement, Jaccard Similarity, scores an average of only **7.27%** across the 50 projects. The scores for each project are shown in Fig. 5.

For the cosine similarity measurements, the similarity scores cover a wide range of average results. Table 1 shows the average score for each word vectorisation method combined with cosine similarity. The cosine similarity with FastText Gensim Wrapper Wiki model scored the highest with an average score of **94.21%** across the 50 GitHub projects sampled.

The results for each method are shown in Fig. 6.

---

[7] https://radimrehurek.com/gensim.
[8] https://fasttext.cc.
[9] https://spacy.io.
[10] https://stackoverflow.com.
[11] https://catalog.ldc.upenn.edu/LDC2013T19.
[12] https://nlp.stanford.edu/projects/glove.

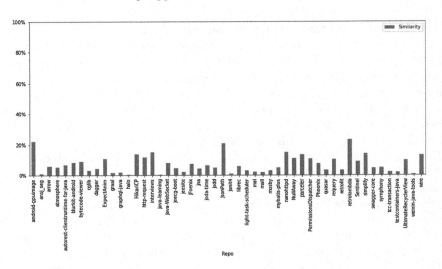

Fig. 5. Jaccard similarity score

Table 1. Average similarity scores across the sampled projects

Method	Score
TFIDFVectorizer	13.38%
StackOverflow Averaged Word2Vec	60.58%
SpaCy en_core_web_md	84.09%
FastText Wiki	94.21%

6   Discussion

It is interesting to note how the degree of similarity can be very different for the various word vectorisation methods. For Jaccard similarity, we can see how the lexical similarity in this scenario does not bode well with concepts extracted from the projects.

For cosine similarity on word vectors, we have a wide range of averages. If we look closely into the trend of projects that consistently score below average among its peers, we can see that this can be attributed to the concepts being non-English, which were disregarded in this paper. This resulted in a lower similarity score across the different measures. When we excluded these projects and only considered those that were fully written in English (42 projects in total), we could see a *noticeable* increase across all measurements. Table 2 shows the new results.

6.1   Limitations

The approach that we have shown has some limitations that we have identified, and we discuss them below:

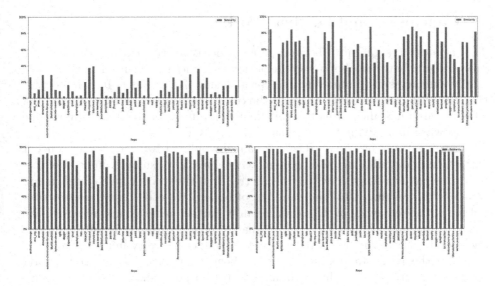

Fig. 6. Cosine similarity scores (the sequence of projects in the X axis is the same as in Fig. 5)

Table 2. Average similarity scores across projects fully in English

Method	Score
TFIDFVectorizer	15.49%
StackOverflow Averaged Word2Vec	66.61%
SpaCy en_core_web_md	88.39%
FastText Wiki	95.36%

1. *Structure of documentation is non-uniform.* Most of the *ReadMe* files are a mixture of code examples and explanations of how to use the system, alongside its options and features. This lack of structure affects the extraction's capability in retrieving the necessary information.
2. *Documentation not in English.* A few of the projects that we analysed are documented in a language different from English (e.g., Chinese). Currently, the extraction of concepts from source code is not affected even in those cases (i.e., the source code is still written with English syntax). On the other hand the non-English documentation cannot be parsed, since our approach ignores non-English characters. This is a limitation because the non-English free text may provide value to the concepts in the code.
3. *Non uniform identifiers.* From the analysis of the systems, we observed that the structure of code has some degree of non-uniformity in the way identifiers are used to represent meaning. In some cases, the classes are explicitly describing some core functionality of the code, while in other cases, they hide the core concepts behind generic descriptors.

7   Conclusion and Further Work

The results of similarity shown between the keywords extracted from the projects' artefacts corroborate on the notion that these terms need to be semantically similar as they represent the constituents of a product. It is important that the terms are similar as they are the core ingredients to software semantics. This research has allowed us to explore this and it has enabled us to move one step forward in applying to the semantics of software.

Bringing this work forward, much can be explored further for the concept extraction and similarity measurement techniques. For example, instead of disregarding the non-English words, we can translate these concepts to English and compare the results. Also, the handling of boilerplate code that does not contribute to the semantics of the software can be further detected and removed. We can also expand our research scope to analyse projects from different languages and whether the structure of the concepts extracted are very much different.

From the NLP standpoint, there has been an increase in research efforts in transfer learning with the introduction of deep pre-trained language models (ELMO, BERT, ULMFIT, Open-GPT, etc.). It is no wonder that deep learning and neural networks will continue to dominate the NLP research field and provide us with an even more effective avenue to further apply to the semantics of software, such as deriving the concepts as a basis of proof for software domain classification, among many others.

References

1. Alqaimi, A., Thongtanunam, P., Treude, C.: Automatically generating documentation for lambda expressions in Java. In: 2019 IEEE/ACM 16th International Conference on Mining Software Repositories (MSR), pp. 310–320 (2019)
2. Antoniol, G., Canfora, G., Casazza, G., De Lucia, A., Merlo, E.: Recovering traceability links between code and documentation. IEEE Trans. Softw. Eng. **28**(10), 970–983 (2002)
3. Bojanowski, P., Grave, E., Joulin, A., Mikolov, T.: Enriching word vectors with subword information. Trans. Assoc. Comput. Linguist. **5**, 135–146 (2017)
4. Capiluppi, A., et al.: Using the lexicon from source code to determine application domain. In: Proceedings of the Evaluation and Assessment in Software Engineering, pp. 110–119 (2020)
5. Efstathiou, V., Chatzilenas, C., Spinellis, D.: Word embeddings for the software engineering domain. In: 2018 IEEE/ACM 15th International Conference on Mining Software Repositories (MSR), pp. 38–41 (2018)
6. Ferreira, R., Lins, R.D., Simske, S.J., Freitas, F., Riss, M.: Assessing sentence similarity through lexical, syntactic and semantic analysis. Comput. Speech Lang. **39**, 1–28 (2016). https://doi.org/10.1016/j.csl.2016.01.003, http://www.sciencedirect.com/science/article/pii/S0885230816000048
7. Haefliger, S., Von Krogh, G., Spaeth, S.: Code reuse in open source software. Manag. Sci. **54**(1), 180–193 (2008)
8. Iyer, S., Konstas, I., Cheung, A., Zettlemoyer, L.: Mapping language to code in programmatic context. CoRR abs/1808.09588 (2018). http://arxiv.org/abs/1808.09588

9. Jaccard, P.: Nouvelles recherches sur la distribution florale. Bull. Soc. Vaud. Sci. Nat. **44**, 223–270 (1908)

10. Kuhn, A., Ducasse, S., Gírba, T.: Semantic clustering: Identifying topics in source code. Inf. Softw. Technol. **49**(3), 230–243 (2007)

11. Kusner, M.J., Sun, Y., Kolkin, N.I., Weinberger, K.Q.: From word embeddings to document distances. In: Proceedings of the 32nd International Conference on International Conference on Machine Learning (ICML 2015), vol. 37, pp. 957–966. JMLR.org (2015)

12. Marcus, A., Maletic, J.I.: Recovering documentation-to-source-code traceability links using latent semantic indexing. In: Proceedings of the 25th International Conference on Software Engineering, 2003, pp. 125–135. IEEE (2003)

13. Marcus, A., Maletic, J.I., Sergeyev, A.: Recovery of traceability links between software documentation and source code. Int. J. Softw. Eng. Knowl. Eng. **15**(05), 811–836 (2005)

14. Pedregosa, F., et al.: Scikit-learn: machine learning in Python. J. Mach. Learn. Res. **12**, 2825–2830 (2011)

15. Righolt, C.H., Monchka, B.A., Mahmud, S.M.: From source code to publication: Code diary, an automatic documentation parser for SAS. SoftwareX **7**, 222–225 (2018). https://doi.org/10.1016/j.softx.2018.07.002, http://www.sciencedirect.com/science/article/pii/S2352711018300669

16. Vilariño, D., Pinto, D., Mireya, T., Leon, S., Castillo, E.: BUAP: lexical and semantic similarity for cross-lingual textual entailment, pp. 706–709, June 2012

Using Kullback-Leibler Divergence to Identify Prominent Sensor Data for Fault Diagnosis

Rodrigo P. Monteiro[1]([✉])[ID] and Carmelo J. A. Bastos-Filho[2]([✉])[ID]

[1] Federal University of Pernambuco, Recife, PE, Brazil
`rodrigo.paula@ufpe.br`
[2] University of Pernambuco, Recife, PE, Brazil
`carmelofilho@ieee.org`

Abstract. The combination of machine learning techniques and signal analysis is a well-known solution for the fault diagnosis of industrial equipment. Efficient maintenance management, safer operation, and economic gains are three examples of benefits achieved by using this combination to monitor the equipment condition. In this context, the selection of meaningful information to train machine learning models arises as an important issue, since it influences the model accuracy and complexity. Aware of this, we propose to use the ratio between the interclass and intraclass Kullback-Leibler divergence to identify promising data for training fault diagnosis models. We assessed the performance of this metric on compressor fault datasets. The results suggested a relation between the model accuracy and the ratio between the average interclass and intraclass divergences.

Keywords: Deep learning · Machine learning · Kullback-Leibler divergence

1 Introduction

Using signal analysis to perform the fault diagnosis on rotating machines is a well-known solution in the literature. Such a solution is widely used in equipment maintenance by companies [9] and plays an essential role in economic and safety issues [5]. The early fault diagnosis allows the maintenance management to be more efficient and leads to safer operation of monitored systems, *e.g.* industrial rotating machines [9] and wind turbines [12].

In this context, machine learning became an important and widely used tool [15]. Such a tool provides fast and accurate responses to the fault diagnosis problem, allowing a more efficient real-time monitoring of the manufacturing line equipment. Thus, machine learning algorithms are commonly trained to learn meaningful patterns on datasets consisting of signals collected by sensors. Those signals may belong to different domains, *e.g.*, time- [4], frequency- [17] and time-frequency domains [18].

© Springer Nature Switzerland AG 2020
C. Analide et al. (Eds.): IDEAL 2020, LNCS 12489, pp. 136–147, 2020.
https://doi.org/10.1007/978-3-030-62362-3_13

However, machine learning-based models show satisfactory results only if the input information is appropriate [3]. An accurate fault diagnosis model depends on data that allows the correct and unique characterization of each fault modality. Selecting the most relevant information to train those models is not a trivial task, and there are different ways to deal with this problem.

Given a scenario in which different sensors provide information about the operation of rotating machines, selecting the most appropriate data to perform the fault diagnosis has three main ways. The first one is establishing a metric to infer the data quality and choose the most promising data according to such a metric [6]. The second way is to train one model for each sensor data and then ensemble the trained models [22]. The results of each model are combined according to a predefined strategy, *e.g.*, majority voting, to perform the fault diagnosis. The last way is to train models to perform the fault diagnosis by assessing and fusing multimodal data [10], *i.e.*, data from multiple sensors.

The selection of the most appropriate data for a given task remains essential despite the ensembles and multimodal classifiers capable of handling information from different input modalities. By identifying the most meaningful data, we can discard information that is not so useful. In this way, we can reduce the amount of training data, the training time, the model complexity, the model size, the fault diagnosis time, among other aspects.

Aware of this, we propose to use the Kullback-Leibler (KL) divergence to select the most promising input data for the fault diagnosis on a two-stage reciprocating compressor. In other words, we use the KL divergence to infer the sensor data related to the trained models with the best accuracy values. Our contribution is how to use the KL divergence for this purpose. For example, given two datasets, *i.e.*, the data of two sensors, we calculate the ratio between the average inter- and intraclass divergences of each one. Then, we select the most promising dataset by comparing the values of those ratios. We noticed that datasets presenting the highest ratios were related to the most accurate models.

We organized the remaining of this work as follows: Sect. 2 presents the theoretical background of the work developed. Section 3 explains how we performed the experiments, *i.e.*, the Methodology. Section 4 presents and discusses the results achieved in this work. Section 5 is the Conclusion we draw from this work.

2   Theoretical Background

In this section, we present the definition of the Kullback-Leibler Divergence and some basic concepts regarding the machine learning techniques deployed in the experiments.

2.1   Kullback-Leibler Divergence General Definition

The Kullback-Leibler (KL) divergence [13,14] measures the difference between two probability distributions. Given two distributions P_1 and P_2 of a continuous random variable x, the KL divergence is defined by the Eq. (1).

$$D_{KL}(P_1(x)||P_2(x)) = \int_{-\infty}^{\infty} p_1(x) \log\left(\frac{p_1(x)}{p_2(x)}\right) dx \qquad (1)$$

in which p_1 and p_2 denote the probability densities of P_1 and P_2, respectively. A Kullback–Leibler divergence equal to 0 suggests that P1 and P2 are identical distributions. Also known as relative entropy and information gain, the KL divergence is a common tool in information theory and machine learning applications.

2.2 Multilayer Perceptron

The Multilayer Perceptron (MLP) [2] is a class of feedforward artificial neural networks. The processing units of those networks are called neurons. Those units are arranged in layers and connect to form a directed graph. The MLPs present three types of layers, *i.e.*, input, hidden, and output layers. The units of a given layer are connected to the units of the subsequent one but do not connect to other units that belong to the same layer. The Multilayer Perceptron can distinguish non-linearly separable patterns. The MLP is a universal approximator, *i.e.*, an MLP with one hidden layer, and enough neurons can approximate any given continuous function [16]. Those networks are frequently used in classification and regression problems [7].

2.3 Convolutional Neural Networks

The Convolutional Neural Networks (CNN) are models inspired by biological processes. They are successful at performing several tasks, *e.g.*, object detection [1] and classification [21]. The fault diagnosis [17] and disease detection [19] are two examples of applications that CNNs can perform. Their basic structure consists of input, convolutional, pooling fully-connected, and output layers [21]. Modifications in this structure may occur, depending on the application. CNNs are often used in problems where the inputs have a high spatial or temporal correlation. We explain the role of each layer in the following:

Input Layer. This layer receives and stores raw input data. It also specifies the width, height, and number of channels of the input data.

Convolutional Layers. They learn feature representations from a set of input data and generate feature maps. Those maps are created by convolving the layer inputs with a set of learned weights. An activation function, *e.g.*, the ReLU function, is applied to the convolution step's output. Equation (2) shows the general formulation of a convolutional layer.

$$x_j^l = f\left(\sum_{i \in M_j} x_i^{l-1} * k_{ji}^l + b_j^l\right) \qquad (2)$$

in which l refers to the current layer, i and j are the element indices of the previous and current layers, respectively, and M_j is a set of input maps. k is the weight matrix of the ith convolutional kernel of the lth layer, applied to the jth input feature map, and b is the bias.

Pooling Layers. They reduce the spatial resolution of feature maps, also improving the spatial invariance to input distortions and translations. Most of the recent works employ a variation of this layer called the max-pooling. It propagates to the next layers the maximum value from a neighborhood of elements. This operation is defined by Eq. (3).

$$y_{rjs} = max_{(p,q)\epsilon R_{rs}} x_{kpq} \qquad (3)$$

in which y_{jrs} is the output of the pooling process for the jth feature map, and x_{kpq} is the element at location (p,q) contained by the pooling region R_{rs}. The pooling process is also known as subsampling.

Fully Connected and Output Layers. They interpret the feature representations and perform high-level reasoning. They also compute the scores of each output class. The number of output nodes depends on the number of classes.

3   Methodology

3.1   Dataset

The dataset contains the frequency spectra of measurements (time-domain signals) provided by nine sensors placed on a two-stage reciprocating compressor, as shown in Table 1. We used a frequency-domain representation because it allows us to analyze the signals in terms of their frequency components. We obtained the spectra by using the Fast Fourier Transform [20] on the time-domain signals. The length of each frequency spectrum was 1,024.

In this work, we analyze two experimental scenarios. Each one contains data collected by nine sensors, *i.e.*, there are nine datasets per scenario. In the first scenario, the signals are divided into four classes, as presented in Table 2. The class P1 is related to the regular operation of the compressor. The remaining classes are related to three modalities of bearing faults. Thus, considering each sensor data, there are 1,500 frequency spectra per class. As we have four classes and nine sensors, the total number of spectra is 54,000.

On the other hand, the second scenario divides the signals into thirteen classes. Class P1 is also related to the regular operation. The remaining ones are related to twelve modalities of multiple faults, *i.e.*, bearing and valve faults. Those classes were listed in Table 3. In this scenario, considering each sensor data, we also have 1,500 frequency spectra per class. As we have thirteen classes, and nine sensors, the total number of spectra is 175,500.

Table 1. List of sensors.

Code	Sensor
MC1	Microphone
MC2	Microphone
CVC1	Current Sensor
CVC2	Current Sensor
CVC3	Current Sensor
A1	Accelerometer
A2	Accelerometer
A3	Accelerometer
A4	Accelerometer

Table 2. The classes regarding the first scenario (bearing faults).

Bearing faults	
Fault code	Fault type
P1	No fault
P2	Inner race crack
P3	Roller element crack
P4	Outer race crack

3.2 The Ratio Between the Average Inter- and Intraclass Kullback-Leibler Divergences

For each sensor data, we calculate the ratio between the average inter- and intraclass KL divergences. Those divergences are calculated between pairs of frequency spectra. We use this ratio to infer how accurate a model trained on those data will perform. This is a comparative analysis. In other words, we compare the ratios obtained from each sensor data and infer which one is the most appropriate to perform the fault diagnosis.

If this ratio is close to one, the inter- and intraclass divergences are close to each other. So, we expect more difficulty to train accurate models. On the other hand, the higher the ratio is, the larger is the interclass divergence concerning the intraclass. So, we expect to be easier to train accurate models.

3.3 Classification Models

We used artificial neural networks to perform the fault diagnosis on the compressor data, *i.e.*, a classification process. We assessed the MLP and CNN. About the MLP, we used a ANN with one hidden layer architecture. Also, we evaluated hidden layers of different sizes, *e.g.*, 512, 1024, and 2048 neurons.

Regarding the CNN, we used an architecture with convolutional, max pooling, flattening, densely-connected, and output layers. We evaluated how the number of convolutional and max-pooling layers influence the results, as well as the number of neurons in the densely-connected layer. We describe the CNN layers in Table 4.

We used the ReLU activation function for both network configurations. This function allows the neural networks to learn more complex patterns because of its non-linearity, and also improves the training process due to the non-saturation of its gradient [8].

The complete description of the experimental setup and signal acquisition process is seen in [4].

Table 3. The classes regarding the second scenario (multiple faults)

Multi faults	
Fault code	Fault type
P1	No fault
P2	Bearing inner race crack/Valve seat wear
P3	Bearing inner race crack/Corrosion of the valve plate
P4	Bearing inner race crack/Fracture of the valve plate
P5	Bearing inner race crack/Spring break
P6	Bearing roller element crack/Valve seat wear
P7	Bearing roller element crack/Corrosion of the valve plate
P8	Bearing roller element crack/Fracture of the valve plate
P9	Bearing roller element crack/Spring break
P10	Bearing outer race crack/Valve seat wear
P11	Bearing outer race crack/Corrosion of the valve plate
P12	Bearing outer race crack/Fracture of the valve plate
P13	Bearing outer race crack/Spring break

3.4 Training the Fault Diagnosis Models

We divided the frequency spectra into training (80%) and test (20%) sets. Also, we trained 10 classifiers for each combination of model parametric configuration and dataset.

The classification models were trained for 50 epochs. We used a computer with the following configuration to perform the training process: OS Windows 10 Home, 64 bits, Memory (RAM) 15.9 GB, Processor Intel® Corte™ i7-6500 CPU @ 2.50 GHz x 2, AMD Radeon™ T5 M330 (No CUDA support). All the scripts were written in Python [23] 3.7, on the Jetbrains PyCharm [11] Community Edition 2019.2.

Table 4. CNN Layers

Layers	Quantity
Convolutional layer with filters (3×1) Max pooling layer with filters (2×1)	Models with 1, 2, 3, and 4 pairs, depending on the configuration
Flattening layer	1
Densely connected layer	1 layer with 32, 64, 128, and 256 neurons, depending on the configuration
Output layer	4 neurons for bearing faults, and 13 neurons for multi faults

4   Results

4.1   Results for Multilayer Perceptron

The first analysis is about bearing faults, whose classes we listed in Table 2. Figure 1 shows the fault diagnosis results obtained by MLPs. Those results belong to the MLP configuration with the best results (the network with 1024 neurons in the hidden layer). Each dot on the chart is related to one sensor dataset, *i.e.*, there are nine dots on the chart since we have nine sensors. They represent the average accuracy of ten fault diagnosis models.

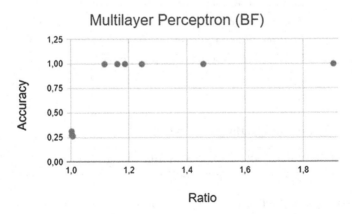

Fig. 1. Ratio x Accuracy for the scenario with bearing faults, and using MLPs to perform the faults diagnosis.

The sensor data with the lowest divergence ratios led to the least accurate models. On the other hand, the sensor data related to ratios with values above 1.1 led to fault diagnosis models with accuracy values close to 1, that is, 100%.

Figure 2 shows the results of MLPs for multiple faults. The classes of this scenario are listed in Table 3. We observe a trend similar to the one seen in Fig. 1. In other words, the model accuracy values are low for the lowest ratio values but tend to increase as the divergence ratio increases.

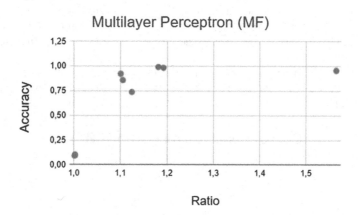

Fig. 2. Ratio x Accuracy for the scenario with multiple faults, and using MLPs to perform the faults diagnosis.

The main difference between the results presented in Figs. 1 and 2 is how fast accuracy increases as the ratio between the inter- and intraclass KL divergences increases. In Fig. 1 the transition is sudden, while in Fig. 2 it is smoother. Probably the reason is the different number of classes. The higher number in the second scenario, *i.e.*, thirteen against four in the first one, may hinder the performance of the faults diagnosis models. So, the difference between the inter- and intraclass divergences must be even greater for datasets to be capable of training accurate classifiers. Table 5 lists the results presented in Figs. 1 and 2. Also, we included the accuracy values concerning different proportions of train/test sets to assess the robustness of our proposal and the occurrence of overfitting. Those proportions were (train/test): 80%/20%, as originally defined in Subsect. 3.4, and 70%/30%, 60%/40%, and 50%/50%. We observe that, despite the fluctuation of accuracy values due to different amounts of train data, the most accurate models remained the ones related to the highest ratios.

4.2   Results for Convolutional Neural Networks

We also used Convolutional Neural Networks in this analysis. CNNs are more powerful classification algorithms than MLPs since they can extract meaningful features from the input data, *e.g.*, the frequency spectra, improving the faults

Table 5. Ratio and accuracy values for MLPs with 1024 neurons in the hidden layer regarding bearing (BF) and multiple (MF) faults. The accuracy values concern only the test sets, and four scenarios regarding different train/test data percentages: 80%/20%, 70%/30%, 60%/40%, and 50%/50%.

Sensor	Ratio (BF)	Acc (BF)				Ratio (MF)	Acc (MF)			
		80/20	70/30	60/40	50/50		80/20	70/30	60/40	50/50
MC1	1.16	1	1	1	1	1.1	0.93	0.93	0.90	0.91
MC2	1.19	1	1	1	1	1.18	0.99	0.99	0.98	0.98
CVC1	1	0.32	0.30	0.30	0.30	1	0.09	0.09	0.09	0.09
CVC2	1	0.28	0.27	0.28	0.27	1	0.11	0.11	0.1	0.1
CVC3	1.01	0.26	0.26	0.26	0.26	1	0.09	0.09	0.09	0.09
A1	1.46	1	1	1	1	1.12	0.74	0.74	0.71	0.70
A2	1.25	0.99	1	1	1	1.19	0.99	0.97	0.97	0.96
A3	1.91	1	1	1	1	1.56	0.95	0.91	0.89	0.87
A4	1.12	1	1	1	1	1.1	0.87	0.86	0.86	0.84

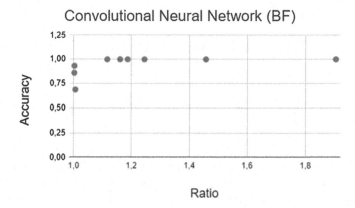

Fig. 3. Ratio x Accuracy for the scenario with bearing faults, and using CNNs to perform the faults diagnosis.

diagnosis. The results presented in Figs. 3 and 4 regard the most accurate CNN configuration (3 pairs of convolutional and max-pooling layers and 256 neurons in the densely connected layer). Figure 3 shows the results related to bearing faults, while Fig. 4 shows the ones related to multiple faults.

We observe similar behavior in Figs. 1 and 3. The sensor data with the lowest ratios led to the least accurate models. Also, the data related to ratios above 1.1 led to models with accuracy values close to 1. The main difference relies on the accuracy results for datasets with a low ratio between inter- and intraclass KL divergences. As already mentioned, CNNs are more powerful algorithms than MLPs, explaining their superior performance even in this case.

The same discussion is valid for Figs. 2 and 4, including the smoother accuracy increase from the lower to the higher ratios. The results presented in Figs. 3 and 4 are listed in Table 6, which also includes accuracy values concerning dif-

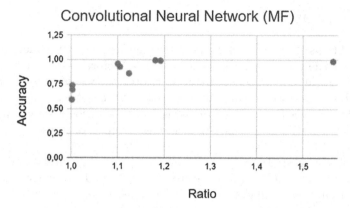

Fig. 4. Ratio x Accuracy for the scenario with multiple faults, and using CNNs to perform the faults diagnosis.

ferent proportions of train/test sets, as in Table 5. We observe that the most accurate models remained the ones related to the highest ratios, despite the variations due to different amounts of train data.

Table 6. Ratio and accuracy values for CNNs with 3 pairs of convolutional and max-pooling layers, and 256 neurons in the densely connected layer. The results regard bearing (BF) and multiple (MF) faults. The accuracy values concern only the test sets, and four scenarios regarding different train/test data percentages: 80%/20%, 70%/30%, 60%/40%, and 50%/50%.

Sensor	Ratio (BF)	Acc (BF)				Ratio (MF)	Acc (MF)			
		80/20	70/30	60/40	50/50		80/20	70/30	60/40	50/50
MC1	1.16	1	1	1	1	1.1	0.96	0.94	0.92	0.89
MC2	1.19	1	1	1	1	1.18	1	0.99	0.98	0.98
CVC1	1	0.86	0.75	0.71	0.66	1	0.74	0.30	0.24	0.18
CVC2	1	0.93	0.71	0.82	0.65	1	0.70	0.27	0.24	0.31
CVC3	1.01	0.69	0.53	0.41	0.38	1	0.59	0.24	0.27	0.25
A1	1.46	1	1	1	1	1.12	0.86	0.65	0.65	0.51
A2	1.25	1	1	1	0.97	1.19	0.99	0.91	0.97	0.84
A3	1.91	1	1	1	1	1.56	0.96	0.90	0.90	0.88
A4	1.12	1	0.96	1	1	1.1	0.93	0.69	0.77	0.67

The results achieved by the MLP and CNN corroborate our hypothesis. The ratios between the average inter- and intraclass KL divergences seem to be related to the model accuracy. When the average values of both divergences are close, *i.e.*, the ratio tends to 1. So, the model tends to have difficulty in identifying the correct classes. On the other hand, a higher ratio means a more significant interclass divergence concerning the intraclass one. In this case, the model can learn the patterns of each class more easily.

5   Conclusions

We proposed to use the Kullback-Leibler divergence to select the most promising input data for the fault diagnosis on a two-stage reciprocating compressor. We evaluated the technique on sensor datasets related to two groups of compressor faults. In this sense, we calculated the ratio between the average inter- and intraclass divergences for each sensor dataset. Those ratios were compared and analyzed together with the accuracy of models trained on these datasets. The models were MLPs and CNNs.

The results suggested a relation between the model accuracy and the ratio between the average inter- and intraclass divergences. Datasets related to the most accurate models were the ones with the highest ratios. Moreover, the data related to the least accurate models presented the lowest ratios. Such a behavior corroborates the expected results, which we discussed in Subsect. 3.2.

Selecting appropriate data is an essential issue for the fault diagnosis, even when we can use data from multiple sensors, *e.g.*, by using network ensembles and algorithms that fuse multimodal data. By discarding irrelevant information for the model learning process, we can reduce the training data, training time, and model complexity, among other benefits.

Future works include assessing the influence of sensor data selection on the performance of network ensembles with the proposed metric. Besides, this analysis may be extended to other datasets to check the robustness of this metric.

Acknowledgments. This study was financed in part by the Coordenação de Aperfeiçoamento de Pessoal de Nível Superior - Brasil (CAPES) - Finance Code 001.

References

1. Abbas, S.M., Singh, S.N.: Region-based object detection and classification using faster R-CNN. In: 2018 4th International Conference on Computational Intelligence & Communication Technology (CICT), pp. 1–6. IEEE (2018)
2. Aggarwal, Charu C.: Neural Networks and Deep Learning. Springer, Cham (2018). https://doi.org/10.1007/978-3-319-94463-0
3. Back, A.D., Trappenberg, T.P.: Input variable selection using independent component analysis. In: IJCNN 1999. International Joint Conference on Neural Networks. Proceedings (Cat. No. 99CH36339), vol. 2, pp. 989–992. IEEE (1999)
4. Cabrera, D., et al.: Bayesian approach and time series dimensionality reduction to lstm-based model-building for fault diagnosis of a reciprocating compressor. Neurocomputing **380**, 51–66 (2020)
5. Chinniah, Y.: Analysis and prevention of serious and fatal accidents related to moving parts of machinery. Safety Sci. **75**, 163–173 (2015)
6. Eriksson, D., Frisk, E., Krysander, M.: A method for quantitative fault diagnosability analysis of stochastic linear descriptor models. Automatica **49**(6), 1591–1600 (2013)
7. Géron, A.: Hands-on Machine Learning with Scikit-Learn and TensorFlow: Concepts, Tools, and Techniques to Build Intelligent Systems. O'Reilly Media, Inc., Newton (2017)

8. Goodfellow, I., Bengio, Y., Courville, A., Bengio, Y.: Deep Learning, vol. 1. MIT press Cambridge (2016)
9. Hashemian, H.M.: State-of-the-art predictive maintenance techniques. IEEE Trans. Instrument. Meas. **60**(1), 226–236 (2010)
10. He, J., Yang, S., Papatheou, E., Xiong, X., Wan, H., Gu, X.: Investigation of a multi-sensor data fusion technique for the fault diagnosis of gearboxes. Proc. Inst. Mech. Eng. Part C: J. Mech. Eng. Sci. **233**(13), 4764–4775 (2019)
11. Islam, Q.N.: Mastering PyCharm. Packt Publishing Ltd., Birmingham (2015)
12. Jiang, G., He, H., Xie, P., Tang, Y.: Stacked multilevel-denoising autoencoders: a new representation learning approach for wind turbine gearbox fault diagnosis. IEEE Trans. Instrument. Meas. **66**(9), 2391–2402 (2017)
13. Kullback, S.: Information theory and statistics. Courier Corporation (1997)
14. Kullback, S., Leibler, R.A.: On information and sufficiency. Ann. Math. Stat. **22**(1), 79–86 (1951)
15. Liu, R., Yang, B., Zio, E., Chen, X.: Artificial intelligence for fault diagnosis of rotating machinery: a review. Mech. Syst. Signal Process. **108**, 33–47 (2018)
16. Mohammed, M., Khan, M.B., Bashier, E.B.M.: Machine Learning: Algorithms and Applications. CRC Press, Boca Raton (2016)
17. Monteiro, R.P., Bastos-Filho, C.J.: Detecting defects in sanitary wares using deep learning. In: 2019 IEEE Latin American Conference on Computational Intelligence (LA-CCI), pp. 1–6. IEEE (2019)
18. Monteiro, R.P., Cerrada, M., Cabrera, D.R., Sánchez, R.V., Bastos-Filho, C.J.: Using a support vector machine based decision stage to improve the fault diagnosis on gearboxes. Comput. Intell. Neurosci. **2019**, (2019)
19. Moran, M.B., et al.: Identification of thyroid nodules in infrared images by convolutional neural networks. In: 2018 International Joint Conference on Neural Networks (IJCNN), pp. 1–7. IEEE (2018)
20. Rao, K.R., Kim, D.N., Hwang, J.J.: Fast Fourier transform-algorithms and applications. Springer, Heidelberg (2011). https://doi.org/10.1007/978-1-4020-6629-0
21. Rawat, W., Wang, Z.: Deep convolutional neural networks for image classification: a comprehensive review. Neural Comput. **29**(9), 2352–2449 (2017)
22. Sharkey, A.J., Chandroth, G.O., Sharkey, N.E.: Acoustic emission, cylinder pressure and vibration: a multisensor approach to robust fault diagnosis. In: Proceedings of the IEEE-INNS-ENNS International Joint Conference on Neural Networks. IJCNN 2000. Neural Computing: New Challenges and Perspectives for the New Millennium, vol. 6, pp. 223–228. IEEE (2000)
23. VanRossum, G., Drake, F.L.: The python Language Reference. Python Software Foundation, Amsterdam (2010)

An Analysis of Protein Patterns Present in the Saliva of Diabetic Patients Using Pairwise Relationship and Hierarchical Clustering

Airton Soares[1]([✉]) [iD], Eduardo Esteves[2,3] [iD], Nuno Rosa[2] [iD],
Ana Cristina Esteves[2] [iD], Anthony Lins[4] [iD], and Carmelo J. A. Bastos-Filho[1] [iD]

[1] Universidade de Pernambuco, Recife, Brazil
assj@ecomp.poli.br
[2] Faculty of Dental Medicine, Center for Interdisciplinary Research in Health (CIIS),
Universidade Católica Portuguesa, Viseu, Portugal
[3] Faculdade de Ciências da Saúde (UBI), Universidade da Beira Interior,
Covilhã, Portugal
[4] Universidade Católica de Pernambuco, Recife, Brazil
http://upe.br/
https://www.ucp.pt/
https://ubi.pt
https://www1.unicap.br/

Abstract. Molecular diagnosis is based on the quantification of RNA, proteins, or metabolites whose concentration can be correlated to clinical situations. Usually, these molecules are not suitable for early diagnosis or to follow clinical evolution. Large-scale diagnosis using these types of molecules depends on cheap and preferably noninvasive strategies for screening. Saliva has been studied as a noninvasive, easily obtainable diagnosis fluid, and the presence of serum proteins in it enhances its use as a systemic health status monitoring tool. With a recently described automated capillary electrophoresis-based strategy that allows us to obtain a salivary total protein profile, it is possible to quantify and analyze patterns that may indicate disease presence or absence. The data of 19 persons with diabetes and 58 healthy donors obtained by capillary electrophoresis were transformed, treated, and grouped so that the structured values could be used to study individuals' health state. After Pairwise Relationships and Hierarchical Clustering analysis were observed that amplitudes of protein peaks present in the saliva of these individuals could be used as differentiating parameters between healthy and unhealthy people. It indicates that these characteristics can serve as input for a future computational intelligence algorithm that will aid in the stratification of individuals that manifest changes in salivary proteins.

Supported by Universidade de Pernambuco (UPE) and Universidade Católica Portuguesa (UCP).

C. Analide et al. (Eds.): IDEAL 2020, LNCS 12489, pp. 148–159, 2020.
https://doi.org/10.1007/978-3-030-62362-3_14

Keywords: Data mining · Clustering algorithms · Saliva · Capillary electrophoresis · Diagnosis

1   Introduction

Molecular diagnosis is based on the quantification of RNA [3], proteins [17], or metabolites, whose concentration can be correlated to clinical situations. Usually, these molecules alone are not suitable for early diagnosis or to follow the clinical evolution. Therefore, strategies to evaluate the complete molecular scenario – early diagnosis, diagnosis, and clinical evolution – are necessary.

The potential of proteins for a large-scale diagnosis depends on cheap and preferably non-invasive strategies for screening. A good approach involves bioinformatics strategies and solutions to work with different types of data, from biological-related data to personal and clinical information. Data integration is an asset to predict the pathological status before clinical outcomes.

In the last decade, saliva has been studied as a non-invasive, easily obtainable diagnosis fluid [14]. It is composed of the secretions of the three largest salivary glands (parotid, submandibular and sublingual), smaller salivary glands, crevicular fluid, and contains serum components, transported by blood capillaries, and subsequently transferred by diffusion, transport and/or ultrafiltration. The presence of serum proteins in saliva enhances its use as a systemic health status monitoring tool [2,10,11,19].

Data on salivary proteins associated with disease or health status is already extensive. Our group has studied salivary proteins and produced the SalivaTecDB database (http://salivatec.viseu.ucp.pt/salivatec-db) [1,15], which is relevant for the identification of proteins that may potentially be associated with specific signatures. SalivaTecDB has currently stored more than 3,500 human salivary proteins.

We recently described an automated capillary electrophoresis-based strategy that allows one to obtain a salivary protein profile – the SalivaPrint Toolkit [4,7]. Since proteins are separated according to their molecular mass, changes in peak morphology or fluorescence intensity (translated by changes in peak height) correspond to fluctuations in the proteins' concentration or the type of proteins being expressed. The association of saliva protein signatures to different health/disease situations allows us to build a cheap and robust framework for the development of a monitoring tool.

The use of machine learning algorithms on risk disease prediction is already a reality [8,12,13,18]. Clinical data patients integrated with laboratory results can contribute to health/disease monitoring, building the foundations for the development of a risk assessment tool for diagnosis.

In this article, we propose a methodology for the analysis of salivary protein patterns that reflects patients' health status. Using a database of healthy and diabetic individuals protein patterns, we analyzed the association of protein peaks with these patients' health status. The goal is to understand the relationships between the protein profiles and the individual's state of health. This analysis can influence the choice of the learning method for the recognition process.

The remainder of the paper is organized as follows. In Sect. 2, we explain the methodology deployed in the project, describing the data acquisition process, graphical representation of the obtained data, pre-analysis, the peaks detection process, the pairwise relationship, and the hierarchical clustering. In Sect. 3, we have the analysis of the results obtained using the described methodology. In Sect. 4, we present proposals for future works in the scope of the project. In Sect. 5, we give our conclusions.

2   Methodology

Aiming at the future development of a computational intelligence algorithm that can differentiate healthy individuals from unhealthy ones, we performed several transformations and analysis procedures on data salivary protein patterns.

2.1   Data Acquisition and Description

All data used in this study were acquired through capillary electrophoresis, using the Experion™Automated Electrophoresis System (BioRad®) in standard protein chips (Experion™Pro260 Analysis Kit[1]). Total protein concentration was normalized to 500 μg/mL, and samples were analyzed in duplicate.

Protein profiles were obtained using the Experion™Software, version 3.20, and exported as an XML file. Once the output file was generated, a Python script extracts the data in the file and generates a CSV (Coma Separated Values) file with 399 signals for each sample. These signals correspond to the fluorescence on each molecular weight measured on the capillary electrophoresis system: 10 kDa to 121 kDa.

The resulting data set was obtained from 77 individuals, 58 are healthy, and 19 have diabetes. Table 1 shows some examples of the data set structure. The first column contains the identifiers. The last column includes the individual's health status, and the columns between them represent protein weights in kDa, ranging from 10 to 121. The values for the columns are the fluorescence returned from the Experion™Software.

Table 1. Examples of data set entries. The first column contains the identifiers, the last column contains the individual's health status and the columns between them represent protein molecular weights

Sample ID	10.0	10.1	10.2		120.3	120.8	121.0	Health status
d1122	−25.914	−23.452	−21.871	...	−1.056	−1.204	−1.332	Healthy
d1127	30.009	25.470	21.091	...	0.756	0.260	−0.472	Healthy
d1132	1.189	0.311	−0.650	...	−0.341	−0.600	−1.043	Healthy
d52	−11.405	−12.425	−13.341	...	−1.025	−1.096	−1.096	Diabetes
d56	−6.595	−6.886	−7.056	...	0.760	0.950	0.950	Diabetes
d59	38.839	25.473	12.131	...	−12.617	−12.682	−12.682	Diabetes

---

[1] http://www.bio-rad.com/webroot/web/pdf/elsr/literature/10000975C.pdf.

2.2   Data Visualization

Figure 1 shows the protein profiles for healthy and unhealthy individuals. Fluorescence values have a small variation for healthy individuals than for unhealthy individuals. Also, unhealthy individuals present higher values on the entire scale and a more significant change than healthy individuals. Both groups show the peaks in similar molecular weights. This behavior indicates that the peak height values identified in the signature of an individual's proteins may be used to characterize his health status.

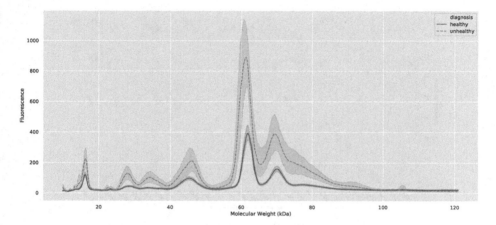

Fig. 1. Representation of the distribution of fluorescence values for each molecular weight (kDa). The average values of the molecular weight's fluorescence for the healthy individuals (n = 58) are represented by the blue line and for the unhealthy individuals (n = 19) represented by the dashed orange line. The shaded areas around the lines represent the standard deviation of the fluorescence values. (Color figure online)

2.3   Data Preparation

As can be seen in Table 1, some fluorescence values are negative. Therefore, each row fluorescence values were normalized by adding the absolute value of the smallest value on each row.

The total number of points in each electropherogram is 395. Since the goal of this analysis is to generate valuable information for the selection and calibration of a future machine learning algorithm, the number of features should be as lower as possible without losing significant information [6]. Therefore, we adopted two simplification procedures.

First, we truncated all the values of the fluorescence towards zero, and we calculated the average of the results for each integer weight (Table 2).

After that, we grouped the weights in sets with a 4 kDa interval, and the value that represents the set is the maximum value in the range (Fig. 2, Table 3). We did this due to the granularity of the measured molecular weights. In this case, the fluorescence values are very close in the neighborhood, and the deployed hardware may not be as accurate, possibly generating lags of some few kDa. The maximum value in the range was used to represent the interval. It helped not to create false peaks, what could be the case if we have used the sum of the values in the ranges, and not to flat some peaks, what could happen if we have used the average value of the ranges.

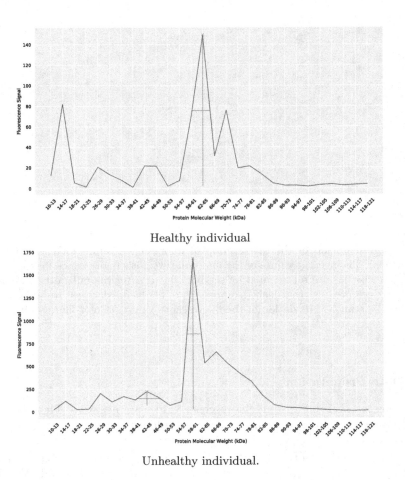

Fig. 2. Graphical representation of peaks detection over samples from Table 3. The "x" axis represents the grouped molecular weights labels. The "y" axis represents the maximum fluorescence value in the molecular weight group. The vertical lines represent the peak prominence and the horizontal ones represent the average width.

Table 2. Examples of data set entries after the fluorescence's values normalization.

Sample ID	10.0	10.1	10.2	...	120.3	120.8	121.0	Health state
d1122	0.0	2.461	4.043	...	24.857	24.709	24.581	Healthy
d1127	33.766	29.228	24.848	...	4.513	4.018	3.284	Healthy
d1132	22.624	21.746	20.784	...	21.093	20.834	20.392	Healthy
d52	6.176	5.156	4.240	...	16.556	16.485	16.485	Diabetes
d56	6.236	5.945	5.775	...	13.592	13.782	13.782	Diabetes
d59	63.151	49.785	36.443	...	11.693	11.629	11.629	Diabetes

Table 3. Examples of data set entries after grouping molecular weights.

Sample ID	10–13	14–17	18–21	...	110–113	114–117	118–121	Health state
d1122	9.455	49.811	24.317	...	24.882	25.244	25.003	Healthy
d1127	12.523	81.814	5.542	...	2.671	3.394	3.915	Healthy
d1132	17.075	28.868	6.731	...	19.325	21.037	20.976	Healthy
d52	5.461	173.141	18.129	...	17.530	16.543	16.609	Diabetes
d56	11.512	44.826	11.464	...	11.888	12.236	13.235	Diabetes
d59	26.700	119.878	27.789	...	9.265	8.962	11.186	Diabetes

2.4 Peak Detection

After the data set was pre-processed, we carried out a peak detection strategy. A peak or local maximum is defined as any entry whose two direct neighbors have a smaller value. Various parameters like prominence, width, and height can be used as thresholds to select specific types of peaks.

We used an algorithm to automatically detect peaks over each sample using a height threshold of 100. We chose this value because it approximates the average amplitude of the relevant lower peaks in unhealthy individuals, as seen in Fig. 1.

With all the peaks detected for all individuals, a new data set is generated. The resulting table contains every height of the distinct peaks detected through the process as features for every sample, 9 in total. If an individual does not present a specific peak, the height for that will be considered 0 (Table 4).

Table 4. Examples of peaks data set entries.

Sample ID	26–29	34–37	42–45	46–49	58–61	62–65	66–69	70–73	74–77	Health state
d1122	0.000	0.000	0.000	0.000	0.000	275.493	0.000	0.000	0.000	Healthy
d1127	0.000	0.000	0.000	0.000	0.000	0.000	0.000	149.120	0.000	Healthy
d1132	161.880	0.000	0.000	0.000	0.000	0.000	0.000	0.000	0.000	Healthy
d52	358.429	0.000	0.000	0.000	0.000	0.000	0.000	0.000	0.000	Diabetes
d56	422.497	0.000	0.000	0.000	0.000	0.000	0.000	0.000	0.000	Diabetes
d59	1680.983	0.000	0.000	0.000	0.000	0.000	0.000	222.072	0.000	Diabetes

2.5 Pairwise Relationships

We performed pairwise relationships to identify relationships between the molecular weights of the identified peaks and also the influence that each of them has on the classification of individuals.

Pairwise relationships can be understood as any process of comparing entities in pairs to judge which of each entity is preferred or has a more significant amount of some quantitative property, or whether or not the two entities are identical [5].

2.6 Hierarchical Clustering

We performed the hierarchical clustering to identify if the given characteristics extracted were sufficient to generate a grouping by the individuals' health status.

Hierarchical clustering is a type of unsupervised machine learning algorithm used to cluster unlabeled data points, grouping the data points with similar characteristics [9]. The calculation that defines the similarity between data points can be different depending on the type of data and how you want to do the grouping.

Because of this grouping property, and because we were looking to explore the data structure to understand emerging profiles, we used hierarchical clustering with the following similarity calculation methods:

1. **average**: Uses the average of the distances of each observation of the two sets.
2. **complete**: Uses the maximum distances between all observations of the two sets.
3. **single**: Uses the minimum of the distances between all observations of the two sets.
4. **ward**: Minimizes the sum of squared differences within all clusters. It is a variance-minimizing approach [20].

The peak heights were treated as coordinates of Euclidean space to calculate the distances.

Another study made over the hierarchical clustering was the silhouette analysis [16]. It consists of calculating the Silhouette Coefficient. It uses the mean intra-cluster distance (a) and the mean nearest-cluster distance (b) for each given sample. The coefficient for a sample is $(b - a)/max(a, b)$. To clarify, "b" is the distance between a sample and the nearest cluster that the sample is not part of. Silhouette coefficients near 1 indicate that the sample is far away from the neighboring clusters. Values around 0 indicate that the sample is on or very close to the decision boundary between two neighboring clusters. Negative values indicate that those samples might have been assigned to the wrong cluster.

3   Results

Figure 3 is a graphical representation of the resulting pairwise relationship, a grid of axes such that each variable will be shared in the "y" axis across a single row and the "x" axis across a single column. The diagonal is treated differently, drawing a plot to show the univariate distribution of the data for the variable in that column.

This representation shows that the peaks "26–29", "42–45", "46–49," and "58–61" are present practically only in unhealthy individuals, making them good candidates for use in the process of differentiation. Furthermore, it shows that in the peaks "34–37", "62–65", "70–73," and "74–77," healthy individuals are concentrated in lower height values, while the unhealthy individuals are better distributed. Finally, it is noticed that the peak "42–45" has a very similar distribution for both profiles, making it not an exciting feature to be used to classify the health of the base entries.

Also, the other graphs show that no binary combination of attributes could separate healthy and unhealthy individuals well. It indicates that the nature of the attributes requires three or more features to classify the health state of the presented examples.

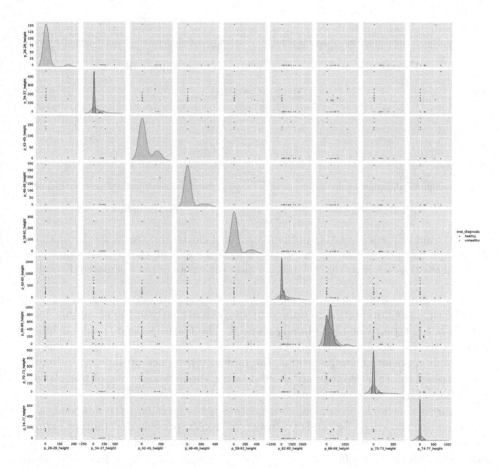

Fig. 3. Pairwise relationships plot of the heights of the peaks to the patient health. Blue represents healthy individuals, while orange represents unhealthy individuals. The graphs in the diagonal axis represent the distribution of individuals for each peak, the remaining represent a binary combination of two peaks in "x" and "y" axis trying to separate the samples. (Color figure online)

The hierarchical clustering results, although different distance calculation types were applied, were very similar, presenting a classification of almost all the unhealthy individuals right at the beginning of the formation of the groups, around distance 250, meaning that this is a reasonable distance for the separation of the categories. Figure 4 depicts a graphical representation of the hierarchical clustering performed using every cited method of calculation.

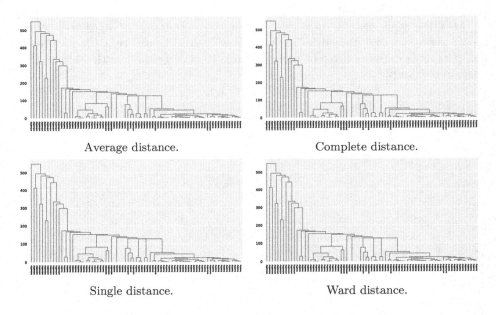

Average distance.   Complete distance.

Single distance.   Ward distance.

Fig. 4. Graphical representation of the hierarchical clustering using different methods of distance calculation. On the "x" axis, we have the classifications' final values, the leaves of the tree, while on the "y" axis, we have the values of the calculated distances.

The rapid agglomeration of unhealthy individuals early in the groupings indicates what the pairwise plot already showed. The features used (peaks heights) manage to characterize well diabetic individuals. Also, as diabetes is a disease with many associated complications, the minority presence of diabetic individuals scattered in other groups may be evidencing the heterogeneity of phenotypes that characterize diabetic patients.

Figure 5 shows the graphical representation of the silhouette analysis. The areas next to the labels "1" and "0" in the "y" axis represent the samples clumping together in cluster "unhealthy" and "healthy," respectively. The dashed lines mark the silhouette coefficient, which is 0.746 for average, complete, and ward methods, showing that they have a reasonable separation distance for the individuals. The coefficient value for the single method is only 0.592.

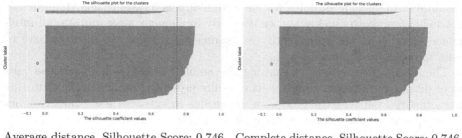

Average distance. Silhouette Score: 0.746   Complete distance. Silhouette Score: 0.746

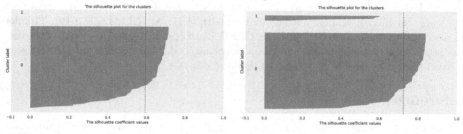

Single distance. Silhouette Score: 0.592   Ward distance. Silhouette Score: 0.746

Fig. 5. Graphical representation of the silhouette analysis in the first level of the hierarchical clusterization, with only two clusters. The label "1" represents the unhealthy group and "0" the healthy group. The dashed vertical line is the silhouette score marker.

4   Future Work

As it could be seen in the analyzes made, the data obtained is promising as input for a learning algorithm. Some computational intelligence algorithms are being tested on this basis to identify an individual health state, the results of that study will be presented in a future work. With positive results in this next study, it would be possible to create an automation for the diagnosis process from the data extracted from a person's saliva using the Experion™Automated Electrophoresis System.

Also, a base with a more significant number of individuals with a greater diversity of diseases is currently being set up. Once this base is ready, we will have more statistical confidence as to whether the peak patterns identified with the current database are sufficient for the health classification and discover new ones if they exist. Besides, we will be able to increase the range of possible classifications, differentiating individuals between healthy and unhealthy (with diabetes), and identifying the specific illness.

5   Conclusions

This article presents an analysis of the saliva protein profiles of diabetic and healthy individuals. The study identified characteristic patterns of variations in the number of specific proteins for these individuals' classification. It indicates that it is possible to quickly and consistently implement a computational intelligence algorithm that can identify a person's health status and automate or assist in a diagnostic process.

The database is limited concerning the number of individuals and the variety of diseases presented. However, the results presented indicate differentiating characteristics between the groups. It is possible to extract these characteristics in a simple way to use them in the process of classification.

Ackownledgments. Thanks are due to FCT/MCTES, for the financial support of the Center for Interdisciplinary Research in Health (UID/MULTI/4279/2019). Thanks are also due to FCT and UCP for the CEEC institutional financing of AC Esteves. This study was financed in part by the Coordenação de Aperfeiçoamento de Pessoal de Nível Superior - Brasil (CAPES) - Finance Code 001.

References

1. Arrais, J.P., et al.: Oralcard: a bioinformatic tool for the study of oral proteome. Arch. Oral Biol. **58**(7), 762–772 (2013)
2. Castagnola, M., et al.: Salivary biomarkers and proteomics: future diagnostic and clinical utilities. Acta Otorhinolaryngol. Ital. **37**(2), 94 (2017)
3. Conde, J., de la Fuente, J.M., Baptista, P.V.: RNA quantification using gold nanoprobes-application to cancer diagnostics. J. Nanobiotechnol. **8**(1), 5 (2010)
4. Cruz, I., et al.: SalivaPRINT toolkit-protein profile evaluation and phenotype stratification. J. Proteomics **171**, 81–86 (2018)
5. David, H.A.: The Method of Paired Comparisons, vol. 12, London (1963)
6. Domingos, P.: A few useful things to know about machine learning. Commun. ACM **55**(10), 78–87 (2012)
7. Esteves., E., Cruz., I., Esteves., A.C., Barros., M., Rosa., N.: SalivaPRINT as a non-invasive diagnostic tool. In: Proceedings of the 13th International Joint Conference on Biomedical Engineering Systems and Technologies, HEALTHINF, vol. 5, pp. 677–682. INSTICC. SciTePress (2020). https://doi.org/10.5220/0009163506770682
8. Ferreira, A.V., Bastos Filho, C.J., Lins, A.J.: An unsupervised analysis of an Alzheimer's disease patient population using subspace search and hierarchical density-based clustering. In: 2019 IEEE Latin American Conference on Computational Intelligence (LA-CCI), pp. 1–6. IEEE (2019)
9. Frigui, H., Krishnapuram, R.: Clustering by competitive agglomeration. Pattern Recogn. **30**(7), 1109–1119 (1997)
10. Kaczor-Urbanowicz, K.E., Martin Carreras-Presas, C., Aro, K., Tu, M., Garcia-Godoy, F., Wong, D.T.: Saliva diagnostics-current views and directions. Exp. Biol. Med. **242**(5), 459–472 (2017)
11. Kaushik, A., Mujawar, M.A.: Point of care sensing devices: better care for everyone (2018)
12. Lins, A., Muniz, M., Bastos-Filho, C.J.: Comparing machine learning techniques for dementia diagnosis. In: 2018 IEEE Latin American Conference on Computational Intelligence (LA-CCI), pp. 1–6. IEEE (2018)
13. Lins, A., Muniz, M., Garcia, A., Gomes, A., Cabral, R., Bastos-Filho, C.J.: Using artificial neural networks to select the parameters for the prognostic of mild cognitive impairment and dementia in elderly individuals. Comput. Methods Programs Biomed. **152**, 93–104 (2017)

14. Loo, J., Yan, W., Ramachandran, P., Wong, D.: Comparative human salivary and plasma proteomes. J. Dent. Res. **89**(10), 1016–1023 (2010)
15. Rosa, N., et al.: From the salivary proteome to the oralome: comprehensive molecular oral biology. Arch. Oral Biol. **57**(7), 853–864 (2012)
16. Rousseeuw, P.J.: Silhouettes: a graphical aid to the interpretation and validation of cluster analysis. J. Comput. Appl. Math. **20**, 53–65 (1987)
17. Sabbagh, B., Mindt, S., Neumaier, M., Findeisen, P.: Clinical applications of MS-based protein quantification. PROTEOMICS-Clin. Appl. **10**(4), 323–345 (2016)
18. Uddin, S., Khan, A., Hossain, M.E., Moni, M.A.: Comparing different supervised machine learning algorithms for disease prediction. BMC Med. Inform. Decis. Mak. **19**(1), 1–16 (2019). https://doi.org/10.1186/s12911-019-1004-8
19. Wang, X., Kaczor-Urbanowicz, K.E., Wong, D.T.W.: Salivary biomarkers in cancer detection. Med. Oncol. **34**(1), 1–8 (2016). https://doi.org/10.1007/s12032-016-0863-4
20. Ward Jr., J.H.: Hierarchical grouping to optimize an objective function. J. Am. Stat. Assoc. **58**(301), 236–244 (1963)

Improving Adversarial Learning with Image Quality Measures for Image Deblurring

Jingwen Su[✉][iD] and Hujun Yin[✉][iD]

Department of Electrical and Electronic Engineering, The University of Manchester,
Manchester M13 9PL, UK
{jingwen.su,hujun.yin}@manchester.ac.uk

Abstract. Generative adversarial networks (GANs) have become popular and powerful models for solving a wide range of image processing problems. We introduce a novel component based on image quality measures in the objective function of GANs for solving image deblurring problems. Such additional constraints can regularise the training and improve the performance. Experimental results demonstrate marked improvements on generated or restored image quality both quantitatively and visually. Boosted model performances are observed and testified on three test sets with four image quality measures. It shows that image quality measures are additional flexible, effective and efficient loss components to be adopted in the objective function of GANs.

Keywords: Generative adversarial networks · Image deblurring · Image quality measures

1 Introduction

Recently, deep neural networks with adversarial learning have become a prevalent technique in generative image modelling and have made remarkable advances. In topics such as image super-resolution, in-painting, synthesis, and image-to-image translation, there are already numerous adversarial learning based methods demonstrating the prominent effectiveness of GANs in generating realistic, plausible and conceptually convincing images [8,14,21,25–27,29].

In this paper we address image enhancement problems such as blind single image deblurring by casting them as a special case for image-to-image translation, under the adversarial learning framework. A straightforward way for realising quality improvement in image restoration is to involve image quality measure as constraints for training GANs. As it is known, the objective function of GANs defines gradient scale and direction for network optimization. Adversarial loss in GANs, an indispensable component, is the foundation that encourages the generation of images to be as realistic as possible. However, details and textures in the generated images are unable to be fully recovered and they are critical for the

© Springer Nature Switzerland AG 2020
C. Analide et al. (Eds.): IDEAL 2020, LNCS 12489, pp. 160–171, 2020.
https://doi.org/10.1007/978-3-030-62362-3_15

human visual system to perceive image quality. Thus, image quality measures that compensate the overlooked perceptual features in images are necessary to take a part in guiding gradient optimization during the training.

An image quality based loss is proposed and added to the objective function of GANs. There are three common quality measures that can be adopted. We investigate their effects on generated/restored image quality, compared with the baseline model without any quality loss. The rest of this paper is structured as follows. Section 2 describes related work. Section 3 introduces the proposed method, followed by experimental settings, results and discussion in Sect. 4. Section 5 concludes the findings and suggests possible future work.

2 Related Work

2.1 Generative Adversarial Networks

GAN consists of a generative model and a discriminative model. These two models are trained simultaneously by the means of adversarial learning, a process that can significantly contribute to improving the generation performance. Adversarial learning encourages competition between the generator and the discriminator. The generator is trained to generate better fake samples to fool the discriminator until they are indistinguishable from real samples.

For a standard GAN (a.k.a the vanilla GAN) proposed by Goodfellow et al. [6], the generator G receives noise as the input and generates fake samples from model distribution p_g, the discriminator D classifies whether the input data is real. There are a great number of variants of GAN proposed afterwards, such as conditional GAN (cGAN) [17], least squares GAN (LSGAN) [16], Wasserstein GAN (WGAN) [1], and Wasserstein GAN with gradient penalty (WGAN-GP) [7].

2.2 Image Deblurring

Image deblurring has been a perennial and challenging problem in image processing and its aim is to recover clean and sharp images from degraded observations. Recovery process often utilises image statistics and prior knowledge of the imaging system and degradation process, and adopts a deconvolution algorithm to estimate latent images. However, prior knowledge of degradation models is generally unavailable in practical situations - the case is categorized as blind image deblurring (BID). Most conventional BID algorithms make estimations according to image statistics and heuristics. Fergus et al. [5] proposed a spatial domain prior of a uniform camera blur kernel and camera rotation. Li et al. [24] created a maximum-a-posterior (MAP) based framework and adopted iterative approach for motion deblurring. Recent approaches have turned to deep learning for improved performances. Xu et al. [23] adapted convolutional kernels in convolutional neural networks (CNNs) to blur kernels. Schuler et al. [20] built stacked CNNs that pack feature extraction, kernel estimation and image estimation modules. Chakrabarti [3] proposed to predict complex Fourier coefficients of motion kernels by using neural networks.

2.3   Image Quality Measures

Image quality assessment (IQA) is a critical and necessary step to provide quantitative objective measures of visual quality for image processing tasks. IQA methods have been an important and active research topic. Here we focus on four commonly used IQA methods: PSNR, SSIM, FSIM and GMSD.

Peak signal-to-noise ratio (PSNR) is a simple signal fidelity measure that calculates the ratio between the maximum possible pixel value in the image and the mean squared error (MSE) between distorted and reference images.

Structural similarity index measure (SSIM) considers image quality degradation as perceived change of structural information in image. Since structural information is independent of illumination and contrast [22], SSIM index is a linear combination of these three relatively independent terms, luminance $l(x,y)$, contrast $c(x,y)$ and structure comparison function $s(x,y)$. Besides, the measure is based on local patches of two aligned images because luminance and contrast vary across the entire image. To avoid blocking effect in the resulting SSIM index map, 11×11 circular-symmetric Gaussian weighing function is applied before computation. Patch based SSIM index is defined as in Eq. 1, while for the entire image, it is common to use mean SSIM (MSSIM) as the evaluation metric for the overall image quality (Eq. 2).

$$
\begin{aligned}
SSIM(x,y) &= l(x,y) \cdot c(x,y) \cdot s(x,y) \\
&= \frac{(2\mu_x\mu_y + c_1)(2\sigma_{xy} + c_2)}{(\mu_x^2 + \mu_y^2 + c_1)(\sigma_x^2 + \sigma_y^2 + c_2)}
\end{aligned} \tag{1}
$$

$$
MSSIM(X,Y) = \frac{1}{M} \sum_{m=1}^{M} SSIM(x_m, y_m) \tag{2}
$$

where x and y are two local windows from two aligned images X and Y. μ_x is the mean of x, μ_y is the mean of y, σ_x^2 is the variance of x, σ_y^2 is the variance of y, σ_{xy} is the covariance of x and y, constants c_1 and c_2 are conventionally set to 0.0001 and 0.0009 to stabilize the division. M is the total number of windows.

Feature similarity index measure (FSIM) is based on similarity of salient low-level visual features, i.e. the phase congruency (PC). High PC means the existence of highly informative features, where the Fourier waves at different frequencies have congruent phases [28]. To compensate the contrast information that the primary feature PC is invariant to, gradient magnitude is added as the secondary feature for computing FSIM index.

First, PC map computation of an image is conducted by generalizing the method proposed in [12] from 1-D signal to 2-D grayscale image, by the means of applying the spreading function of Gaussian. 2-D log-Gabor filter extracts a quadrature pair of even-symmetric filter response and odd-symmetric filer response $[e_{n,\theta_j}(a), o_{n,\theta_j}(a)]$ at pixel a on scale n in the image. Transfer function is formulated as follows,

$$
G(\omega, \theta_j) = exp\left(-\frac{(\log(\frac{\omega}{\omega_0}))^2}{2\sigma_r^2}\right) \cdot exp\left(-\frac{(\theta - \theta_j)^2}{2\sigma_\theta^2}\right) \tag{3}
$$

where ω represents the frequency, $\theta_j = \frac{j\pi}{J}$ ($j = \{0, 1, \ldots, J - 1\}$) represents the orientation angle of the filer, J is the number of orientations. ω_0 is the filter center frequency, σ_r is the filter bandwidth, σ_θ is the filter angular bandwidth. And the PC at pixel a is defined as,

$$PC(a) = \frac{\sum_j E_{\theta_j}(a)}{\epsilon + \sum_n \sum_j A_{n,\theta_j}(a)} \tag{4}$$

$$E_{\theta_j}(a) = \sqrt{F_{\theta_j}(a)^2 + H_{\theta_j}(a)^2} \tag{5}$$

$$F_{\theta_j}(a) = \sum_n e_{n,\theta_j}(a), \; H_{\theta_j}(a) = \sum_n o_{n,\theta_j}(a) \tag{6}$$

$$A_{n,\theta_j}(a) = \sqrt{e_{n,\theta_j}(a)^2 + o_{n,\theta_j}(a)^2} \tag{7}$$

where $E_{\theta_j}(a)$ is the local energy function along orientation θ.

Gradient magnitude (GM) computation follows the traditional definition that computes partial derivatives $G_h(a)$ and $G_v(a)$ along horizontal and vertical directions using gradient operators. GM is defined as $G(a) = \sqrt{G_h(a)^2 + G_v(a)^2}$.

For calculating FSIM index between X and Y, PC and GM similarity measure between these two images are computed as follows,

$$S_{PC}(a) = \frac{2PC_X(a) \cdot PC_Y(a) + T_1}{PC_X^2(a) + PC_Y^2(a) + T_1} \tag{8}$$

$$S_G(a) = \frac{2G_X(a) \cdot G_Y(a) + T_2}{G_X^2(a) + G_Y^2(a) + T_2} \tag{9}$$

$$S_L(a) = S_{PC}(a) \cdot S_G(a) \tag{10}$$

where T_1 and T_2 are positive constants depending on dynamic range of PC and GM values respectively. Based on similarity measure $S_L(a)$, the FSIM index is defined as,

$$FSIM(X, Y) = \frac{\sum_a^\Omega S_L(a) \cdot PC_m(a)}{\sum_a^\Omega PC_m(a)} \tag{11}$$

where $PC_m(a) = max(PC_X(a), PC_Y(a))$ is to balance the importance between similarity between X and Y, Ω is the entire spatial domain of image. Introduced in [28], $FSIM_c$ is for colour images by incorporating chormatic information.

Gradient magnitude standard deviation (GMSD) mainly utilizes feature properties in image gradient domain to derive quality measure. GMSD metric calculates the standard deviation of gradient magnitude. Prewitt filter is commonly adopted as the gradient operator. Similar to FSIM index, GM similarity measure is firstly computed using Eq. 9. The difference is the Eq. 12. So the smaller GMSD the higher image perceptual quality.

$$GMSD(X, Y) = \sqrt{\frac{1}{N} \sum_{a \in \Omega} (S_G(a) - mean(G(a)))^2} \tag{12}$$

where N is the total number of pixels in image, $mean(G(a)) = \frac{1}{N} \sum_{a \in \Omega} S_G(a)$.

3   The Proposed Method

We propose modified GAN models that are able to blindly restore sharp latent images with better quality from single blurred images. Quality improvement of restored images is realized by adding a quality loss into the training objective function. We compare three image quality measure based losses, which are based on SSIM, FSIM and MSE. We apply these quality losses to two types of GAN models, LSGAN and WGAN-GP, respectively.

3.1   Loss Function

For simplicity, we first define variables and terms as follows. Batch size is m, input blurred image samples $\{I_B^{(i)}\}_{i=1}^m$, restored image samples $\{I_R^{(i)}\}_{i=1}^m$, and original sharp image samples $\{I_S^{(i)}\}_{i=1}^m$. The adversarial loss $\mathcal{L}_{\text{ad}}$, content loss $\mathcal{L}_X$, quality loss $\mathcal{L}_Q$ are as follows.

Adversarial Loss. For LSGAN,

$$G : \mathcal{L}_{\text{ad}} = \frac{1}{m}\sum_{i=1}^m \frac{1}{2}\left(D(G(I_B^{(i)})) - 1\right)^2 \tag{13}$$

$$D : \mathcal{L}_{\text{ad}} = \frac{1}{m}\sum_{i=1}^m \frac{1}{2}\left[(D(I_S^{(i)}) - 1)^2 + D(G(I_B^{(i)}))^2\right] \tag{14}$$

For WGAN-GP,

$$\mathcal{L}_{\text{ad}} = \frac{1}{m}\sum_{i=1}^m D(I_S^{(i)}) - D(G(I_B^{(i)})) + \lambda\left[(\|\nabla_{\tilde{x}} D(\tilde{x})\| - 1)^2\right] \tag{15}$$

Content Loss. $\mathcal{L}_X$ is a L_2 loss based on the difference between the VGG-19 feature maps of generated image and sharp image. As proposed in [9], the VGG19 network is pretrained on ImageNet [4]. $\mathcal{L}_X$ is formulated as,

$$\mathcal{L}_X = \frac{1}{W_{j,k}H_{j,k}}\sum_{x=1}^{W_{j,k}}\sum_{y=1}^{H_{j,k}}(\phi_{j,k}(I_S^{(i)})_{x,y} - \phi_{j,k}(G(I_B^{(i)}))_{x,y})^2 \tag{16}$$

where $\phi_{j,k}$ is the feature map of the k-th convolution before j-th maxpooling layer in the VGG19 network. $W_{j,k}$ and $H_{j,k}$ are the dimensions of feature maps.

Quality Loss. Based on SSIM and FSIM, quality loss functions are defined as in Eqs. 17 and 18. In addition we experiment a MSE based quality loss (Eq. 19) that computes between $I_R^{(i)}$ and $I_S^{(i)}$ and name this quality loss as Pixel Loss.

$$SSIM Loss : \mathcal{L}_Q = 1 - SSIM(I_R^{(i)}, I_S^{(i)}) \tag{17}$$

$$FSIM Loss : \mathcal{L}_Q = 1 - FSIM(I_R^{(i)}, I_S^{(i)}) \tag{18}$$

$$Pixel Loss : \mathcal{L}_Q = MSE(I_R^{(i)}, I_S^{(i)}) \tag{19}$$

Combining the adversarial loss $\mathcal{L}_{ad}$, content loss $\mathcal{L}_X$ and image quality loss $\mathcal{L}_Q$, the overall loss function is formulated as,

$$\mathcal{L} = \mathcal{L}_{ad} + 100\mathcal{L}_X + \mathcal{L}_Q \tag{20}$$

3.2 Network Architecture

We adopted the network architecture proposed in [13]. The generator has two strided convolution blocks, nine residual blocks, two transposed convolution blocks. The residual block was formed by one convolution layer, an instance normalization layer and ReLU activation. Dropout regularization with rate of 50% was adopted. Besides, global skip connection learned a residual image, which was added with the output image to constitute the final restored image I_R. The discriminator was a 70×70 PatchGAN [8], containing four convolutional layers, each followed by BatchNorm and LeakyReLU with $\alpha = 0.2$ except for the first layer.

4 Experiments

4.1 Datasets

The training dataset was sampled from the train set of the Microsoft Common Object in COntext (MS COCO) dataset [15], which contains over 330,000 images covering 91 common object categories in natural context. We adopted the method in [2] to synthesize motion blur kernels. Kernel size was set as 31×31, motion parameters followed the default setting in the original paper. In total, we generated 250 kernels to randomly blur MS COCO dataset images. We randomly selected 6000 images from the MS COCO train set for training and 1000 from the test set for evaluation. Besides, trained models were tested on two other datasets, the GoPro dataset [18] and the Kohler dataset [11].

GoPro dataset has 3214 pairs of realistic blurry images and their sharp version at 1280×720 resolution. Images are 240 fps video sequences captured by GoPro Hero 4 camera in various daily or natural scenes. Blurry images are averaged from a varying number of consecutive frames, in order to synthesize motion blur of varying degrees. This is a common benchmark for image motion deblurring. We randomly select 1000 pairs for evaluation.

Kohler dataset contains four original images, 48 blurred images that are generated by applying 12 approximations of human camera shakes on original images respectively. The dataset is also considered as a benchmark for evaluation of blind deblurring algorithms.

4.2   Implementation

We performed experiments using PyTorch [19] on a Nvidia Titan V GPU. All images were scaled to 640 × 360 and randomly cropped to patches of size 256 × 256. Networks were optimized using the Adam solver [10]. Initial learning rate was 10^{-4} for both generator and critic. For LSGAN models, learning rate remained unchanged for the first 150 epochs and linearly decayed to zero for the rest 150 epochs, and it took around 6 days to finish the training. For WGAN-GP models, learning rate was maintained for 50 epochs and then linearly decreased to zero for another 50 epochs. Training took around 3 days to converge.

4.3   Results and Analysis

We name the model without quality loss as the baseline model. Evaluation metrics include PSNR, SSIM, FSIM and GMSD. Quantitative performances are given in Tables 1, 2 and 3. Examples of resulting images are shown in Figs. 1, 2 and 3.

MS COCO Dataset. From Table 1, we can observe that WGAN-GP model with SSIM loss function has the best performance on all four measures. The WGAN-GP model with FSIM loss function has a comparable performance with subtle differences in values. But significant improvements from the baseline model that does not include quality losses demonstrate the usefulness of SSIM or FSIM loss. From the examples shown in Fig. 1, restored images by SSIM loss and FSIM loss contain more details visually and also have better quantitative evaluation results than their counterparts.

Table 1. Model performance evaluation measures averaged on 1000 images of MS COCO dataset.

Measures	Baseline		With SSIM Loss		With FSIM Loss		With Pixel Loss	
	LSGAN	WGAN-GP	LSGAN	WGAN-GP	LSGAN	WGAN-GP	LSGAN	WGAN-GP
PSNR(dB)	18.14	19.57	17.50	**20.16**	19.91	19.98	16.50	19.52
SSIM	0.7447	0.8032	0.7513	**0.8218**	0.8175	0.8196	0.7412	0.8022
FSIM	0.8178	0.8207	0.8170	**0.8217**	0.8208	0.8216	0.8172	0.8205
GMSD	0.1460	0.1419	0.1438	**0.1391**	0.1404	0.1393	0.1439	0.1413

GoPro Dataset. We can find similar performances on MS COCO dataset, although the training was solely based on synthetic blurred images from MS COCO dataset. Still WGAN-GP is the model that gives better performance. In terms of PSNR and SSIM metrics, performance of WGAN-GP model with SSIM loss function is ranked the first. And FSIM loss function encourages the model to produce better results with regard to FSIM and GMSD metrics.

Fig. 1. Results generated by WGAN-GP model with various loss functions on MS COCO dataset.

Table 2. Model performance evaluation measure averaged on 1000 images of GoPro dataset.

Measures	Baseline		With SSIM Loss		With FSIM Loss		With Pixel Loss	
	LSGAN	WGAN-GP	LSGAN	WGAN-GP	LSGAN	WGAN-GP	LSGAN	WGAN-GP
PSNR(dB)	18.58	20.01	19.06	**20.99**	20.42	20.86	19.06	20.17
SSIM	0.8017	0.8446	0.8047	**0.8650**	0.8557	0.8627	0.8160	0.8480
FSIM	0.8266	0.8379	0.8188	0.8379	0.8337	**0.8386**	0.8263	0.8379
GMSD	0.1181	0.1029	0.1195	0.0988	**0.0962**	0.0990	0.1117	0.0991

Fig. 2. Results generated by WGAN-GP model with different loss functions on GoPro dataset.

Table 3. Model performance evaluation measure averaged on 48 images of Kohler dataset.

Measures	Baseline LSGAN	WGAN-GP	With SSIM Loss LSGAN	WGAN-GP	With FSIM Loss LSGAN	WGAN-GP	With Pixel Loss LSGAN	WGAN-GP
PSNR(dB)	15.94	17.81	16.50	18.19	**18.37**	17.79	13.18	17.73
SSIM	0.6042	0.6775	0.5689	0.6818	**0.6908**	0.6886	0.5444	0.6810
FSIM	0.7562	0.7565	**0.7607**	0.7559	0.7555	0.7539	0.7512	0.7564
GMSD	0.1984	0.2040	**0.1971**	0.2055	0.2048	0.2044	0.2002	0.2031

Kohler Dataset. Compared to results of above two datasets, results given in Table 3 are generally low. Considering images from Kohler dataset are approximations of human camera shake, models trained by synthetic blurred images have limited generalization on tackling such real blurry images. But SSIM and FSIM loss still demonstrate their effectiveness in improving image quality as shown in Table 3 and Fig. 3, although the example in Fig. 3 is a challenging one to restore.

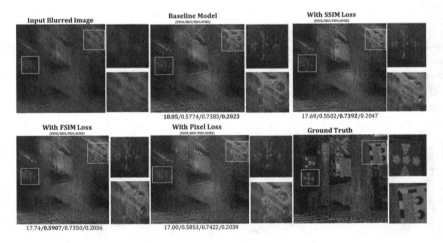

Fig. 3. Results generated by WGAN-GP model with different loss functions on Kohler dataset.

As one can observe from Tables 1, 2 and 3, quantitative results show that image quality measure based loss functions are effective components for GANs to further improve generated image quality. Among the three loss functions, models trained with the SSIM loss and the FSIM loss have comparable performances and generate the best results, compared to the baseline model and model trained with the pixel loss. Experimentation on three different datasets in two different types of GAN model demonstrates effectiveness of inclusion of such quality loss functions.

For visual comparison, models with the SSIM or FSIM loss function restore images with better texture details and edges. However, if we carefully observe the details in generated image patches, we can find that with the SSIM loss function, the patches have window artifacts while models with FSIM loss produce smoother details when zoomed in. It is because the SSIM loss is computed by basing on local windows of images while the FSIM loss is computed pixel by pixel. And in the result images generated by models trained with pixel loss function, details are still blurred, illustrating that L2 loss in the spatial domain has little contribution to image quality improvement. In general, compared to FSIM loss, SSIM loss has the advantage of computation efficiency and performance stability in various quantitative evaluation metrics and visual quality.

It is also noted that WGAN-GP generates better results than LSGAN and converges faster. But training WGAN-GP model is more difficult; during the experimentation, model training diverges more often than LSGAN. Parameter tuning becomes a crucial step in experiment setting for training WGAN-GP, and it is very time-consuming to find a feasible model structure and network parameters.

5   Conclusion

In this paper, we tackled the problem of image deblurring with the framework of adversarial learning models. Losses based on image quality measures were proposed as additional components in the training objective function of the GAN models. Experimental results on various benchmark datasets have demonstrated the effectiveness of adding such image quality losses and their potential in improving quality of generated images.

For future work, training data could include more diverse datasets to improve generalization ability of the network. So far the weightings of these various losses in the overall objective function have not been fine-tuned; further experiments could be conducted on further improving the performance by fine-tuning these parameters. Besides, considering flexibility and adaptability of these image quality losses, applications for solving other image enhancement and restoration tasks would also be worth investigating in the future.

References

1. Arjovsky, M., Chintala, S., Bottou, L.: Wasserstein GAN. arXiv preprint arXiv:1701.07875 (2017)
2. Boracchi, G., Foi, A.: Modeling the performance of image restoration from motion blur. IEEE Trans. Image Process. **21**(8), 3502–3517 (2012)
3. Chakrabarti, A.: A neural approach to blind motion deblurring. In: Leibe, B., Matas, J., Sebe, N., Welling, M. (eds.) ECCV 2016. LNCS, vol. 9907, pp. 221–235. Springer, Cham (2016). https://doi.org/10.1007/978-3-319-46487-9_14
4. Deng, J., Dong, W., Socher, R., Li, L.J., Li, K., Fei-Fei, L.: ImageNet: a large-scale hierarchical image database. In: Proceedings of the IEEE International Conference on Computer Vision and Pattern Recognition, pp. 248–255. IEEE (2009)

5. Fergus, R., Singh, B., Hertzmann, A., Roweis, S.T., Freeman, W.T.: Removing camera shake from a single photograph. In: ACM SIGGRAPH 2006 Papers, pp. 787–794. Association for Computing Machinery (2006)
6. Goodfellow, I., et al.: Generative adversarial nets. In: Advances in Neural Information Processing Systems, pp. 2672–2680 (2014)
7. Gulrajani, I., Ahmed, F., Arjovsky, M., Dumoulin, V., Courville, A.C.: Improved training of Wasserstein GANs. In: Advances in Neural Information Processing Systems, pp. 5767–5777 (2017)
8. Isola, P., Zhu, J.Y., Zhou, T., Efros, A.A.: Image-to-image translation with conditional adversarial networks. In: Proceedings of the IEEE International Conference on Computer Vision and Pattern Recognition, pp. 1125–1134 (2017)
9. Johnson, J., Alahi, A., Fei-Fei, L.: Perceptual losses for real-time style transfer and super-resolution. In: Leibe, B., Matas, J., Sebe, N., Welling, M. (eds.) ECCV 2016. Lecture Notes in Computer Science, vol. 9906, pp. 694–711. Springer, Cham (2016). https://doi.org/10.1007/978-3-319-46475-6_43
10. Kingma, D.P., Ba, J.: Adam: a method for stochastic optimization. arXiv preprint arXiv:1412.6980 (2014)
11. Köhler, R., Hirsch, M., Mohler, B., Schölkopf, B., Harmeling, S.: Recording and playback of camera shake: benchmarking blind deconvolution with a real-world database. In: Fitzgibbon, A., Lazebnik, S., Perona, P., Sato, Y., Schmid, C. (eds.) ECCV 2012. LNCS, vol. 7578, pp. 27–40. Springer, Heidelberg (2012). https://doi.org/10.1007/978-3-642-33786-4_3
12. Kovesi, P., et al.: Image features from phase congruency. Videre J. Comput. Vis. Res. 1(3), 1–26 (1999)
13. Kupyn, O., Budzan, V., Mykhailych, M., Mishkin, D., Matas, J.: DeblurGAN: blind motion deblurring using conditional adversarial networks. In: Proceedings of the IEEE International Conference on Computer Vision and Pattern Recognition, pp. 8183–8192 (2018)
14. Ledig, C., et al.: Photo-realistic single image super-resolution using a generative adversarial network. In: Proceedings of the IEEE International Conference on Computer Vision and Pattern Recognition, pp. 4681–4690 (2017)
15. Lin, T.-Y., et al.: Microsoft COCO: common objects in context. In: Fleet, D., Pajdla, T., Schiele, B., Tuytelaars, T. (eds.) ECCV 2014. LNCS, vol. 8693, pp. 740–755. Springer, Cham (2014). https://doi.org/10.1007/978-3-319-10602-1_48
16. Mao, X., Li, Q., Xie, H., Lau, R.Y., Wang, Z., Paul Smolley, S.: Least squares generative adversarial networks. In: Proceedings of the IEEE International Conference on Computer Vision, pp. 2794–2802 (2017)
17. Mirza, M., Osindero, S.: Conditional generative adversarial nets. arXiv preprint arXiv:1411.1784 (2014)
18. Nah, S., Hyun Kim, T., Mu Lee, K.: Deep multi-scale convolutional neural network for dynamic scene deblurring. In: Proceedings of the IEEE International Conference on Computer Vision and Pattern Recognition, pp. 3883–3891 (2017)
19. Paszke, A., et al.: Automatic differentiation in PyTorch. In: NIPS-W (2017)
20. Schuler, C.J., Hirsch, M., Harmeling, S., Schölkopf, B.: Learning to deblur. IEEE Trans. Pattern Anal. Mach. Intell. 38(7), 1439–1451 (2015)
21. Sønderby, C.K., Caballero, J., Theis, L., Shi, W., Huszár, F.: Amortised map inference for image super-resolution. arXiv preprint arXiv:1610.04490 (2016)
22. Wang, Z., Bovik, A.C., Sheikh, H.R., Simoncelli, E.P.: Image quality assessment: from error visibility to structural similarity. IEEE Trans. Image Process. 13(4), 600–612 (2004)

23. Xu, L., Ren, J.S., Liu, C., Jia, J.: Deep convolutional neural network for image deconvolution. In: Advances in Neural Information Processing Systems, pp. 1790–1798 (2014)
24. Xu, L., Zheng, S., Jia, J.: Unnatural l0 sparse representation for natural image deblurring. In: Proceedings of the IEEE International Conference on Computer Vision and Pattern Recognition, pp. 1107–1114 (2013)
25. Yeh, R., Chen, C., Lim, T.Y., Hasegawa-Johnson, M., Do, M.N.: Semantic image inpainting with perceptual and contextual losses 2(3) (2016). arXiv preprint arXiv:1607.07539
26. Zhang, H., et al.: Stackgan: text to photo-realistic image synthesis with stacked generative adversarial networks. In: Proceedings of the IEEE International Conference on Computer Vision, pp. 5907–5915 (2017)
27. Zhang, H., et al.: Stackgan++: realistic image synthesis with stacked generative adversarial networks. IEEE Trans. Pattern Anal. Mach. Intell. 41(8), 1947–1962 (2018)
28. Zhang, L., Zhang, L., Mou, X., Zhang, D.: FSIM: a feature similarity index for image quality assessment. IEEE Trans. Image Process. 20(8), 2378–2386 (2011)
29. Zhu, J.Y., Park, T., Isola, P., Efros, A.A.: Unpaired image-to-image translation using cycle-consistent adversarial networks. In: Proceedings of the IEEE International Conference on Computer Vision, pp. 2223–2232 (2017)

On Random-Forest-Based Prediction Intervals

Aida Calviño[✉][iD]

Department of Statistics and Data Science,
Complutense University of Madrid, Madrid, Spain
aida.calvino@ucm.es

Abstract. In the context of predicting continuous variables, many proposals in the literature exist dealing with point predictions. However, these predictions have inherent errors which should be quantified. Prediction intervals (PI) are a great alternative to point predictions, as they permit measuring the uncertainty of the prediction. In this paper, we review Quantile Regression Forests and propose five new alternatives based on them, as well as on *classical* random forests and linear and quantile regression, for the computation of PIs. Moreover, we perform several numerical experiments to evaluate the performance of the reviewed and proposed methods and extract some guidelines on the method to choose depending on the size of the data set and the shape of the target variable.

Keywords: Quantile · Regression · Coverage probability · Skewness · Kurtosis · Machine learning

1 Introduction and Motivation

Predictive modeling is the process of using known data to create, process and validate a model that can be used to forecast future (or simply unknown) outcomes. It is a very useful tool as it permits reducing the uncertainty about future processes or observations. Some areas of applications include, but are not limited to, prediction of the selling price of a house, contamination level of a city in the following days, etc. The required input consists of a data set that contains information on a set of observations (e.g., houses, days, people, etc.) regarding the variable to be predicted (called target variable) as well as other variables that might help in the prediction (usually referred to as input variables).

Although predictive modeling has been studied very thoroughly in the literature (see [5] for a review on the most important predictive models), many of the methods proposed only provide point predictions, that is, they provide a single value that represents the expected value of the target variable, under the circumstances given by the input variables. Nevertheless, models are rarely perfect and, thus, point estimates should go together with an uncertainty measure that allows to evaluate the certainty of the prediction. In this paper we deal with prediction intervals (PIs), that is, instead of providing a unique forecast,

© Springer Nature Switzerland AG 2020
C. Analide et al. (Eds.): IDEAL 2020, LNCS 12489, pp. 172–184, 2020.
https://doi.org/10.1007/978-3-030-62362-3_16

the models studied here provide a (hopefully narrow) interval of values that will contain the real value with a prespecified (and usually high) probability.

PIs are not new in the literature and many methods exist for producing them, from the simplest ones based on classical linear and quantile regression models (see [12] and [7], respectively) to more sophisticated ones, as the ones in [10,11]. However, many of the methods that can be used to generate PIs (such as linear regression or quantile regression forest) have not been designed specifically for the generation of PIs and, thus, if analyzed with PIs metrics (as we do in this paper), they might underperform. An interesting reference where several PI methods are applied in the particular case of energy consumption and production is [2], where the authors conclude that both quantile regression forest [8] and quantile boosting-based methods [14] provide high quality results.

In this paper we propose and analyze methods to produce PIs based on random forests (RF), proposed in [3], as they have been proved to have high predictive power, are robust to the presence of missing and outlier values and do not require previous variable selection [5]. Furthermore, the building process of random forest is very intuitive, making them understandable for a wider audience than more complex models. The reason not to include boosting-based methods, despite its known good performance, is three fold. First, their building process is not so straightforward, reducing its potential applicability due to its added difficulty. Second, their computation is not easily parallelizable (as opposed to RF) and, therefore, their processing times might be larger. Finally, as mentioned in [13], simulation studies have shown that quantile boosting-based methods fail at predicting extreme quantiles, which are the ones used when producing PIs.

Our aim is to evaluate several procedures to generate PIs based on RFs (more specifically on its philosophy) to discover the circumstances, if any, that make each method preferable. Thus, our main contribution is the proposal of five new random-forest-based methods, as well as the analysis and comparison of those methods with previously proposed ones, using specific metrics for evaluating PIs.

The paper is organized as follows. In Sect. 2, we review some methods previously proposed in the literature to generate PIs, whereas in Sect. 3 we present five new procedures inspired by the previous ones. Section 4 is devoted to test the performance of the reviewed and proposed methods by means of an experimental study, where the methods are applied to a wide variety of data sets. Finally, some conclusions are given in Sect. 5.

2   Previous Prediction Interval Methods

As already stated, in this paper we aim at evaluating new and already proposed methods to obtain PIs. A PI is a range of values where the unknown value of a certain observation will fall, given enough samples, a certain percentage of the times. We now introduce some notation and define formally PIs.

For the remaining of the paper we assume that we have a set of n observations for which we know the value of m variables (called input variables) and the variable to be predicted (called target). Let Y be a vector of size n containing the target variable and X, a $n \times m$ matrix containing the input variables. The objective is, for a new observation k for which the vector of input variables x_k is known, to find a narrow interval $(\hat{y}_k^\ell, \hat{y}_k^u)$ such that there is a high probability $1 - \alpha$ (called confidence) that the real value y_k lies inside that interval.

Many methods have been proposed in the literature to produce PIs (such as [10,11]) based on different prediction tools (e.g., linear and quantile regression, cluster analysis, neural networks, etc.). In this paper we restrict ourselves to methods based on random forest. However, our proposals are inspired by two different classical approaches, which are now reviewed briefly.

Linear regression: In linear regression (see [12]), PIs are computed as

$$\hat{y}_k^\ell = \hat{y}_k - z_{\alpha/2}\hat{\sigma}_k \qquad \hat{y}_k^u = \hat{y}_k + z_{\alpha/2}\hat{\sigma}_k, \qquad (1)$$

where $\hat{y}_k$ is the predicted value of y_k with the estimated linear regression, z_α is the value of the standard normal with cumulative probability level of α and $\hat{\sigma}_k = \frac{1}{n-m-1} \sum_{i=1}^{n} (y_i - \hat{y}_i)^2$.

Quantile regression: In quantile regression (see [7]), the aim is not to compute the conditional expectation, but the conditional quantiles. In other words, the aim is to predict as closely as possible the α-quantile of the target variable, denoted $Q_\alpha(y)$, based on the values of the input variables.

In this context, PIs can be computed considering the corresponding conditional quantiles as follows:

$$\hat{y}_k^\ell = \widehat{Q}_{\alpha/2}(y_k) \qquad \hat{y}_k^u = \widehat{Q}_{1-\alpha/2}(y_k), \qquad (2)$$

where $\widehat{Q}_\alpha(y_k)$ is the predicted α-quantile given the vector of input variables x_k.

As it can be seen, two different approaches arise from the previous definitions. The first one (from now on referred to as *mean-variance*) consists of predicting separately the conditional mean and variance, and then use Eq. (1) to compute PIs, whereas the second approach (from now on referred to as *quantile-based*) consists of predicting conditional quantiles (not necessarily by means of quantile regression), and then apply Eq. (2) to compute PIs.

A third alternative to the ones mentioned above (from now on referred to as *mean-quantile*) is based on [6], where the authors suggest that, in the context of mixed linear models (which are a generalization of linear regression models), a good way to obtain PIs is by means of the following expression:

$$\hat{y}_k^\ell = \hat{y}_k + \widehat{Q}_{\alpha/2}(e) \qquad \hat{y}_k^u = \hat{y}_k + \widehat{Q}_{1-\alpha/2}(e), \qquad (3)$$

where $\widehat{Q}_\alpha(e)$ is the α-quantile of the residuals vector of the estimated linear regression, given by: $e = (y_i - \hat{y}_i)_{i=1,\ldots,n}$. Note that, in this case, the procedure to generate PIs consists of predicting the conditional mean (i.e., obtaining the usual point prediction) and, then, estimate the uncertainty of the prediction by means of the quantiles of the residuals.

All previous alternatives have advantages and disadvantages. In particular, they are all very flexible regarding the specific models to be chosen (linear ones, neural networks, random forest, etc.). However, the first partially relies on the assumption of normality of the target variable, which might not always hold. However, taking into account Chebyshov inequality and the Central limit theorem, even if Y does not resemble *gaussian*, this approach can still provide results of good quality. Regarding the second one, its main disadvantage is that specific models need to be train for each value of α, which might be time consuming. Finally, in the case of the alternatives based on linear regression, PI lengths are assumed common for all observations, which might seem a strong assumption. However, as it will be shown in Sect. 4, *homogeneous* methods provide PIs of good quality.

Considering the three previous approaches, in this paper we propose five new PI models based on them, as well as on random forests. For that reason, we now review the basic concepts of random forests.

2.1   Random Forests

Random forests (RFs), proposed by [3], are a combination (or *ensemble*) of decision tree predictors such that each tree is trained with an independent random *bootstrap* sample of the original data set and a random selection of input variables. In other words, RFs first compute a large set of decision trees (say T), based on different subsets of the original data set, make then predictions based on those trees and finally combine all predictions through the predictions average. For a review on random forests and decision trees, please refer to [5].

For the ease of clarity, we now introduce some notation on how predictions are obtained by means of RFs. Again, we assume that we want to predict y_k for a new observation k for which its input vector x_k is known. Decision trees provide a partition of the space generated by the input variables (the subsets of the partition are usually referred to as leafs) in such a way that the values of the target variable of the observations in the same leaf are similar. Considering a specific tree t, let S_k^t be the set of observations lying in the same leaf as observation k. Under this scenario, $\hat{y}_k$ can be computed as:

$$\hat{y}_k = average\left(\{\hat{y}_k^t\}_{t=1,...,T}\right), \ \ \text{with} \ \ \hat{y}_k^t = average\left(\{y_i\}_{i \in S_k^t}\right), \tag{4}$$

that is, predictions are obtained by means of the average of the predictions of each tree which, in turn, are obtained by means of the average of the target values of the observations in the same leaf as k.

It is important to highlight that RFs have some parameters that need to be tuned. The three most important ones are: *ntree*, which is the number of trees in the forest (previously referred as T); *mtry*, which is the proportion of input variables randomly selected; and *nodesize*, which is the minimum number of observations on a leaf.

2.2   Quantile Regression Forests (QRF)

QRFs were proposed in [8] as an extension to the classical RF [3] when the aim is not to compute the conditional expectation of the target variable, but the conditional quantiles. For that reason, when applied for computing PIs, QRFs correspond to the quantile-based approach previously defined.

Based on the previous definition of RF, QRFs, as originally proposed in [8], also build a large set of decision trees, but the prediction procedure is as follows: instead of computing the average of the target of the observations in the leafs, QRFs compute the empirical cumulative distribution function (ecdf) and, then, the ecdfs from all trees are combined through its average. Finally, the average ecdf can be used to compute any quantile by means of its inverse.

[8] showed that QRFs provide consistent estimators for conditional quantiles. However, this approach is memory consuming as the ecdfs of all leafs in all trees need to be stored. For that reason, the authors proposed and implemented a new QRF version in [9] where one observation per leaf is randomly selected and, then, the empirical desired quantile of the random observation of all trees is obtained. The advantage of this version is that a single observation per node needs to be retained and not the whole ecdf. This is the QRF approach that will be evaluated in the remaining on the paper, which will be referred to as QRF.

Formally, QRFs provide quantiles predictions $\widehat{Q}_\alpha(y_k)$ in the following way:

$$\widehat{Q}_\alpha(y_k) = \alpha\text{-}quantile\left(\{\hat{y}_k^t\}_{t=1,\dots,T}\right), \quad \text{with} \quad \hat{y}_k^t = random\left(\{y_i\}_{i \in S_k^t}\right).$$

As QRFs are based on RFs, they rely on the same parameters. In the experiments conducted in this paper we assume the default parameters proposed by the authors: 500, 33% and 5, for *ntree*, *mtry* and *nodesize*, respectively. Moreover, in order to make *fair* comparisons, from now on we assume the same default parameters when computing RFs.

3   Proposed Prediction Interval Methods

Appart from QRF, in this paper we evaluate five newly proposed methods, which are based on the three approaches mentioned previously and are briefly explained in the following subsections.

3.1   Mean-Variance Methods

In this subsection we deal with models that aim at predicting means and variances separately and then apply Eq. (1) in order to compute PIs.

Homogeneous Random Forest. As in the classical linear regression model, in this method we assume that the variance of all observations is equal (homogeneous) and, thus, it does not need to be predicted individually. Following the ideas of the classical linear regression, this method consists of applying Eq. (1), where $\hat{y}_k$ is obtained by means of the classical random forest in Eq. (4) and $\hat{\sigma}_k$ is given by the square root of $\hat{\sigma}_k^2 = \frac{1}{n}\sum_{i=1}^{n}(y_i - \hat{y}_i)^2$. Note that, in the case of RF, as opposed to linear regression, the number of parameters is not obvious and, thus, we resort to the classical mean square error as an estimator of the variance. We will refer to this method as homogeneous random forest (HRF).

Mean-Variance Random Forest. In the mean-variance random forest (MVRF) we make use of a unique random forest to estimate both $\hat{y}_k$ and $\hat{\sigma}_k$. The idea is to take advantage of the predictions made inside the RF by each of the trees to estimate the variance, assuming that the difference among them can be used for that purpose. The procedure is as follows:

1. Build a classical RF and compute $\hat{y}_k$ using Eq. (4).
2. Based on the same RF, compute $\hat{\sigma}_k$ as the squared root of the following predicted variances:

$$\hat{\sigma}_k^2 = variance\left(\{\hat{y}_k^t\}_{t=1,\ldots,T}\right), \quad \text{with} \quad \hat{y}_k^t = average\left(\{y_i\}_{i \in S_k^t}\right).$$

Chained Random Forest. In this method, two different RFs are applied for predicting mean and variance separately. We called this method *chained random forest* (CRF) as it requires training first a random forest for computing $\hat{y}_k$ and, then, using the residuals of that model as input, training another RF for the computation of $\hat{\sigma}_k$. The procedure is as follows:

1. As previously done, consider a set of n observations $\mathcal{D}$. Then, divide it randomly into two subsets $\mathcal{D}_1$ and $\mathcal{D}_2$, each of them containing half of the observations.
2. Build a classical RF using only the observations in $\mathcal{D}_1$, as in Eq. (4), and compute $\hat{y}_k$ for all the observations in $\mathcal{D}_2$.
3. Compute the squared residuals of the first RF for the observations in $\mathcal{D}_2$: $\hat{e}_k = (y_k - \hat{y}_k)^2$, $\forall k \in \mathcal{D}_2$.
4. Build a second RF to predict the variances by means of the residuals:

$$\hat{\sigma}_k^2 = average\left(\{\hat{e}_k^t\}_{t=1,\ldots,T}\right), \quad \text{with} \quad \hat{e}_k^t = average\left(\{\hat{e}_i\}_{i \in S_k^t}\right).$$

In this case, it is important to build both models independently (and not with the same observations) in order to avoid overfitting. The underlying assumption of this method is that the uncertainty of a prediction (which can be measured by means of its squared error) can also be estimated using the input variables, and not only the prediction itself.

3.2 A Quantile-Based Method: Quantile Random Forest

In this subsection we deal with models that aim at predicting quantiles, and then apply Eq. (2) in order to compute PIs.

As it will be shown in the experiments, in the context of PIs, QRFs do not always achieve satisfactory results, unless the confidence is very high (larger than 99%). This is not a surprising fact as, according to [13], simulation studies have shown that the weakness of QRF lies in predicting central quantiles. In order to overcome its deficiencies, we propose a modification of the QRF model proposed in [9], which will be referred to as quantile random forest (QRDF).

The basic idea is to compute the mean of the leafs empirical quantiles. More formally, QRDF provides quantiles predictions $\widehat{Q}_\alpha(y_k)$ in the following way:

$$\widehat{Q}_\alpha(y_k) = average\left(\{\hat{y}_k^t\}_{t=1,\dots,T}\right), \quad \text{with} \quad \hat{y}_k^t = \alpha\text{-}quantile\left(\{y_i\}_{i\in S_k^t}\right).$$

Contrary to the previous RF-based methods, a remarkable feature of QRDF is the fact that it needs a large value of *nodesize* (we remind the reader that this parameter imposes a minimum number of observations per leaf) in order to produce meaningful quantiles in all leafs. This, in turn, implies smaller computing times than QRF, as the trees in the forest have fewer leafs.

Experimental analysis carried out in order to evaluate the impact of the parameter *nodesize* (not shown here for the sake of conciseness), point out that the best value should be around 10% of the number of observations in the data set used for training. However, regarding *ntree* and *mtry*, default values are kept as in QRF.

3.3   A Mean-Quantile Method: Homogeneous Quantile Random Forest

In this subsection we deal with models that aim at predicting means and residual quantiles, and then apply Eq. (3) in order to compute PIs. As it is done in HRF, we assume that the uncertainty of all predictions is the same, that is, we assume again that the PI length is homogeneous. For that reason, we called this method homogeneous quantile random forest (referred to as HQRF).

However, unlike HRF, HQRF does not need to estimate the variance of the predictions, but the quantiles of the residuals. More formally, HQRF provides PIs by means of expression (3), where $\hat{y}_k$ is obtained by means of the classical random forest in Eq. (4) and $\widehat{Q}_\alpha(e)$ is obtained as the α-quantile of the residuals vector of the RF used to obtain $\hat{y}_k$, given by: $e = (y_i - \hat{y}_i)_{i=1,\dots,n}$.

4   Benchmarking Experiments

To compare the previous methods, we adopt a common experimental procedure in the Machine Learning literature (see [10], among others), consisting of running the experiments on ten open-access data sets, which can be found in [4]. The aim of the experiments is to evaluate the performance of the methods in different scenarios. In particular, we analyze the effect of the size of the data set and the shape of the target variable (specifically, its skewness and kurtosis). It is important to take into account these last features in the experiments as mean-variance methods in Sect. 3.1 rely partially on the normality of the target variable and may underperform when this hypothesis does not hold. Unfortunately, the ten usual data sets considered in this common experimental procedure do not cover all the desired combinations of data set size and shape of the target variable that would allow evaluating most of the scenarios that can arise in real world applications. In order to overcome this drawback, we have considered another

Table 1. Summary of the datasets used in Sect. 4.

	Minimum	Q_1	Median	Q_3	Maximum
Number of obs.	205	1041	5337	15818	45730
Number of variables	5	9	14	30	82
Skewness	-0,9831	0,1647	0,68875	1,7496	4,0235
Kurtosis	-1,2498	-0,5501	0,3108	2,5098	37,0648

ten open-access data sets, that can be found in [1]. Table 1 gives a summary of the twenty data sets, including their sizes, as well as the skewness and kurtosis of the corresponding target variables.

For the data sets considered, 90, 95 and 99% PIs are computed using the previously explained methods. As it is usually done in the Machine Learning community, the data sets are randomly partitioned (80% train and 20% test) using twenty different random seeds. This approach allows us to analyze the methods average performance, as well as its variability. The results shown in the comparison below refer always to the test partition.

Along with the methods explained in Sect. 2, we include in the comparison what we call "classical methods", that is, linear and quantile regressions (referred to as LR and QR, respectively) explained at the beginning of Sect. 2. The objective is to verify if the more complex methods proposed in this paper outperform simpler ones. We note that we have not considered expression (3) for the case of linear regression as, according to [6], this alternative leads to similar results to Eq. (1) in the context of linear regression.

In order to evaluate PI methods with specific metrics, we need to take into account two aspects: the first one refers to the fact that PIs should contain a proportion of observations close to the nominal confidence level; and the second one is related to their length, as very wide intervals are little informative. For that reason, when computing PIs using the previous methods, we calculate the proportion of observations that indeed lie in their corresponding PI (called Coverage Probability (CP)), as well as their average relative length. PI relative lengths are computed as: $(\hat{y}_k^u - \hat{y}_k^\ell)/\bar{y}$, where $\bar{y}$ is the mean value of vector $\boldsymbol{Y}$.

Table 2 contains a summary of the numerical experiments of the eight PI methods. In particular, mean and variance (in brackets) of the coverage probabilities and mean lengths of the *test* partition for the twenty data sets and twenty repetitions is shown. For the sake of better visualization, a color scale (specific for each nominal level) has been added, where lighter colors indicate better results (CPs close to the nominal levels and shorter PIs). As it can be seen, regarding CPs, the best methods depend on the nominal confidence level. For 90 and 95%, LR, HRF, HQRF and QRDF lead to the best results, while QRF and MVRF give CPs much larger than the nominal levels. On the other hand, for 99%, all methods lead to CPs closer to the nominal levels. An interesting fact is that QR always leads to CPs much smaller than the nominal levels.

Table 2. Prediction intervals coverage probabilities and mean lengths. The values shown in the tables correspond to the mean and standard deviation (in brackets).

		0,9	0,95	0,99
QRF	Cov. Prob.	0,9543 (0,0347)	0,9753 (0,0224)	0,9912 (0,0119)
	Mean length	0,7705 (0,6163)	0,9763 (0,8187)	1,4135 (1,3118)
QRDF	Cov. Prob.	0,9208 (0,0389)	0,9526 (0,0325)	0,9769 (0,0278)
	Mean length	1,0048 (0,9808)	1,1891 (1,2069)	1,4973 (1,6213)
HRF	Cov. Prob.	0,9197 (0,0247)	0,9469 (0,0194)	0,9748 (0,0151)
	Mean length	0,8013 (0,8178)	0,9549 (0,9745)	1,2549 (1,2807)
HQRF	Cov. Prob.	0,8973 (0,0238)	0,9415 (0,0239)	0,9783 (0,0246)
	Mean length	0,7529 (0,8411)	0,9986 (1,0727)	1,6284 (1,7835)
MVRF	Cov. Prob.	0,9539 (0,0435)	0,9720 (0,0313)	0,9873 (0,0196)
	Mean length	0,8115 (0,7539)	0,9669 (0,8983)	1,2702 (1,1802)
CRF	Cov. Prob.	0,9313 (0,0254)	0,9634 (0,0185)	0,9875 (0,0104)
	Mean length	0,7721 (0,7971)	0,9200 (0,9498)	1,2090 (1,2482)
LR	Cov. Prob.	0,9079 (0,0307)	0,9421 (0,0264)	0,9787 (0,0179)
	Mean length	1,2324 (1,3983)	1,4685 (1,6661)	1,9299 (2,1897)
QR	Cov. Prob.	0,8740 (0,0596)	0,9175 (0,0684)	0,9477 (0,0789)
	Mean length	1,0302 (1,0839)	1,1818 (1,2628)	1,4423 (1,6332)

As to mean lengths, we remind that wide PIs are little informative, even if their CPs are close to the nominal levels. In this sense, note that LR provides wider PIs than HRF or HQRF, even when they have similar CPs. This is not a surprising fact as both LR and HRF obtain PI widths by means of its mean squared errors (by means of linear regression and random forest, respectively) and RFs usually lead to more accurate results. Furthermore, we highlight that CRF shows short PIs while having CPs not far from the nominal ones.

Moreover, we highlight the fact that, according to Table 2, standard deviations are not very different among the studied methods and, for that reason, they are not mentioned previously. Nevertheless, we do consider the variance of the methods performance into consideration in the following analysis.

As already highlighted, the shape of the target variable can be decisive when selecting a PI method. In this sense, we have performed two different analysis regarding skewness and kurtosis, separately. We point out that all of the skewed data sets (those showing a skewness index larger than one in absolute value) used in this experiment are leptokurtic (kurtosis index larger than one) and, thus, the analysis regarding kurtosis only considers symmetric data sets. After a careful review of many open-access data sets, we realized that real skewed data sets with kurtosis indexes smaller than one are very rare and, thus, although not considering this type of target variable, our study does consider the vast majority of situations that arise in real cases.

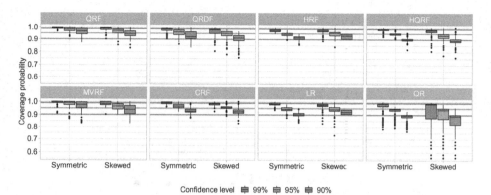

Fig. 1. Coverage probabilities boxplots by model, confidence level and symmetry of the target variable.

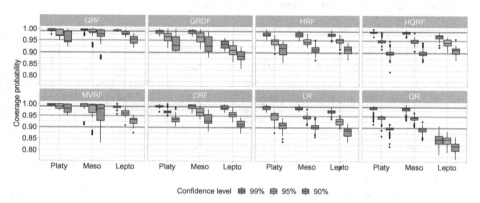

Fig. 2. Coverage probabilities boxplots by model, confidence level and kurtosis of the target variable (only data sets with symmetric target variables are considered here).

Regarding skewness, Fig. 1 contains information on the CPs of the eight methods plotted against their symmetry level and nominal confidence level. As it can be seen, when used for symmetric target variables, LR, QR, HRF and HQRF lead to the best results. More precisely, HQRF shows mean CPs very close to the corresponding nominal level, with very low variability. On the contrary, QRF and MVRF show CPs larger than the nominal levels if the target variable is symmetric. Concerning skewed target variables, QRDF, HQRF and CRF lead to the best results, whereas QR shows a very unstable performance. All in all, it seems that CRF and HQRF are the methods least affected by the symmetry level and the ones with the smallest variability.

On the other hand, Fig. 2 contains information on the CPs of the eight methods plotted against their kurtosis level and nominal confidence level. We remind that those data sets with skewed target variables are not considered in this analysis and, thus, conclusions drawn here apply only to data sets with symmetric

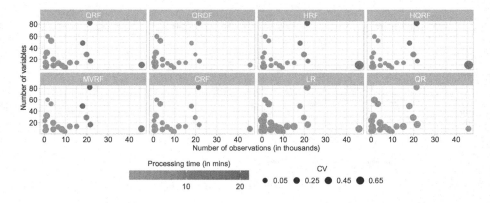

Fig. 3. Processing times of the eight methods analyzed for the twenty data sets by their size (number of variables and observations). Average times is shown by means of a color scale, whereas size represents time variability.

target variables. As it can be seen, data sets with platykurtic and mesokurtic target variables lead to similar results in all methods, except in the case of QRF, where worse CPs arise in the case of mesokurtic target variables. On the other hand, data sets with leptokurtic target variables show a different behavior. For instance, QR leads to CPs much smaller than the nominal level in the case of leptokurtic target variables, but close ones otherwise. QRDF, HQRF and LR also show worse results in the leptokurtic scenario. On the contrary, MVRF and CRF exhibit CPs closer to the nominal levels. Finally, HRF does not seem to be affected significantly by the kurtosis level.

Another key factor when comparing methods is processing times. In this sense, Fig. 3 shows, by means of a color scale, the average processing time of the eight methods for the twenty data sets, considering their size. Moreover, point size represents time variability. As it can be seen, for small data sets (up to 10,000 observations), all analyzed methods require similar times, whereas the effect of larger data sets is very different. As it can be seen, LR, QR and QRDF seem independent of the number of observations and variables, whereas QRF, HRF, HQRF and MVRF seem to depend exponentially on the combination of the number of variables and observations. Finally, CRF is more affected by the size than the former methods, but less than the last ones.

Finally, we would like to highlight that more analyses have been conducted to evaluate the effect of the target variable shape on PI lengths and the effect of the data set size in CPs and PI lengths showing no influence. For that reason, and for the sake of brevity, results on those analyses are skipped here.

5   Conclusions

Finally, based on the analysis performed in Sect. 4, we draw some conclusions on the comparison of the methods explained in Sects. 2 and 3:

1. Despite its good name, QRF does not seem a good choice (except if the nominal level is 99% or higher) when computing PIs as CPs are much larger than the corresponding nominal levels and processing times are highly affected by data set size. Similar conclusions apply to MVRF, but slightly better results are obtained for symmetric and leptocurtic target variables.

2. LR and QR lead to quite good coverage probabilities (except for confidence levels equal to 99%) when the target variable is symmetric and not leptokurtic and their processing times are not affected by data set size. However, they provide very wide PIs and, specially QR, lead to poor results if the target variable is skewed and/or symmetric but leptokurtic.

3. Although CRF is not the "best" method in any of the scenarios analyzed, it seems to be a good default one, as it is the method least affected by the target variable shape and its processing times are not heavily affected by data set sizes.

4. The indirect assumption of normality of the target variable of mean-variance methods does not affect its quality. In fact, mean-variance methods lead to better results for skewed and symmetric-but-leptocurtic target variables.

5. *Homogeneous* methods achieve promising results indicating that assuming common PI length for all observations is not a strong assumption.

6. Taking into consideration all the characteristics analyzed previously, we can extract the following generic guidelines:
   - If the data set has less than $10,000$ observations, HQRF should be chosen, except if the target variable is symmetric and leptocurtic, where HRF is preferable.
   - If the data set has more than $10,000$ observations, QRDF should be chosen, except if the target variable is symmetric and leptocurtic, where CRF is preferable.
   - For confidence levels close to 99% and *small* data sets, independently of the shape of the target variables, QRF seems the best option.

References

1. Kaggle (2017). https://www.kaggle.com/. Accessed Mar 2020
2. Bogner, K., Pappenberger, F., Zappa, M.: Machine learning techniques for predicting the energy consumption/production and its uncertainties driven by meteorological observations and forecasts. Sustainability **11**(12), 1–22 (2019)
3. Breiman, L.: Random forests. Mach. Learn. **45**, 5–32 (2001)
4. Dua, D., Graff, C.: UCI Machine Learning Repository (2017). http://archive.ics.uci.edu/ml. Accessed Mar 2020
5. Hastie, T., Tibshirani, R., Friedman, J.: The Elements of Statistical Learning, 2nd edn. Springer, Heidelberg (2009)
6. Jiang, J., Zhang, W.: Distribution-free prediction intervals in mixed linear models. Stat. Sin. **12**, 537–553 (2002)
7. Koenker, R.: Quantile Regression. Econometric Society Monographs. Cambridge University Press (2005). https://doi.org/10.1017/CBO9780511754098
8. Meinshausen, N.: Quantile regression forests. J. Mach. Learn. Res. **7**, 983–999 (2006)

9. Meinshausen, N., Michel, L.: quantregForest: Quantile Regression Forests (2017). https://CRAN.R-project.org/package=quantregForest, R package version 1.3-7
10. Pearce, T., Brintrup, A., Zaki, M., Neely, A.: High-quality prediction intervals for deep learning: a distribution-free, ensembled approach. In: Proceedings of Machine Learning Research, Stockholm, Sweden, vol. 80, pp. 4075–4084 (2018)
11. Shrestha, D.L., Solomatine, D.P.: Machine learning approaches for estimation of prediction interval for the model output. Neural Netw. **19**(2), 225–235 (2006)
12. Wooldridge, J.M.: Introductory Econometrics: A Modern Approach, 5th edn. South-Western CENGAGE Learning (2013)
13. Yuan, S.: Random gradient boosting for predicting conditional quantiles. J. Stat. Comput. Simul. **85**(18), 3716–3726 (2015)
14. Zheng, S.: QBoost: predicting quantiles with boosting for regression and binary classification. Expert Syst. Appl. **39**(2), 1687–1697 (2012)

A Machine Learning Approach to Forecast the Safest Period for Outdoor Sports

João Palmeira$^{(\boxtimes)}$ (ID), José Ramos (ID), Rafael Silva (ID), Bruno Fernandes (ID),
and Cesar Analide (ID)

Department of Informatics, ALGORITMI Centre, University of Minho,
4704-553 Braga, Portugal
joopalmeira@gmail.com, zramosbg@gmail.com, silvarafael029@gmail.com,
bruno.fmf.8@gmail.com, analide@di.uminho.pt

Abstract. Vulnerable Road Users (VRUs) are all to whom danger may befall on the road, such as pedestrians and cyclists. The research here conducted aims to reduce such vulnerability by providing every VRU with a new tool to help them know when it is the safest period to do sports during the day. To achieve the proposed goal, the focus was set on Machine Learning models, such as Random Forest and Multilayer Perceptrons (MLP), trained and tuned with data received from multiple sources, including the weather, pollution levels, traffic flow and traffic incidents. A mobile application was designed and conceived, making use of the model with the best performance, an MLP, to provide forecasts to VRUs. Our results show promising prospects regarding the use of Machine Learning models to quantify the best period for sports during the day.

Keywords: Deep learning · Machine learning · Multilayer perceptrons · Vulnerable Road Users · Road safety

1 Introduction

Road safety is a major health issue of today's society. In 2018 only, as seen in PORDATA database [1], more than five thousand people were run over on Portuguese roads. Most of them are categorized as Vulnerable Road Users (VRUs), i.e., users with an increase vulnerability at the road. Following recent developments in information technologies, the increasing search for applications which integrate both Artificial Intelligence (AI) and Machine Learning (ML) has led to the possibility of developing artefacts to advise VRUs about the current and future safety conditions outside, allowing VRUs to adapt and prevent themselves from facing dangerous situations. The research here presented makes use of multiple ML candidate models, which were fed with hourly environmental and traffic

---

J. Palmeira, J. Ramos and R. Silva—The authors contributed equally to this work.

data, to forecast when the safest period for a VRU for outdoor sports is. The development of the predictive process followed the CRISP-DM methodology, with Random Forests and Multilayer Perceptrons (MLP) as the best candidates for the predictive model. The remainder of this paper is structured as follows, viz. Section 2 briefly describes the current state of the art, while Sect. 3 provides an overview of the user materials and methods. Section 4 sets the experimental setup, with results being gathered and discussed in Sect. 5. Finally, Sect. 6 presents our conclusions and outlines future work.

2   State of the Art

The use of ML and AI methods to aid our everyday activities has been growing exponentially recently. In fact, the literature shows that ML is already being used to increase VRU's safety on the road.

Studies have attempted to reach this goal by imbuing measures in the cars that travel the road [2]. The goal is to prevent incidents by monitoring the environment around the vehicles, providing a technological link between the information captured with the use of appropriate sensors and a driver warning system that implements safety measures should some eventual and possibly dangerous situation appear [3].

Another study followed the possibility of using VRUs' smartphones as mobile sensors in active security systems. Taking advantage of the GPS capabilities of such devices, it is possible to circumvent the need to install sensors in vehicles and, with geo-positioning, alarm VRUs if any danger or vehicle is approaching them, alerting them to be cautious [4].

In 2012, David Shinar's study focused on accident analysis and prevention, gathering knowledge about the disparity of resources invested in traffic safety over safe mobility for VRUs [5].

In 2010, the U.N. General Assembly, recommended "strengthening road safety management and paying particular attention also to the needs of vulnerable road users, such as pedestrians, cyclists and motorcyclists".

This project places more emphasis on the dangers to which VRUs are more susceptible, not only accounting vehicular problems, but also striving to gather information about weather conditions and atmosphere particle concentrations that should not be neglected when assessing VRUs safety.

3   Materials and Methods

The datasets used in this study have been collected since July 2018 and comprise data about the city of Braga, in Portugal, containing features related to traffic flow and incidents, pollution levels, and weather conditions.

3.1   Data Exploration

The first dataset, the one on pollution, holds data regarding the concentration values of multiple particles in the atmosphere (43845 observations). Focusing on the particles extracted from sensors that may present danger to VRUs, this study centred itself on ascertain particle concentrations that may indicate a dangerous environment. The used dataset includes data about the Ultraviolet index (UV), Particle matter 10 (PM_{10}), Carbon Monoxide (CO), Sulfur dioxide (SO_2), Nitrogen Dioxide (NO_2), and Ground-level Ozone (GLO). With well-known scales, it is possible to set the boundaries that make a particular parameter threatening to the human being. For example, UV values of 8 to 10.9 are considered of very high risk while above 11 represented extreme danger [6]. Another example corresponds to CO values at concentrations of 1.28%, which can cause loss of consciousness after 3 breaths [7].

Ultra-Violet Radiation (UV) can very harmful, at times when it is unusually strong it is customary for national Television to broadcast a safety warning. Excessive ultraviolet radiation causes skin diseases, such as skin cancer[6]. More preemptive safety thresholds were researched, as can be seen in Table 1, to prepare our data to better account for this health risk.

Table 1. Risk of damage from exposure to Ultra-violet radiation for an average adult

UV range	Risk
0.0–2.9	Low
3.0–5.9	Moderate
6.0–7.9	High
8.0–10.9	Very high
11.0+	Extreme

Particle matter 10 (PM_{10}) diameter enables them to bypass our body's defenses[7]. Exposure can affect the lungs and the heart. In Table 2 it is indicated the safety threshold values researched for this substance.

Table 2. The referenced limit values of PM_{10} established.

Reference period	Threshold	Tolerance
1 day	50 mg/m3̂	50%
1 civil year	40 mg/m3̂	20%

Carbon Monoxide (CO) when inhaled, prevents blood cells from linking with the oxygen molecule, not allowing for the efficient oxygenation of vital organ tissue. Over a prolonged interval of time it can cause serious problems even ulti-mately resulting in death. Intoxication may happen when the concentrations of carbon monoxide in the blood surpass a certain threshold as exposed in Table 2. Most cases of intoxication originate from the inhalation of CO produced by motor vehicles powered by fossil fuels in direct proximity of VRUs[7] (Table 3).

Table 3. Effects of different amounts of CO on the body of an average adult human.

CO concentration	Symptoms
0.0035%	Headaches and dizziness within 6 to 8 h of continuous exposure
0.01%	Light headache within 2 a 3 h of exposure
0.02%	Mild headache within 2 to 3 h of exposure Loss of judgment
0.04%	Frontal headache within 1 to 2 h
0.08%	Nausea and seizures in 45 min In-sensitiveness within 2 h
0.16%	Headache, increased heart rate, dizziness and nausea in 20 min. Death in less than 2 h
0.32%	Headache, dizziness and nausea in 5 to 10 min Death in 30 min
0.64%	Headache and dizziness in 1 to 2 min. Convulsions, respiratory arrest and death in less than 20 min
1.28%	Unconsciousness after 2–3 breaths Death in less than 3 min

80% of Sulfur dioxide (SO_2) in the atmosphere is due to the incomplete burning of fossil fuels[9]. When inhaled it affects the mucous membranes found on the nose, throat, and airways. The simultaneous presence of SO_2 and PM_{10} in the atmosphere can aggravate cardiovascular problems. In Table 4 is indicated the safety threshold values researched for this substance.

Table 4. Air quality depending on the level of sulfur dioxide.

SO_2 concentration (mg/m$\hat{3}$)	Air quality
0–19	Good
20–39	Moderate
40–364	Slightly unhealthy
365–800	Unhealthy
800+	Dangerous

Nitrogen Dioxide (NO_2) is mainly a residue of car fossil fuel consumption. Its inhalation may aggravate respiratory illnesses. Over prolonged exposure increases our body's susceptibility to dangerous respiratory infections[8]. In the Table 5 we indicate the values at which NO_2 concentration becomes prejudicial to humans.

Table 5. Effects on humans of nitrogen dioxide in air. [7]

NO_2concentration (ppb)	Air quality	Considerations
0–50	Good	No health impacts are expected when air quality is in this range
51–100	Moderate	Individuals extraordinarily sensitive to nitrogen dioxide should consider limiting prolonged effort outdoors
101–150	Unhealthy for sensitive groups	The following groups should limit prolonged effort outdoors:
		People with lung diseases, such as asthma
		Children and older adults
151-20	Unhealthy	The following groups should avoid prolonged efforts outdoors:
		People with lung diseases, such as asthma
		Older children and adults
		All others should limit prolonged effort outdoors
201–300	Dangerous	The following groups should avoid all effort outdoors:
		People with lung diseases, such as asthma
		Older children and adults
		All others should limit the effort outdoors

Ground-level ozone (GLO) is produced by chemical reactions between nitrogen oxides and volatile organic compounds in the presence of sunlight[7]. Inhaling this gas can be harmful to cells that line your lungs. Table 6 presents values thresholds values for VRUs safety.

Table 6. The referenced limit values of Ground-level Ozone established.

GLO reference period	Threshold
8 h a day 5 days a week	0,1 ppm
15 min (short time exposure)	0,3 ppm

A second dataset, the one on *traffic flow*, is composed only of three features, i.e., the *city_name* (as Braga), *speed_diff*, and *creation_date* (21049 observations). From these, the most important one is *speed_diff*, which provides information linked to the traffic flow status at the corresponding timestamp. It communicates the difference between the maximum speed that cars can reach in a traffic-free scenario and the speed at which cars are travelling at that moment [8]. Higher values suggest the presence of traffic, while near zero values indicate no traffic at all. Figure 1 expresses the *speed_diff* variation throughout a day of the dataset.

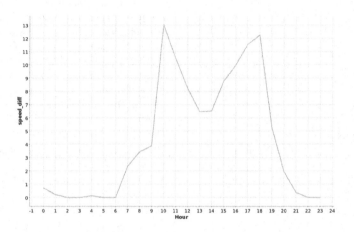

Fig. 1. Average *speed_diff* during one day of the dataset.

The third dataset, the *traffic incidents* one, contains hourly information about traffic incidents in Braga (83347 observations). Its features consist of the incident *description, impact category, magnitude of the delay* and *delay in seconds* it caused as well the *length in meters* the incident is causing.

It contains an overwhelming amount of unneeded information and may deviate the learning process of any ML algorithm. Thus reviewing and setting aside a few features became paramount. The features composing our selection, allowed for an efficient and in context treatment of the information this dataset contained. Description, indicates the type of incident presenting high importance for risk assessment, as it may indicate different levels of danger for the VRUs. Incident_category_desc, enables understanding of the impact the incident road. Magnitude_of_delay_desc, indicates the time delay that an accident causes. Length_in_m, stands for the distance of the traffic line the incident caused. Delay_in_s represents a calculated number for the time delay an incident causes.

Lastly, the *weather* dataset provides information about the weather conditions in the same city (6821 observations). The used dataset includes data about the weather *description, temperature, atmospheric pressure, humidity, wind speed, precipitation* and *luminosity*, among others. Likewise to the other datasets there were a few features which did not need an in depth review, such as city_name, creation_date and seq_num. The remaining data could be divided in two sub-groups, nominal and numerical. Weather_description provides nominal information about weather conditions in Braga. A few of it's values presented were similar meanings such as for the values of "mist" and "haze", "loosely clouded" and "a few clouds", finally "drizzle and light rain" and "low precipitation". Temperature designates the outside general temperature level. Atmospheric pressure as it tends do lower the amount of oxygen particles in the environment also lowers becoming harder to breath. Humidity varies along the year and at extreme values can cause respiratory diseases. Wind speed may difficult VRUs movement outside at high enough levels. Rain lowers drivers awareness of VRUs increasing the possibility incidents occurring. Current_luminosity indicates luminosity values in the city. It presents high correlation with sunrise and sunset features. These two indicate the time of the that the actual sun-setting and sunrise occurred.

3.2   Data Treatment and Pre-processing

The *creation_date* feature, which had milliseconds precision, was normalized to an hourly one and used to join all datasets into one ultimate dataset. This final dataset is the result of joining each individual dataset through each observation's *creation_date* up to the hour. Subsequently, the few existing missing values were replaced by the values of the previous hour.

The *pollution* features were then binned into five bins, with each features ranging from posing *very_low_risk* to *very_high_risk*. This allows us to arrange the risk values for each given pollution type. Since both the SO_2 as CO_2 features failed to present meaningful data, they were discarded from the final dataset.

From the *traffic flow* dataset, only *speed_diff* Hence, with the extraction process completed, the labelling was done in accordance to the information assimilated in the data exploration so that risk values could be correctly inferred.

Using a correlation matrix as well as intuition, it was found that some features from the *traffic incidents* dataset presented no value to the main objective. Such features were discarded from the final dataset in order do improve its quality.

From the *weather* dataset, features considered as relevant were the *temperature, weather_description* and *humidity*. The features in the *weather_description* were then grocategoriseduped by similarity and labeled into different levels of danger that allowed a better discrimination of weather events and their possible influence in the safety of VRUs.

3.3   Data Labelling

The final step to prepare the dataset consisted in labelling the dataset in order to create the target column (danger criteria) for each single hour, i.e., observation. Table 7 depicts the weights defined for the calculation of danger criteria per feature per original dataset. The weight value distribution allowed us to guarantee that in case of a life level threatening scenario, a greatest level of risk would be assured. As can be seen from Table 7, each dataset contains new data with the danger criteria for several variables. Taking this in account, it was possible to create a final criterion for each dataset as the criteria final pollution that we call criteria pollution is an accumulation of the dangers criteria of that dataset. The final criteria is created based on all datasets final criteria based on the final weights, obtaining the value of the final criteria.

The risk value of each feature is summed and the final overall danger is assessed. There are two cases that are not shown in Table 7, *temperature* and *humidity*, which depict a variation in weights between 1 and 5. I.e., *temperature* values between 10 and 15 °C are considered to be of very low risk, while values above 20 °C or less than 5 °C are considered to be of very high risk. *Humidity* values between 45 and 60 are of very low risk, while values above 90 and below 20 are of very high risk.

The final dataset consists of 6820 observations with 16 features (*hour, Day of month, Month, criteria_ozone, criteria_nitrogen_dioxide, criteria_particulate, criteria_ultraviolet, temperature, criteria_weather, count_incidents, criteria_incidents, criteria_pollution, criteria_final_incidents, criteria_speed_diff, criteria_final_weather, criteria_final*), including the target one. The feature *criteria_final* is the target, ranging between 0 and 3. The target feature, which was not used as training input, was further encoded using a one-hot encoding technique. The final target value has a shape of 4, with the correct value marked as 1.

Table 7. Criteria used to label the dataset. The following acronyms were used: VLR - Very Low Risk, LR - Low Risk, MR - Medium Risk, HR - High Risk, and VHR - Very High Risk.

Dataset	Feature	VLR	LR	MR	HR	VHR
Pollution	*criteria_particulate*	–	2	–	20	–
Pollution	*criteria_ozone*	1	2	3	6	8
Pollution	*criteria_nitrogen_dioxide*	2	4	7	12	–
Pollution	*criteria_ultraviolet*	2	4	8	20	–
Weather	*criteria_weather*	1	2	3	6	8
Traffic incidents	*criteria_incidents*	2	4	7	12	–
Traffic flow	*speed_dif*	1	2	3	4	5

3.4   Technologies

As mentioned, KNIME was used for data exploration and data processing, yet for the designing of the models, tools such as Python 3.7.6 and TensorFlow 2.0 were used. In order to apply the model Google Colab's platform GPU was used as it presented greater processing power and was capable of delivering the final product faster and with more concise information regarding time and performance.

4   Experiments

Two distinct algorithms were implemented, in particular, Random Forests and
MLPs. In both cases, results were evaluated using cross-validation, with k = 5
for the Random Forest and k = 10 for the MLP. The dataset was further split
into training (75%) and evaluation data (25%) prior to cross-validation (holdout-
validation). The shape of the input, without the target feature, is of (5115, 15)
for training and (1705, 15) for evaluation.

4.1   Random Forest

The first step was to prepare both the nominal and the numeric parameters
to acquiesce with the model's requirements and specifications. Tuning was performed for the quality measure, the number of models, the number of levels of
each tree, and finally the number of nodes (Table 2).

4.2   Multi Layer Perceptron

Features were prepared according to the data treatment phase converting the
categorical data into a format compatible with our neuronal network. In order to
improve performance values, tuning was performed for the following parameters:
batch size, number of neurons per layer, epochs, dropout, and the number of deep
layers of the model. Call-backs methods were implemented in order to prevent
overfitting allowing the training to end pre-emptively. Table 8 describes the
search space for each hyper-parameter of the candidate models.

Table 8. Search space for each hyper-parameter of the candidate models.

Model	Hyper-parameter	Search space
Random forest	Quality measure	[Gain ratio, Gini index, Gini]
	Number of levels	[10, 40] with step size of 10
	Number of models	[100, 500] with step size of 100
	Minimum number of records	[1, 12] with a unitary step size
MLP	Batch size	[32, 64]
	Number of neurons	[32, 128] with step size of 32
	Epochs	[5, 25] with step size of 5
	Dropout	[0.1, 0.4] with step size of 0.1

5   Results and Discussion

With the tuning process finished, the technique used to choose the best candidate
models was based on the attained accuracy. The next lines describe the main
results obtained from the conducted experiments.

Table 9. Summary results for the conceived models.

Method	Accuracy (%)	Error (%)	Runtime	Hyper-parameters
Random forest	95.234	4.766	8 h (w/ tuning)	Gain Ratio 500 trees 20 levels
MLP	99.41	0.59	3 min	32 neurons 50 epochs 64 Batch size 0.2 Dropout Adam

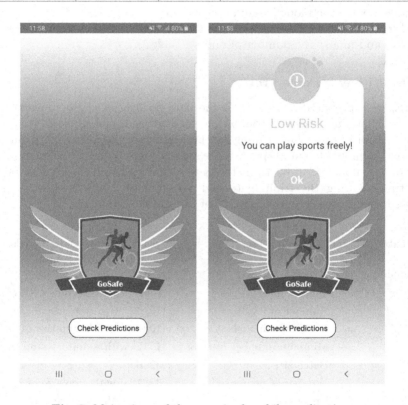

Fig. 2. Main views of the conceived mobile application.

5.1   Model Results

The best candidate, Random Forest, used the hyper-parameters presented in Table 8. It saw an accuracy of 95,234%. On the other hand, the best candidate MLP presented two dense layers, one layer of dropout (of 0.2), and again used the hyper-parameters that are in Table 8 and presented an accuracy of 99%.

It is our assumption that such high accuracies emerge from the fact that both models were able to learn the rules that were followed to label the dataset. If the models are able to learn such rules, it means they are able to accurately categorize the presence of dangerous situations. Both models present satisfactory accuracy, being able to successfully categorise an hour in respect to its safety to go outside. The best MLP behaved slightly better than the best Random Forest. It also presented more stable results and had a quick execution time. Table 9 presents the summarised results for the conceived models.

5.2 Platform

A platform was developed to work as a front-end of information for VRUs. The first stage to conceive the platform focused on deploying the best MLP model in an online server. Afterwards, a mobile application was developed using React Native. The mobile application queries the server to attain forecasts and presents the results to the user. Figure 2 portrays the main view of the platform, which is available for both Android and iOS.

6 Conclusions

From data investigation to its treatment the research and work conducted focused in guaranteeing VRUs safety by predicting if a certain period of the day was safe for sports. The process of transforming the data fed by the sensors across the city assures that safety thresholds for each given feature are obeyed. With these methods it was then possible to infer the danger a VRU incurred as the output of the data fed to the prediction model developed in this project. The application is able to link itself to the web server that holds the predictive model receiving the result. Both selection of the best ML model to implement and ensuring that no incompatibilities and inaccuracies regarding the theoretical thresholds for each feature existed were difficulties found. However, after preparation and careful investigation, by employing the knowledge acquired they were possible. In addition, although this performance is still calculated with the data at hand, it is expected that after entering circulation in Braga it will be precise. It is also expected to further tune the networks' performance, either by training with future data acquired or even by adding different features that may present alternative types of danger to the ones covered in this article.

Acknowledgments. This work has been supported by FCT – *Fundação para a Ciência e Tecnologia* within the R&D Units Project Scope: UIDB/00319/2020. It was also partially supported by a doctoral grant, SFRH/BD/130125/2017, issued by FCT.

References

1. PORDATA. https://www.pordata.pt/Municipios/Pe~oes+atropelados+total+e+m ortos-233. Accessed 18 June 2020

2. Tahmasbi-Sarvestani, A., Nourkhiz Mahjoub, H., Fallah, Y.P., Moradi-Pari, E., Abuchaar, O.: Implementation and evaluation of a cooperative vehicle-to-pedestrian safety application. In: IEEE Intelligent Transportation Systems Magazine, vol. 9, no. 4, pp. 62–75 (2017). https://doi.org/10.1109/MITS.2017.2743201
3. Moxey, E., Johnson, N., McCarthy, M.G., McLundie, W.M., Parker, G.A.: The advanced protection of vulnerable road users (APVRU) - project information sheet. In: IEE Target Tracking 2004: Algorithms and Applications, Brighton, UK, p. 83 (2004). https://doi.org/10.1049/ic:20040056
4. Liebner, M., Klanner, F., Stiller, C.: Active safety for vulnerable road users based on smartphone position data. In: 2013 IEEE Intelligent Vehicles Symposium (IV), Gold Coast, pp. 256–261 (2013). https://doi.org/10.1109/IVS.2013.6629479
5. Shinar, D.: Safety and mobility of vulnerable road users: pedestrians, bicyclists, and motorcyclists. Accid. Anal. Prev. 44(1), 1–2 (2012)
6. Religi, A., et al.: Prediction of anatomical exposure to solar UV: a case study for the head using SimUVEx v2. In: 2016 IEEE 18th International Conference on e-Health Networking, Applications and Services (Healthcom), Munich, pp. 1–6 (2016). https://doi.org/10.1109/HealthCom.2016.7749513
7. Bierwirth, P.: Carbon dioxide toxicity and climate change: a major unapprehended risk for human health. Emeritus Faculty, Australian National University (2020). https://doi.org/10.13140/RG.2.2.16787.48168
8. Lu, F., Chen, X.: Analyzing the speed dispersion influence on traffic safety. In: 2009 International Conference on Measuring Technology and Mechatronics Automation, Zhangjiajie, pp. 482–485 (2009). https://doi.org/10.1109/ICMTMA.2009.566
9. Tomás, V.R., Pla-Castells, M., Martínez, J.J., Martínez, J.: Forecasting adverse weather situations in the road network. IEEE Trans. Intell. Transp. Syst. 17(8), 2334–2343 (2016). https://doi.org/10.1109/TITS.2016.2519103

One-Shot Only Real-Time Video Classification: A Case Study in Facial Emotion Recognition

Arwa Basbrain[1,2(✉)] and John Q. Gan[1(✉)]

[1] School of Computer Science and Electronic Engineering, University of Essex, Colchester, UK
{amabas,jqgan}@essex.ac.uk
[2] Faculty of Computing and Information Technology, King Abdul-Aziz University, Jeddah, Kingdom of Saudi Arabia
abasabreen@kau.edu.sa

Abstract. Video classification is an important research field due to its applications ranging from human action recognition for video surveillance to emotion recognition for human-computer interaction. This paper proposes a new method called One-Shot Only (OSO) for real-time video classification with a case study in facial emotion recognition. Instead of using 3D convolutional neural networks (CNN) or multiple 2D CNNs with decision fusion as in the previous studies, the OSO method tackles video classification as a single image classification problem by spatially rearranging video frames using frame selection or clustering strategies to form a simple representative storyboard for spatio-temporal video information fusion. It uses a single 2D CNN for video classification and thus can be optimised end-to-end directly in terms of the classification accuracy. Experimental results show that the OSO method proposed in this paper outperformed multiple 2D CNNs with decision fusion by a large margin in terms of classification accuracy (by up to 13%) on the AFEW 7.0 dataset for video classification. It is also very fast, up to ten times faster than the commonly used 2D CNN architectures for video classification.

Keywords: Video-based facial emotion recognition · Convolutional neural network · Spatial-temporal data fusion

1 Introduction

With the rapid growth of video technology, there is a widespread demand for video recognition systems across different fields, such as visual surveillance, human-robot interaction and autonomous driving vehicle. Recently, a significant amount of literature has been published on computer vision in relation to developing robust recognition systems which can achieve a similar level of performance to that of human agents. With the recent advances in GPUs and machine learning techniques, the performance of image-based recognition systems has been significantly boosted, however what video-based recognition systems can achieve today is still far from what the human perception system can do.

© Springer Nature Switzerland AG 2020
C. Analide et al. (Eds.): IDEAL 2020, LNCS 12489, pp. 197–208, 2020.
https://doi.org/10.1007/978-3-030-62362-3_18

Numerous methods have been proposed to utilize convolutional neural networks (CNNs) for video classification tasks such as action recognition and emotion recognition. Many state-of-the-art methods employ a single frame baseline approach that treats video frames as still images, applies 2D CNNs to classify each frame, and then fuses the individual predictions to obtain the final decision at the video level [1, 2]. This approach does not take the order of frames into account. However, a video is not just a stack of frames; every single frame represents a small portion of the video's story, and it also includes a temporal component which gives vital information to the recognition task.

To address the problem in the single frame baseline approach to video classification, it is natural to use the temporal information in videos to achieve better performance. Some existing methods [3–5] apply feature fusion approaches to aggregate spatial and temporal information. On the other hand, 3D CNNs have been developed to learn spatio-temporal features for various video analysis tasks. As 3D CNNs have more parameters than 2D CNNs, they are more complex and harder to train.

Inspired by the You Only Look Once system for real-time object detection [6], we propose a general model called One-Shot Only (OSO) for video classification, which converts a video-based problem to an image-based one by using frame selection or clustering strategies to form a simple representative storyboard for spatio-temporal video information fusion. The work in this paper is different from that of Jing et al. [7] (Video You Only Look Once for Action Recognition) which uses complex 3D CNNs to learn temporal features from a video and classify the actions it contains in a single process. Using 2D CNNs without losing the temporal information due to the use of the storyboard representation of videos, the OSO methods proposed in this paper can not only meet the requirement for real-time video analysis but also produce competitive video classification accuracy by combatting the overfitting problem existing in commonly used 2D CNN architectures for video classification. The main contributions of this paper are as follows:

- A novel spatio-temporal data fusion approach to video representation is proposed, which speeds up video classification with competitive accuracy to meet the requirements of real-time applications.
- Frame selection and clustering strategies are proposed to handle varied video length and the redundancy in consecutive video frames for effective video representation.
- Two 2D CNN based pipelines for video classification are proposed and evaluated using the AFEW facial emotion video dataset.

The remainder of this paper is organised as follows. Section 2 reviews related work. Section 3 describes the proposed OSO methods for real-time video classification. The experiments are described in Sect. 4, with the results presented and discussed in Sect. 5. Finally, this paper is concluded in Sect. 6.

2   Related Work

2.1   Methods for the EmotiW Challenge

Numerous researchers have made intensive efforts to improve audio-visual emotion recognition based on images and videos [8]. This paper focuses on visual emotion recognition from videos. Based on Ekman's study [9], most emotion recognition systems categorise facial expressions as anger, disgust, fear, happiness, sadness, surprise, and neutrality, which will also be considered in this paper. Many methods for video classification for emotion recognition have been proposed in response to the Emotion Recognition in the Wild Challenge (EmotiW). Kahou et al. [10] combined a number of deep neural networks, with each designed to handle different kinds of data sources. They used 2D CNNs to classify aligned images of faces and a shallow network to extract features related to the mouth. They used support vector machines (SVM) to classify these extracted features and a multilayer perceptron (MLP) for decision fusion to provide the final video classification result. They achieved an accuracy of 41.03% on the test set, the highest among the submissions to the EmotiW 2013 challenge [11].

Sikka et al. [12] employed a feature fusion approach based on multiple kernel learning (MKL), which was used to find an optimal combination of audio and visual features for input into a non-linear SVM classifier. They utilised four different hand-crafted models to extract ten different visual features, including HOG, PHOG, BoW and LPQ-TOP. The classification accuracy they achieved was 37.08% on the EmotiW validation dataset.

Liu et al. [13] represented all the frames of a video clip as an image set and modelled it as a linear subspace to be embedded in Grassmannian manifold. After the features were extracted from each video frame by using a CNN, a class-specific one-to-rest partial least squares (PLS) model was employed to learn, in relation to video and audio features separately, to distinguish between classes. The final accuracy achieved on the EmotiW validation and test set was 35.85% and 34.61% respectively.

Chen et al. [14] proposed a feature descriptor called histogram of oriented gradients from three orthogonal planes (HOG_TOP), which could be used to extract dynamic visual features from video sequences, and adopted the multiple kernel fusion framework method to find an optimal combination of visual and audio features. Their methods achieved an overall classification accuracy of 40.21% and 45.21% on the EmotiW validation set and test set, respectively.

The baseline method for the EmotiW Challenge [15] utilized the local binary pattern from three orthogonal planes (LBP-TOP) method to extract features from nonoverlapping spatial 4×4 blocks and concatenated these features to create a feature vector. A non-linear Chi-square kernel-based SVM was used for classification.

Jing et al. [7] proposed a 3D CNN model to learn extracting visual and temporal information from a whole video at one pass by selecting a subset of frames. They designed 8 types of 3D CNNs to handle different lengths of proxy videos. Their work opened up new ways of integrating spatio-temporal information for video classification.

2.2 Key-Frame Selection Strategies

As a short-length video clip of 2–3 s could contain 70–90 frames, processing each frame separately, even in a short-length video, is time-consuming and would usually affect the system accuracy. Many studies proposed key-frame selection strategies to handle this problem. However, most of these strategies are either complicated and computationally expensive, or cannot effectively work. Some of the common key-frame selection strategies are as follows:

- Predefined-frames strategy, which takes specific frames depending on their position in the video, such as the middle and boundary frames [16, 17]. Although this strategy is straightforward and fast, it depends on some knowledge of dataset [16]. In addition, predefined frames are usually not stable and do not detain most of the visual content.

- Motion-analysis strategy, which computes the optical flow for each frame to evaluate the changes on the facial expression [18], using specific points such as the left and inner eyebrow or the corners of the mouth.

- Visual-content strategy, which computes similarities between frames represented by colour histograms or other features [19]. The first frame is chosen as the first key-frame, then the similarity between adjacent frames is computed and the frame which has a significant content change is selected as the next key-frame.

- Clustering strategy, which clusters similar frames by assigning each frame to a corresponding group and then selects the centroid frame of each group as key-frames [19]. Although clustering methods demonstrate good results in general, noise and motion can easily affect their performance. In addition, the key-frames selected may be only from the dominant clusters.

3   Proposed Methods for One-Shot Only Real-Time Video Classification

Two OSO methods for facial emotion recognition based on video classification are proposed in this paper, named Frame Selecting Approach and Frame Clustering Approach, which benefit from the hierarchical representation of spatio-temporal information in video frames. The structures of the proposed OSO approaches are shown in Fig. 1. Both approaches apply three pre-processing steps that detect and track faces across the video frames, and extract the region of interest (ROI) of the detected faces. Then the facial landmark points are used to align the faces of the frames chosen by frame selection or clustering strategies. The pre-processed facial images are combined to create a storyboard in the form of single image, in which spatio-temporal information fusion is conducted at raw data level, i.e., at the level of the ROIs of the selected video frames.

In the frame selecting approach, as shown in Fig. 1(A), the storyboard is created from selected frames and is used as the input to a 2D CNN which predicts the emotion class of the video directly. Video clips have differing lengths or different number of frames. Also, the period of the same emotion may vary when performed by different subjects or by the same subject at different times. When selecting only a small number of frames from a video clip showing the emotion, it is critical to select the frames which are most different from the average frame of the whole original video.

In the frame clustering approach, as shown in Fig. 1(B), the video frames are clustered into groups of frames with certain similarity, and a storyboard is then created for each group respectively. In a video clip, the subject might start by showing one emotion and end up by presenting quite another. In other words, the clip may contain several consecutive emotions: pre-emotion, post-emotion and the main one. For example, the "surprised" emotion may be followed by one of the post-emotions, perhaps "happy" or "fear." By modelling the temporal relationships between consecutive emotions, we can distinguish between the compound and the individual ones. Based on this idea, we

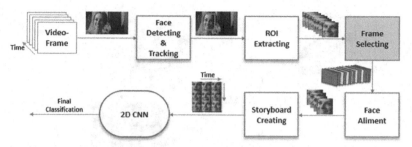

A. OSO Video-based Classification Pipeline for Emotion Recognition Using Frame Selecting Approach

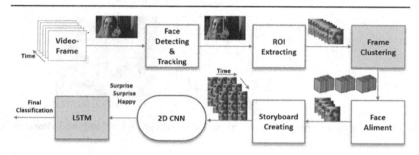

B. OSO Video-based Classification Pipeline for Emotion Recognition Using Frame Clustering Approach

Fig. 1. OSO video-based classification pipelines for emotion recognition.

propose to produce pre-prediction of class for each storyboard using 2D CNNs, and the sequence of these class pre-predictions are sent to a long short-term memory (LSTM) network to obtain the final class prediction of the whole video.

3.1 Spatio-Temporal Information Fusion (Storyboard Creating)

The facial emotion of a subject in a video generates space-time images within a 3D space, which encode both spatial and temporal information related to the subject's emotion. Instead of creating a 3D volume for the space-time information, this paper proposes a storyboard creation technique that conflates video frames into one image based on keyframe selection or clustering.

Figure 2 shows the three dimensions used in presenting the storyboard for video-based classification. Before constructing the storyboard, the selected frames are resized to a fixed size of 224 × 224 in order to reduce the interference caused by the images' boundaries. Then these frames are concatenated to build one image, as illustrated in Fig. 2. After that, the constructed storyboard is resized to 224 × 224 pixels because this size fits most 2D CNN models.

3.2 Frame Selection and Clustering Strategies

The purpose of keyframe selection is to find a set of representative frames from an image sequence while the purpose of frame clustering is to segment the set of sequential frames into subsets based on similarity matching.

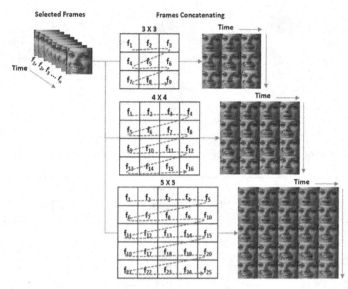

Fig. 2. The three dimensions used in presenting the storyboard for video-based classification.

In this paper, a clustering-based strategy is used to achieve automatic keyframe selection and frame clustering. The proposed clustering-based strategy works, fundamentally, by measuring the dissimilarity or distance between frames using the Euclidean distance:

$$d_t = \sqrt{\sum_{j=1}^{L} \left(\hat{f}_t(j) - \hat{f}_{t+1}(j) \right)^2} \tag{1}$$

where $\hat{f}_t$ denotes a frame feature vector at a specific time and L is the length of the vector. The following steps are followed to assign frames to the most similar cluster:

1. Normalise the ROIs extracted from successive frames by resizing them to 224×224 and converting them to grey images.
2. Represent every frame as a feature vector $\hat{f}$.
3. Compare adjacent frames with each other using Eq. (1) to determine how dissimilar they are. Denote the set of frame vectors as $\hat{F} = \left\{ \hat{f}_1, \hat{f}_2, \hat{f}_3, \dots \hat{f}_N \right\}$, where N is the number of frames in the video clip, and the difference between these frames as $D_{if} = \{d_1, d_2, d_3 \dots d_{N-1}\}$.
4. Determine a boundary-threshold value Ψ by calculating the mean value of the D_{if} set.
5. Use the threshold Ψ to determine the borders of each cluster - where a dissimilarity value higher than Ψ indicates the start or end of a frame cluster. Denote the set of clusters as $C = \{c_1, c_2, c_3 \dots c_M\}$ where M is the number of clusters that consist of similar frames.
6. For the keyframe selection strategy:

a. If the number of clusters M is smaller than the preset storyboard size (9, 16 or 25), decrease the value for Ψ. Alternatively, increase it when M is larger. Then go back to step 5.

b. If M equals the preset storyboard size (9, 16 or 25), select the mid-frame of each cluster as a keyframe.

Or for the frame clustering strategy, preset a cluster-threshold value γ as the maximum number of clusters to be generated (it is set to 3 in this paper):

a. If the number of clusters M is larger than γ, decrease the value for Ψ. Alternatively, increase it when M is smaller. Then go back to step 5.

b. If M equals γ, use all the frames in each cluster to build a storyboard.

(1) If the number of frames in a cluster Q is larger than the preset storyboard size (9, 16 or 25), choose the middle frames of the cluster.

(2) If Q is smaller than the preset storyboard size (9, 16 or 25), duplicate the middle frames of the cluster to compensate.

4 Experiments

In this section, the facial emotion datasets used in the experiments are briefly described first, and the details about the implementation of the proposed methods are explained, including the pre-processing and the settings of the CNNs.

4.1 Datasets

In order to evaluate the proposed methods, experiments were conducted on two facial emotion datasets, AffectNet [20] and AFEW [21]. AffectNet is a large-scale facial emotion dataset containing more than one million images obtained from the Internet. Twelve expert annotators annotated a total of 450,000 images according to the facial expressions in the images using both discrete categorical and continuous dimensional (valence and arousal) models. The Acted Facial Expressions in the Wild (AFEW) dataset consists of short video clips of facial expressions in close to real-life environments. AFEW contains 1,426 video clips of 330 subjects aged 1–77 years. The video clips were annotated as one of the six basic expressions or neutral by two independent labellers.

4.2 Implementation Details

Pre-processing. As the videos in the AFEW 7.0 dataset might be taken from more than one subject, we used MATLAB "Face Detection and Tracking" to track the main subject's face in the videos automatically. A set of feature points in the detected facial region were identified using the standard "good features to track" process [22] and tracked using the Kanade-Lucas-Tomasi (KLT) algorithm [23]. For facial landmark detection and face alignment, we utilised the algorithm proposed by Zhang et al [24]. The generated landmarks (two eyes, nose and mouth corners) were then used to determine the inner area of the face as a ROI. For normalisation, the face is cropped to a 224 × 224 RGB image.

Training the 2D CNN and LSTM Models. In order to increase the generalisation ability of the CNN models and to tackle the overfitting problem, seven well-known pre-trained 2D CNNs (GoogLeNet, VGG16, VGG19, ResNet50, ResNet101, ResNet18, Inceptionv3) were utilised in our experiments and a large number of still images in the AffectNet dataset were used to fine-tune these models via two-fold cross-validation. To find the appropriate size for the storyboard, these models were fine-tuned on three different sizes, 3 × 3, 4 × 4, and 5x5. This resulted in 28 fine-tuned models that were able to classify emotional images. To train the frame clustering OSO model for video classification, each video in the AFEW training dataset was clustered into three groups of frames and classified by the fine-tuned 2D CNNs, producing three emotion-words. Then the produced series of emotion-words were used to train the LSTM whose output is the final classification for each video. The AFEW validation dataset was used to evaluate the trained models.

5   Results and Discussion

In our experiments the proposed OSO video classification pipelines for emotion recognition using frame selecting approach and frame clustering approach respectively (as shown in Fig. 1) were evaluated based on three storyboard sizes (3 × 3, 4 × 4, 5 × 5), with their performance compared to those of the two single-frame baseline (1 × 1) methods. In the first baseline method, a decision level fusion method based on majority voting is used to combine the CNN emotion predictions of all the frames in a video. The second baseline method follows a feature fusion approach where an LSTM model was trained to classify videos using fused features of all the frames extracted by the 2D CNNs. Table 1 and Table 2 show the results on the AFEW dataset in terms of validation accuracy and runtime of the OSO methods using seven 2D CNNs respectively, in comparison with the baseline methods.

Table 1. Validation accuracy of the OSO methods using 2D CNNs for video classification on the AFEW dataset, in comparison with single frame baseline (1 × 1) approaches

Accuracy		1 × 1		3 × 3		4 × 4		5 × 5	
		Decision Fusion	Feature Fusion	Selecting	Clustering	Selecting	Clustering	Selecting	Clustering
CNN	GoogLeNet	33.8	35.1	46.4	48.7	49.0	46.7	41.9	45.0
	VGG16	39.3	34.6	51.1	55.3	47.7	55.6	47.7	50.1
	VGG19	37.7	30.1	49.8	53.8	47.7	51.0	46.7	47.8
	ResNet50	33.8	41.1	50.0	54.9	50.6	53.8	43.0	47.3
	ResNet101	36.9	37.2	48.7	52.8	46.7	51.1	44.6	48.0
	ResNet18	34.5	30.1	46.9	51.1	45.1	46.3	46.9	40.9
	Inceptionv3	36.1	34.8	50.8	54.6	48.0	51.3	41.7	45.7
	Average	36.0	34.71	49.1	53.03	47.83	50.83	44.64	46.40

As shown in Table 1, the OSO approaches outperformed both baseline methods in terms of validation accuracy by 10% to 17%. The frame clustering approach outperformed the frame selecting approach in almost all cases by 1.1% (VGG19, 5 × 5) to

Table 2. Comparison of validation time on the AFEW dataset between the OSO methods using 2D CNNs and the single frame baseline (1 × 1) approaches for video classification

Speed S/Video		1 × 1		3 × 3		4 × 4		5 × 5	
		Decision Fusion	Feature Fusion	Selecting	Clustering	Selecting	Clustering	Selecting	Clustering
CNN	GoogLeNet	0.31	0.33	0.038	0.047	0.055	0.061	0.074	0.091
	VGG16	0.60	0.63	0.037	0.046	0.056	0.059	0.073	0.090
	VGG19	0.69	0.71	0.039	0.047	0.056	0.061	0.075	0.092
	ResNet50	0.40	0.43	0.038	0.047	0.056	0.061	0.075	0.091
	ResNet101	0.52	0.56	0.046	0.050	0.062	0.064	0.078	0.099
	ResNet18	0.23	0.27	0.035	0.047	0.053	0.061	0.075	0.088
	Inceptionv3	0.59	0.62	0.045	0.052	0.063	0.064	0.077	0.099
	Average	0.48	0.51	0.040	0.048	0.057	0.062	0.075	0.093

Table 3. Comparison with the state-of-the-art results on the AFEW 7.0 dataset

		Methods	Val	Test
EmotiW baseline [15]		LBP-TOP-SVM	36.08	39.33
3D CNN	Ouyang et al. [25]	3D CNN	35.20	
	Lu et al. [3]	3D CNN	39.36	
	Fan et al. [26]	3D CNN	39.69	
	Vielzeuf et al. [27]	3D CNN+LSTM	43.20	
Decision/Feature Fusion	Yan et al. [25]	Trajectory+SVM	37.37	
	Fan et al. [28]	MRE-CNN (AlexNet)	40.11	
	Yan et al. [29]	VGG-BRNN	44.46	
	Ding et al. [30]	AlexNet	44.47	
	Fan et al. [26]	VGG16+LSTM	45.43	
	Ouyang et al. [25]	ResNet+LSTM	46.70	
	Ouyang et al. [25]	VGG+LSTM	47.40	
	Fan et al. [28]	MRE-CNN (VGG16)	47.43	
	Vielzeuf et al. [22]	VGG16+LSTM	48.60	
	Proposed OSO 1	FrameSelecting-VGG16-3 × 3	**51.10**	**51.15**
	Proposed OSO 2	FrameClustering-VGG16-4 × 4+LSTM	**55.60**	**52.37**

7.9% (VGG16, 4 × 4) and on average by 3.93%, 3.0% and 1.76%, corresponding to storyboard size 3 × 3, 4 × 4 and 5 × 5 respectively. The highest accuracy was achieved by the OSO method using frame clustering and VGG16 with storyboard size 4 × 4. On average, the OSO method using frame clustering with storyboard size 3×3 achieved the highest accuracy of 53.03%. It can be observed that among the seven 2D CNNs, VGG16 achieved the highest accuracy in almost all cases.

One key advantage of the OSO approaches is their efficiency. To show this, we compared the runtime of OSO approaches with the two baseline methods with the AFEW validation dataset by using a single NVIDIA TITAN X GPU. As shown in Table 2, it is clearly demonstrated that the OSO approaches are about ten times faster than the single frame baseline methods.

Most winners of the EmotiW Challenge (2017–2019) utilised both audio and visual information in their approaches to increase the overall accuracy. To further evaluate the proposed OSO methods, we compared their performance with the EmotiW baseline performance and those of the winner methods reported in the literature that used visual information only. Table 3 shows the comparison results on the AFEW dataset. The proposed OSO methods using frame selecting and frame clustering approaches with the best 2D CNN and storyboard size achieved validation accuracy of 51.10% and 55.60% respectively and test accuracy of 51.15% and 52.37% respectively, much superior to the competition baseline performance and those achieved by the methods reported in the literature that used 2D CNNs or 3D CNNs for video-based emotion recognition without using audio information.

6 Conclusion

This paper proposes fast OSO methods for video-based facial emotion recognition to meet the requirements of real-time applications. Different from other approaches that aggregate temporal information from video frames, the proposed methods take the advantage of spatio-temporal data fusion based on novel frame selection and clustering strategies and use 2D CNN models to predict emotional category from videos with facial expressions. The experimental results show that the proposed OSO methods are not only fast but also capable to achieve competitive accuracy of video classification.

References

1. Kim, B.-K., Roh, J., Dong, S.-Y., Lee, S.-Y.: Hierarchical committee of deep convolutional neural networks for robust facial expression recognition. J. Multimodal User Interfaces **10**(2), 173–189 (2016). https://doi.org/10.1007/s12193-015-0209-0
2. Liu, C., Tang, T., Lv, K., Wang, M.: Multi-feature based emotion recognition for video clips. In: Proceedings of the ACM International Conference on Multimodal Interaction, pp. 630–634. ACM, Boulder (2018)
3. Lu, C., Zheng, W., Li, C., Tang, C., Liu, S., Yan, S., Zong, Y.: Multiple spatio-temporal feature learning for video-based emotion recognition in the wild. In: Proceedings of the ACM International Conference on Multimodal Interaction, pp. 646–652. ACM, Boulder (2018)
4. Knyazev, B., Shvetsov, R., Efremova, N., Kuharenko, A.: Convolutional neural networks pretrained on large face recognition datasets for emotion classification from video. arXiv preprint arXiv:1711.04598 (2017)
5. Bargal, S.A., Barsoum, E., Ferrer, C.C., Zhang, C.: Emotion recognition in the wild from videos using images. In: Proceedings of the ACM International Conference on Multimodal Interaction, pp. 433–436. ACM, Tokyo (2016)

6. Redmon, J., Divvala, S., Girshick, R., Farhadi, A.: You only look once: unified, real-time object detection. In: Proceedings of the IEEE Conference on Computer Vision and Pattern Recognition (CVPR), pp. 779–788 (2016)
7. Jing, L., Yang, X., Tian, Y.: Video you only look once: overall temporal convolutions for action recognition. J. Vis. Commun. Image Representation **52**, 58–65 (2018)
8. Samadiani, N., Huang, G., Cai, B., Luo, W., Chi, C.-H., Xiang, Y., He, J.: A review on automatic facial expression recognition systems assisted by multimodal sensor data. Sensors **19**, 1863 (2019)
9. Ekman, P., Friesen, W.V.: Constants across cultures in the face and emotion. J. Pers. Soc. Psychol. **17**, 124–129 (1971)
10. Kahou, S.E., et al.: Combining modality specific deep neural networks for emotion recognition in video. In: Proceedings of the 15th ACM International conference on multimodal interaction, pp. 543–550. ACM, Sydney (2013)
11. Dhall, A., Goecke, R., Joshi, J., Wagner, M., Gedeon, T.: Emotion recognition in the wild challenge 2013. In: Proceedings of the 15th ACM on International Conference on Multimodal Interaction, pp. 509–516. ACM, Sydney (2013)
12. Sikka, K., Dykstra, K., Sathyanarayana, S., Littlewort, G., Bartlett, M.: Multiple kernel learning for emotion recognition in the wild. In: Proceedings of the 15th ACM on International Conference on Multimodal Interaction, pp. 517–524. ACM, Sydney (2013)
13. Liu, M., Wang, R., Huang, Z., Shan, S., Chen, X.: Partial least squares regression on grassmannian manifold for emotion recognition. In: Proceedings of the 15th ACM on International Conference on Multimodal Interaction, pp. 525–530. ACM, Sydney (2013)
14. Chen, J., Chen, Z., Chi, Z., Fu, H.: Facial expression recognition in video with multiple feature fusion. IEEE Trans. Affect. Comput. **9**, 38–50 (2018)
15. Dhall, A., Murthy, O.V.R., Goecke, R., Joshi, J., Gedeon, T.: Video and image based emotion recognition challenges in the wild: EmotiW 2015. In: Proceedings of the ACM on International Conference on Multimodal Interaction, pp. 423–426. ACM, Seattle (2015)
16. Yang, B., Cao, J., Ni, R., Zhang, Y.: Facial expression recognition using weighted mixture deep neural network based on double-channel facial images. IEEE Access **6**, 4630–4640 (2018)
17. Doherty, A.R., Byrne, D., Smeaton, A.F., Jones, G.J.F., Hughes, M.: Investigating keyframe selection methods in the novel domain of passively captured visual lifelogs. In: Proceedings of the International Conference on Content-based Image and Video Retrieval, pp. 259–268. ACM, Niagara Falls (2008)
18. Guo, S.M., Pan, Y.A., Liao, Y.C., Hsu, C.Y., Tsai, J.S.H., Chang, C.I.: A key frame selection-based facial expression recognition system. In: Proceedings of ICICIC 2006 Innovative Computing, Information and Control, pp. 341–344 (2006)
19. Zhang, Q., Yu, S.-P., Zhou, D.-S., Wei, X.-P.: An efficient method of key-frame extraction based on a cluster algorithm. J. Hum. Kinet. **39**, 5–14 (2013)
20. Mollahosseini, A., Hasani, B., Mahoor, M.H.: Affectnet: a database for facial expression, valence, and arousal computing in the wild. IEEE Trans. Affect. Comput. **10**, 18–31 (2019)
21. Dhall, A., Goecke, R., Lucey, S., Gedeon, T.: Collecting large, richly annotated facial-expression databases from movies. IEEE Multimed. **19**, 34–41 (2012)
22. Shi, J., Tomasi, C.: Good features to track. In: Proceedings of the IEEE Conference on Computer Vision and Pattern Recognition, pp. 593–600 (1994)
23. Tomasi, C., Kanade, T.: Detection and tracking of point features. Technical report, Carnegie Mellon University (1991)
24. Zhang, K., Zhang, Z., Li, Z., Qiao, Y.: Joint face detection and alignment using multitask cascaded convolutional networks. IEEE Signal Process. Lett. **23**, 1499–1503 (2016)

25. Ouyang, X., et al.: Audio-visual emotion recognition using deep transfer learning and multiple temporal models. In: Proceedings of the 19th ACM International Conference on Multimodal Interaction, pp. 577–582. ACM, Glasgow (2017)

26. Fan, Y., Lu, X., Li, D., Liu, Y.: Video-based emotion recognition using CNN-RNN and C3D hybrid networks. In: Proceedings of the 18th ACM International Conference on Multimodal Interaction, pp. 445–450. ACM, Tokyo (2016)

27. Vielzeuf, V., Pateux, S., Jurie, F.: Temporal multimodal fusion for video emotion classification in the wild. In: Proceedings of the 19th ACM International Conference on Multimodal Interaction, pp. 569–576. ACM, Glasgow (2017)

28. Fan, Y., Lam, Jacqueline C.K., Li, Victor O.K.: Multi-region ensemble convolutional neural network for facial expression recognition. In: Kůrková, V., Manolopoulos, Y., Hammer, B., Iliadis, L., Maglogiannis, I. (eds.) ICANN 2018. LNCS, vol. 11139, pp. 84–94. Springer, Cham (2018). https://doi.org/10.1007/978-3-030-01418-6_9

29. Yan, J., et al.: Multi-clue fusion for emotion recognition in the wild. In: Proceedings of the 18th ACM International Conference on Multimodal Interaction, pp. 458–463. ACM, Tokyo (2016)

30. Ding, W., et al.: Audio and face video emotion recognition in the wild using deep neural networks and small datasets. In: Proceedings of the 18th ACM International Conference on Multimodal Interaction, pp. 506–513. ACM, Tokyo (2016)

Integrating a Data Mining Engine into Recommender Systems

Vitor Peixoto[1] , Hugo Peixoto[2] , and José Machado[2(✉)]

[1] University of Minho, Campus Gualtar, Braga 4710, Portugal
a79175@alunos.uminho.pt
[2] Algoritmi Center, University of Minho, Campus Gualtar, Braga 4710, Portugal
{hpeixoto,jmac}@di.uminho.pt

Abstract. History could be epitomised to a handful of events that changed the course of human evolution. Now, we found ourselves amid another revolution: the data revolution. Easily unnoticeable, this new outlook is shifting in every possible way how we interact with the internet and, for the first time in history, how the internet interacts with us. This new kind of interactions is defined by connections between users and consumable goods (products, articles, movies, etc.). And through these connections, knowledge can be found. This is the definition of data mining. Buying online has become mainstream due to its convenience and variety, but the enormous offering options affect negatively the user experience. Millions of products are displayed online, and frequently the search for the craved product is long and tiring. This process can lead to a loss of interest from the customers and, consequentially, losing profits. The competition is increasing, and personalisation is considered the game-changer for platforms. This article follows the research and implementation of a recommender engine in a well-known Portuguese e-commerce platform specialised in clothing and sports apparel, aiming the increase in customer engagement, by providing a personalised experience with multiple types of recommendations across the platform. First, we address the reason why implementing recommender systems can benefit online platforms and the state of the art in that area. Then, a proposal and implementation of a customised system are presented, and its results discussed.

Keywords: Recommender system · E-commerce · Data science · Business intelligence · Machine learning · Data driven decision making

1 Introduction

Commerce has been defined by the term '*mainstream*' since the start of the last century. Society started following fashion trends, renowned brands, or their favourite movie star fashion. To increase profits, commerce stores focused on popular products. Interestingly enough, the Pareto principle (the "80/20" rule)

© Springer Nature Switzerland AG 2020
C. Analide et al. (Eds.): IDEAL 2020, LNCS 12489, pp. 209–220, 2020.
https://doi.org/10.1007/978-3-030-62362-3_19

also applies to commerce [5]. Indeed, while observing products available to recommend in our platform, it showed that approximately 80% of all purchases come from 30% of all products (see Fig. 1), close enough to the "80/20" rule. The other 70% of niche products, are known as the *long tail*.

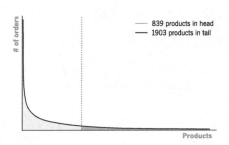

Fig. 1. Distribution of orders on this case study platform.

But then, at the turn of the millennium, the *dot-com bubble* burst, and with it, the beginning of the era of online commerce. This promoted a paradigm shift from the Pareto principle. Given the internet advantage of infinite shelf storage, the price of adding more and more products to platforms' portfolios is close to zero, so companies started increasing the number of available products.

However, according to Anderson, *"simply offering more variety, does not shift demand by itself. Consumers must be given ways to find niches that suit their particular needs and interests"* [4]. In fact, increasing the array of products hinders the task for clients to find what they desire. The solution relies in **recommender systems**. Through the analysis of customers' behaviours: their purchases, reviews and similarities among other customers, several assumptions can be made on the likelihood of a customer buying a certain product.

This solution develops a market of niches and subcultures accountable for a tremendous amount of profits in products outside the hits, which would likely remain unknown to most, affecting user satisfaction, due to the apparent lack of product diversity.

By creating opportunities for unpopular products to be sold, the *long tail* thickens when business drives from hits to niches, increasing overall profitability.

This theory has been explored through the last two decades, achieving remarkable advances in new data mining techniques. *Amazon* was one of the first companies investing in recommender systems, when they released their first patent in 1998 [6], more than two decades ago. By 2003, these recommendations were already widely disseminated across their platform. And it worked! By 2015, a research report estimated that 30% of *Amazon*'s product views originated from recommendations [7].

In another research, *Amazon*'s *long tail* was calculated and compared between 2000 and 2008 (see Fig. 2). During that period, the *tail* has gotten significantly longer, and overall gains from niche books increased five-fold, being accountable for 37% of *Amazon*'s book sales [8].

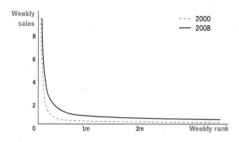

Fig. 2. *Amazon*'s *long tail*: 2000 *vs* 2008.

Several other companies followed *Amazon* and nowadays, recommendations make part of our quotidian in internet sites like *Facebook, Netflix, Spotify, YouTube, TripAdvisor*, but also in some local e-commerce platforms.

The data revolution is shifting society as a whole, from commercials to political campaigns [1], from healthcare [2] to weather forecasts. These data-driven decision-making techniques are helping companies expand their revenues by making targeted sales or improving their business decisions such as predicting available stock ruptures or future trends. In fact, companies that were mostly data-driven had 6% higher profits and 4% higher productivity than average [3]. That being said, the importance of hit products should never be undervalued. Hits are here to stay, and companies must invest in them. However, some consideration should always be directed to niche markets. The increase in supply must be supplemented by a recommendation system in any company that desires to thrive in the current competitive environment of e-commerce, aiming customer engagement and consequent increase in profitability.

2   State of the Art

A recommender system is comprised of one or more types of recommendations. This section gives an overview of the most popular and accepted approaches and those best suited for this case study.

Collaborative Filtering. Often referred as *people-to-people correlation* [9], this approach is based on the principle that two or more individuals sharing similar interests in one kind of product, tend to become interested in similar products from another kind (see Fig. 3).

Content-Based Filtering. This approach depends on the similarity between products. It is built upon the concept "Show me more of what I have liked" [10] (see Fig. 4). The association between such products can be calculated by their related features.

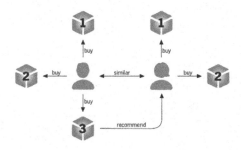

Fig. 3. Logic behind Collaborative filtering.

Fig. 4. Content-based filtering connections.

Hybrid Systems. As the name suggests, hybrid systems combine two or more systems. The idea behind this combination is that the shortcomings of one system are overcome by the other so that the final prediction is more accurate. *Netflix* employs a perfect example of a hybrid recommender system as it couples collaborative and content-based filtering, recommending content based on previous content and similarity with other users [10].

Cart Suggestions. Unlike previous approaches, this method is not personalised to each customer. Instead, it depends only on the product and the connections established between other products through orders, implementing a Market Basket Analysis recommendation.

This approach builds association rules between products through frequent patterns in order items. This helps to build recommendations based on the concept of "Users who buy this product, frequently buy these too" [11].

Product Similarity. A content-based model for product similarities can be an exciting approach for this type of e-commerce platform. Contrary to the content-based filtering described earlier, these recommendations are not personalised to each customer and are solely based on product traits. These suggestions can drive customers that are observing a certain product that it is not precisely what they desired, to a similar product, more fit to the likes of the customer.

Popularity. Suggesting niche products is key in today's online business model, but the demand for hit products is still significant. Hence, building recommendations based on product popularity is always a solid bet and should not be undervalued. These recommendations can be based on product information, such as the number of orders, ordered quantity, product rating, product hits (number of visits on the product page), among others.

3 Proposal

The main goal is to implement a recommender engine, containing multiple types of suggestions that enhances profitability on products available in a specific e-commerce platform. The platform in question sells clothing and sports apparel, specialised in the surf market.

3.1 Data

To aid the process of finding knowledge, data often comes structured into two entities and one or more types of interactions between those two. In this platform, the entities are **products** and **clients**, interacting through **orders** and are stored in *Elasticsearch*. These three entities contain attributes (Table 1) with information on each order, product or client.

Table 1. Dataset details.

Entities	Products	Clients	Orders
Rows	8860	23889	14945
Attributes	product_id	client_id	order_item_id
	parent_id	birth_date	order_id
	name	gender	product_id
	stock	country	client_id
	hits	zip	quantity
	manufacturer	locality	created_at
	categories		ship_locality
	color		ship_country
	size		hits_count

3.2 Applicability

With the available data, these are the recommendations that can be implemented in the platform. This wide range of suggestions is suited for the clothing e-commerce scene and satisfies multiple areas of user activity in the platform:

- **Hybrid:** These recommendations could be implemented in a product list, ordering the displayed products by the probability of purchasing, or in a homepage carousel.

- **Cart suggestions:** As the name suggests, this type of recommendations could be implemented in a shopping cart, suggesting products that are often purchased together with the products already in the cart.
- **Product similarity:** These recommendations could be implemented in product details pages, redirecting customers to similar and perhaps more interesting products.
- **Popularity:** This recommender is often displayed in homepages.

3.3   The 'Update' Dilemma

Some services, like *YouTube*, need their recommendations continually updated due to its regular visitors and the hours of videos uploaded every second [12]. The content consumption is constant, ergo, user taste is in permanent transformation.

However, our platform is visited occasionally when customers want to purchase specific products. In these circumstances, the period of inactivity is tremendously higher than the period of activity for the vast majority of clients. Besides that, the limited computational resources available for this project make this approach seems unattractive.

At this stage, all recommendations will be updated on a specific day and time with the aid of *Crontab*, a time-based job scheduler for *Unix* systems. Given a specific time and day or a time interval, *Crontab* triggers a designated command, executing all *Python* files responsible for generating recommendations.

3.4   Cold-Start Problem

Cold-start is a frequent problem when dealing with recommender systems. This occurs when it cannot draw any inferences for clients that have not yet sufficient information, in this case, clients with no orders.

Orders are the foundation of hybrid recommendations and cart suggestions, ergo, clients without orders will not be able to get this kind of recommendations. Still, similar and popular products do not depend on the orders of a specific client, making these recommendations available for every client. The cold-start problem on hybrid and cart recommendations could be surmounted with an initial questionnaire to build a user profile, but its implementation outruns the scope of this project.

3.5   Architecture

The architecture of the recommender engine comprises not only the engendering of recommendations but also the communication between components. The engine's workflow is complex and can be broken into the subsequent stages:

1. Query data (products, clients and orders) from the *Elasticsearch* search engine necessary to train recommendation models.

2. Get the recommender engine configuration variables from *MongoDB*. These variables add customisation to the recommendations so that the platform can develop recommendations more suited to variable conditions of the business.
3. Begin the process of Exploratory Data Analysis (EDA) and Data Preparation (DP), performing all necessary and suitable operations to improve the quality of the data.
4. Create predictive models for every recommendation featured in the recommender engine and evaluate the accuracy of those models using tested and suitable metrics for recommender systems.
5. Save recommendations in *MongoDB* to be prepared on platform demand.

Furthermore, *Python* was chosen as the programming language to build recommendation models as it offers powerful libraries. Weinberger stated that *"Python is not only the leading way to learn how to build a recommendation system, it is also one of the best ways to build a recommendation system in general"* [13]. *Elasticsearch* was chosen because of its unstructured and adaptable architecture and its speed of querying, essential when dealing with variable and large amounts of data. *MongoDB* is the preferred recommendation storage database due to its high query performance, scalability and availability, essential to retrieve the recommendations to the e-commerce platform.

The recommender engine workflow commented above can be translated to the architectural design in the Fig. 5.

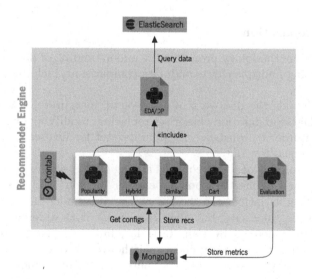

Fig. 5. Recommender engine architecture.

4 Implementation

The implementation of the designed architecture followed a defined data mining process called CRISP-DM. This methodology is independent of both the industry sector and the technology used, and aims to make large data mining projects, less costly, more reliable, and more manageable [14].

4.1 Configurations

In order to make the recommender engine customised to the platform's current business situation, a set of configurations are stored in a *MongoDB* document. This is the first data to be parsed by the recommender engine. The engine can be customised by the following configurations:

1. **evaluate:** Boolean that defines if the models accuracy should be calculated.
2. **min_score:** All recommended products have a score (float 0–1) of its certainty. This variable removes all recommendations below that score (for the 'cart suggestions' model, this variable is called 'min_conf').
3. **last_month_orders:** Use only orders from the last X months. This improves recommendations accuracy as orders from a long time ago usually do not reflect the current user's taste.
4. **bought_products:** Boolean that defines if already purchased products should be recommended to a client.

4.2 Data Preparation

The CRISP-DM methodology promotes the understanding of data. This process is done by plotting multiple charts and data comparisons, including the long tail graph in Fig. 1.

This process aids the discovery of incomplete, inaccurate or irrelevant portions of data. Data preparation comprises replacing, modifying or deleting these portions [15] in order to construct the final dataset for further model construction, from the initial raw data.

While exploring the locality frequency among clients, an incoherence between city names appears. In effect, Lisbon appears as "Lisbon", "Lisboa", "Lx", "lisboa", among other variations. To the human eye, it is understandable that these names refer to the same city, but algorithms treat it as different values. Also, clients' age is only defined by their birthday, but, in order to establish connections and similarities between clients, birthdays are converted into two new attributes: age and age group (children, youth, adults and seniors [16]).

4.3 Recommendations

After the study on recommendation approaches for this e-commerce platform, follows the process of implementing the philosophy of these approaches in *Python*.

- **Popularity:** This recommendation is unvarying for users of the platform. To compute the most popular products in store, a score for each order is estimated based on the following parameters: ordered quantity and order status (delivered, cancelled, etc.). Next, product scores are computed, by grouping orders by product. Product hits are also taken in consideration in product scores. These four parameters have different weights on the score, according to their significance. After computed, scores are normalised (0–1) and products sorted by score.

- **Hybrid:** This recommendation is a compound of content-based and collaborative recommended products. For the **content-based** recommendation, all product features were grouped in a single string, containing its product name, main size (more generic), size (more specific), its categories, colour, manufacturer, and season (when it sells the most). *TF-IDF* is a statistic widely adopted in search engines and was used to compute frequent terms among clients purchases. This technique converts unstructured text into a vector, where each word represents a position in the vector, and the value measures its relevance for a product [17]. When applying cosine similarity to those vectors, it translated to product similarity. This similarity is used to compute product profiles by using scores to define the most similar products. After that, the last step is to build a client profile based on profiles of products purchased in the past and use it as a recommendation. The **collaborative** recommendation is based on the latent factor model *Singular Value Decomposition* (SVD) that creates and compresses a client-product matrix (filled with the scores of orders) into a low-dimensional representation in terms of latent factors. One advantage of this approach is that it deals with a much smaller matrix, instead of a high dimensional matrix containing an abundant number of missing values, increasing computational effort [17]. After the factorisation, the matrix is reconstructed by multiplying its factors. The resulting matrix results in scores (predictions) for products the client have not yet purchased, that will be exploited for recommendations. To compute hybrid recommendations, the lists of recommended products are merged and a mean of both scores is computed and normalised (0–1).

- **Similar Products:** This approach follows the same *TF-IDF* technique described above for content-based recommendations. However, here, client profiles are not computed, using only products profiles to build recommendations based on products with similar frequent terms (attributes).

- **Cart Suggestions:** As explained, this suggestion relies on association rules. Frequent item sets for each product are computed from clients orders, using the *Apriori* algorithm. These item sets are converted into association rules, and consequents of those rules are recommended as products frequently bought simultaneously. Confidence indicates how often the obtained rule was found to be true, and will be used as a normalised score for the recommendation.

- **Save recommendations:** These four models are executed simultaneously, producing recommendations for every client (hybrid and popularity) or every product (similar products and cart suggestions), saving every recommendation by replacing the previous, thus not increasing exponentially storage space. Each model is saved in its own collection named by the platform and the model (i.e. platform_popularity). Furthermore, recommendations are stored as a JSON where the key is the client or product ID, following a list of IDs for recommended products sorted by score.

Every document is saved separately and atomically so there is no downtime in the recommendations. This means that while the recommender engine is updating the recommendations, the platform remains usable.

5 Results

The recommender engine comprises four different and versatile models, that cover a wide range of areas on our platform. Triggered by a *Crontab* job, the engine produces four collections with the following number of recommendations:

Table 2. Statistics on saved recommendations.

Model	Popularity	Hybrid	Similar	Cart
Recommendations	1[a]	5996	787	183
Total size	25.55 kB	62.49 MB	449.06 kB	167.37 kB
Recom. coverage[b]	100% of clients	25% of clients	9% of products	2% of products

[a] Popularity has one recommendation because popular products are the same for every client.
[b] The low coverage of some recommendations is due to some clients that have never ordered products or products that are set as non buyable, therefore, excluded from the recommendation.

5.1 Evaluation

Evaluation is a critical step of CRISP-DM, as it allows to compare algorithms and potentiate continuously improvement of the models, aiming accuracy gain. One common practice is to divide orders into test and train sets, generate predictions based on the train set, and compare with the actual orders in the test set.

Time plays an important role in e-commerce platforms, ergo, orders are sorted by date and then split (80% for training, 20% for testing). Evaluating using hold-out allows emulating a scenario where we are predicting actual customer future behaviour. Additionally, the low computational effort of this technique facilitates the engine, as it generates recommendations for every product and client concomitantly.

Among the most adopted evaluation metrics in recommender systems, we find Recall. Recall@N is the rate of the ordered products in the test set that are among the top-N recommended products. The results vary from 0 to 1 being that 1 means that every ordered product is displayed in the recommendation. This process comprises the following steps:

1. Generate recommendations for every client using orders from the train set.
2. Get ordered products in test set.
3. For every ordered product in test set verify if it's among the top-N recommendations.
4. Compute recall.

This evaluation process was developed to every model in this recommender system. The results for the top-10 recommendations are exposed in Table 3.

Table 3. Evaluation results.

Model	Popularity	Hybrid			Similar	Cart
		Content-based	Collaborative	Hybrid		
Recall@10	0.22	0.37	0.30	0.41	0.04	0.08

These results confirm two theories stated in the Proposal section. First, recommending hit products using the popularity recommender achieves very decent results, as hit products still represent a major portion of sales in e-commerce platforms. Second, merging recommender systems results in greater accuracy, as we can observe in the hybrid recommendation.

The execution of the recommender engine completed with a run time of 10 min and 16 s. This includes all CRISP-DM stages, from the preparation of all data in Table 1, modelling and storing of recommendations in Table 2, and the model accuracy evaluation explained in the section below.

6 Conclusions

As online interactions increase, demand for more customisation and personalised content rises too. Connecting clients to fresh and undiscovered products through recommendations is essential to develop niche markets and increase revenue. The implementation of such strategies should be as hybrid and versatile as possible, combining multiple types of recommendations, striving an increase customer involvement and satisfaction.

This engine comprises four distinct types of recommendations, each suitable to every platform context, employing established algorithms and embracing multiple attributes present in the available dataset. All steps were verified and followed a defined methodology, producing results that comply with the proposal and satisfactory accuracy results of the models. However, the importance of the accuracy results should not be overemphasised, as the true virtue of recommendations is not only to predict what products clients will purchase but to suggest possibly unknown but appealing products.

The personalisation allowed by this recommender engine, combined with its customisation through the available configurations, make this approach essential in an e-commerce platform aiming the increase in client satisfaction through the adoption of a data-driven decision-making strategy.

Acknowledgments. This work has been supported by FCT – Fundação para a Ciência e Tecnologia within the RD Units Project Scope: UIDB/00319/2020.

References

1. Nickerson, D., Rogers, T.: Political campaigns and big data. In: Harvard Kennedy School Faculty Research Working Paper Series (2013)
2. Cruz, M., Esteves, M., Peixoto, H., Abelha, A., Machado, J.: Application of data mining for the prediction of prophylactic measures in patients at risk of deep vein thrombosis. In: Rocha, Á., Adeli, H., Reis, L.P., Costanzo, S. (eds.) WorldCIST'19 2019. AISC, vol. 932, pp. 557–567. Springer, Cham (2019). https://doi.org/10. 1007/978-3-030-16187-3_54
3. McAfee, A., Brynjolfsson, E.: What makes a company good at IT? Wall Street J (2011). https://www.wsj.com/articles/SB10001424052748704547804576260781324726782. Accessed 26 May 2020
4. Anderson, C.: The Long Tail, 1st edn. Hyperion, New York (2006)
5. Kruse, K.: The 80/20 Rule and How It Can Change Your Life. Forbes (2016). https://www.forbes.com/sites/kevinkruse/2016/03/07/80-20-rule/
6. Linden, G., Jacobi, J., Benson, E.: Collaborative recommendations using item-to-item similarity mappings. US Patent 6,266,649 to Amazon Technologies Inc. (1998)
7. Sharma, A., Hofman, J. M., Watts, D. J.: Estimating the causal impact of recommendation systems from observational data. In: Proceedings of the 6th ACM Conference on Economics and Computation, pp. 453–470 (2015)
8. Brynjolfsson, E., Hu, Y. J., Smith, M. D.: The longer tail: the changing shape of Amazon's sales distribution curve. SSRN Electron. J. (2010)
9. Ekstrand, M.D., Riedl, J.T., Konstan, J.A.: Collaborative Filtering Recommender Systems. Foundations and Trends in Human-Computer Interaction, pp. 81–173 (2011)
10. Sharma, R., Singh, R.: Evolution of recommender systems from ancient times to modern era: a survey. Indian J. Sci. Technol. **9**, 1–12 (2016)
11. Tatiana K., Mikhail M.: Market basket analysis of heterogeneous data sources for recommendation system improvement. In: 7th International Young Scientist Conference on Computational Science, pp. 246–254 (2018)
12. Covington, P., Adams, J., Sargin, E.: Deep neural networks for YouTube recommendations. In: Proceedings of the 10th ACM Conference on Recommender Systems, pp. 191–198 (2016)
13. Weinberger, M.: Why you need python machine learning to build a recommendation system. Dataconomy (2018)
14. Wirth, R., Hipp, J.: CRISP-DM: towards a standard process model for data mining. In: Proceedings of the 4th International Conference on the Practical Application of Knowledge Discovery and Data Mining, pp. 29–39 (2000)
15. Mayo, M.: 7 Steps to Mastering Data Preparation for Machine Learning with Python. KDNuggets (2019). https://www.kdnuggets.com/2019/06/7-steps-mastering-data-preparation-python.html
16. Age Categories, Life Cycle Groupings. StatCan: National Statistical Office of Canada(2017). https://www.statcan.gc.ca/eng/concepts/definitions/age2
17. Moreira, G.: Recommender Systems in Python 101. Kaggle (2019). https://www.kaggle.com/gspmoreira/recommender-systems-in-python-101

Free-Floating Carsharing in **SimFleet**

Pasqual Martí$^{(\boxtimes)}$, Jaume Jordán ⓘ, Javier Palanca ⓘ, and Vicente Julian ⓘ

Valencian Research Institute for Artificial Intelligence (VRAIN),
Universitat Politècnica de València, Camino de Vera s/n, 46022 Valencia, Spain
pasmargi@inf.upv.es, {jjordan,jpalanca,vinglada}@dsic.upv.es
http://vrain.upv.es/

Abstract. With the number of people that live in cities increasing every
year, the complexity of urban traffic increased as well, making it more
necessary than ever to find solutions that are good for the citizens,
energy-efficient, and environmentally friendly. One of the systems that
are becoming more popular is carsharing, specifically free-floating carsharing: fleets of cars that are parked around a city that can be temporarily booked for private use within the borders of the city by the system
users. In this work, we implement one of these systems over SimFleet, an
agent-based fleet simulator. We present how the original SimFleet agents
are adapted to our system and how they interact with each other, as
well as the strategies they follow to address the urban traffic problem
efficiently. Our implementation for the simulation of free-floating carsharing scenarios is crucial for companies or municipalities to make the
necessary tests before deploying the systems in real life.

Keywords: Multi-agent system · Simulation · Transportation ·
Electric vehicle · Smart city · Urban fleets · Carsharing

1 Introduction

The use of privately owned vehicles in cities is becoming every time more incon-
venient for the citizens. As the concerns about carbon dioxide emissions increase,
cities adapt by creating more green areas, penalising or completely banning the
use of private vehicles in their city centre, and encouraging both public transport and electric vehicles. Besides that, there are problems that are inherent for
vehicle owners like the lack of parking space, which implies on many occasions
the need for paying for a private space. To such disbursement, one must add fuel
or electricity expenses as well as vehicle maintenance.

Considering all of these inconveniences, a good amount of the people that
live in cities opt by not buying a vehicle. For such users, public transport usually
fulfils their displacement needs. However, there are cities in which, because of
their structures or by the lack of resources, the public transport services are not
enough to comply with the needs of citizens. Besides that, there might be some
trips which entail needs that can not be provided by public transport.

© Springer Nature Switzerland AG 2020
C. Analide et al. (Eds.): IDEAL 2020, LNCS 12489, pp. 221–232, 2020.
https://doi.org/10.1007/978-3-030-62362-3_20

Aiming to solve all of the previously mentioned issues, carsharing systems were proposed [10]. In these systems, private vehicles are owned by an enterprise, which rents them temporarily to their customers. In general, the vehicles are parked in specific locations and must be either returned to the original location or parked in a different predetermined location. Also, carsharing companies are transitioning to the use of electric vehicles, since they offer great performance in urban areas. Europe accounts for about 50% of the global carsharing market and is expected to grow further to 15 million users by 2020 [5].

Taking this into account, the system we will describe in our work is a modification over the original carsharing system: a free-floating carsharing system [7]. These systems are much more flexible for the users since their vehicles can be picked up and parked anywhere within a specified urban area. Customers, that have access to vehicle locations, book a vehicle for a determined amount of time. The company takes care of vehicle relocation and recharge if necessary. For a monthly fee, these companies provide the benefits of owning a private vehicle without its drawbacks.

Free-floating carsharing systems need to be tested and improved to provide better service, for example, reducing the distance users have to walk to a vehicle and minimising waiting times to provide better service. While testing on real cities might be expensive and sometimes completely impossible, the use of simulators can be of help. In this work, we design and implement a carsharing system of this type for SimFleet [14], an agent-based simulator for urban fleets, hoping that it provides a platform to test a carsharing system over real city infrastructures.

The rest of the paper is structured as follows: Sect. 2 analyses previous works in the simulation of systems of this kind; Sect. 3 describes SimFleet, which is an agent-based simulator for urban fleets; Sect. 4 details the proposed extension of SimFleet for free-floating carsharing systems and its implementation aspects, and Sect. 5 illustrates different possible experiments; finally, Sect. 6 shows some conclusions and future works.

2   Related Work

In the last few years, several works have appeared in the line of being able to simulate the behaviour of shared vehicle fleets. These simulations aim to optimise resources by comparing different strategies when allocating vehicles or deciding on their location. In this way, we can find in the literature several proposals of simulators implemented from scratch or implementations made on known simulation software. In any case, most of the proposals make use of Agent-Based Simulation (ABS) techniques, which have several interesting properties which makes it useful for many domains. ABS supports structure-preserving modelling of the simulated reality, simulation of pro-active behaviour, parallel computations, and very dynamic simulation scenarios [4].

At a general level, we can find ABS software that facilitates to some extent the development of simulations of any type of strategy or transport model such as the MATSim tool[1] and the AnyLogic[2] simulation modelling software. Similarly, the system proposed in [15] deals with the problem of relocating unused vehicles to provide a high level of accessibility to carsharing services with few vehicles.

With regard to specific work to model the behaviour of shared vehicle fleets, there are various approaches in recent years such as the work proposed in [2], implemented on the MATSim framework. This work analyses the interaction of two free-floating carsharing companies competing in the city of Zurich (Switzerland). Another example in MATSim is the work proposed in [11], where a framework that integrates different carsharing services with the typical problems of electric vehicles such as power sharing capabilities, smart charging policies, booking services, fleet redistribution, and membership management.

A more general view can be found in the work proposed in [9], where an agent-based model simulates flexible demand responsive shared transport services. The goal is to obtain a platform to simulate strategic planning in a real context. The system has been tested in the city of Ragusa (Italy). Another approach is the proposed [16] where a reputation system is implemented using intelligent software agents. The goal of the system is to identify good driving behaviours which would get discounts in carsharing fees to make it more attractive and to promote its use. In [8], authors explore different model implementations to assess the potential of predicting free-floating cars from the non-habitual user population. On the other hand, authors in [3] face the problem of finding the optimal location of charging stations, and the design of smart car return policies using data from a carsharing system in the city of Turin (Italy). Finally, the work presented in [1] evaluates the free-floating carsharing system through the analysis of the temporal distribution of the main flows. In this sense, the work provides the first spatial analysis of systems of this kind in the city of Madrid (Spain) using rental data collected from the operators.

As this analysis reveals, free-floating carsharing systems are currently a hot topic for researchers. There are different implementations of models or strategies that are difficult to compare and often are adapted to specific problems and cities (difficult to extrapolate). In this work, we propose a simulation environment for this type of systems that allows flexible and adaptable modelling in order to analyse in detail the operation of this type of fleets.

3   SimFleet

SimFleet [14] is an agent-based urban fleet simulator built on the SPADE platform [6]. This fleet simulator was built to allow complex simulations over cities where a large number of agents interact to perform fleet coordination. SimFleet uses the multi-agent systems paradigm to allow the user to get access to autonomous, pro-active, intelligent, communicative entities. SPADE is a multi-agent systems platform that provides these features using a simple but powerful interface, which is why it was chosen for the development of SimFleet.

---

[1] https://www.matsim.org.
[2] https://www.anylogic.com.

SPADE allows the creation of autonomous agents that communicate using the open XMPP instant messaging protocol [17]. This protocol is the standard protocol of the IETF and W3C for instant messaging communication and it (or some variant of it) is used in such important communication platforms as WhatsApp, Facebook or Google Talk. SPADE agents have also a web-based interface to create custom app frontends for agents, which is also used by SimFleet to show how every agent is moving through the city in a map that represents all routes made by agents.

Finally, SimFleet is based on the Strategy design pattern, which allows the user to introduce new behaviours to the SimFleet agents without the need of modifying the simulator. This design pattern is used to introduce new algorithms that follow a common interface. In this case, introducing new coordination algorithms to an agent is as simple as building a *StrategyBehaviour* and loading it at SimFleet startup. SimFleet also provides some common agent classes that can be used (or inherit from them) to create a simulation. These agents represent the entities involved in a fleet simulator: fleet managers, transports, customers, and service directory. Next, we shortly describe these classes and how they interact.

Fleet Manager Agents. SimFleet simulates an environment where there can be different kinds of fleets that provide services in a city. Each fleet has a fleet manager who takes care of the fleet, allows transports to be accepted in the fleet and puts customers and carriers in touch with each other to provide a service. An example of a fleet manager is a taxi company call centre or a goods transport operator with trucks.

Transport Agents. These agents represent the vehicles that are moving through the city providing services. SimFleet supports any kind of city transport such as cars, trucks, taxis, electric cars, skateboards or even drones. However, the user can customise the kind of transport for its simulations. Transport may or may not belong to a fleet, but belonging to a fleet brings them some benefits like being found more easily and having a coordinator for its operations. Transport agents receive transport requests from customers and, if free, they will pick the customer up (or the package) and drive to its destination. However, before attending a request, a transport agent will make sure it has enough autonomy to do the whole trip. If not, the agent drops the request and goes to recharge its batteries or refuel to the nearest station. After serving one request, the agent awaits for more requests until the simulation is finished.

Directory Agent. SimFleet has a place where services provided during the simulation may be registered to be found later. This is done by the directory agent, which offers this subscription and search service. Usually, fleets are registered in this directory to be found by customers and to allow new transports to find them and sign up for the fleet.

Customer Agents. Customers are the entities that want to perform an operation: calling a taxi to move from one place to another, send a package to a destination, etc. This entity is represented by the customer agent. In SimFleet customers do not have the ability to move. They call a transport service which goes to the customer's position, picks up the package (or customer in case of a taxi, a bus, etc.), and transports the goods to a destination. Customer agents depend completely on the transport agents. To get a transport service the customer looks for an appropriate fleet in the directory and contacts to its fleet manager to get a transport service for the customer. The fleet manager broadcasts the requests to some or all of their registered transports (depending on its strategy) and any transport interested in attending it will send a proposal to the customer, who has to accept or refuse it (depending on the customer's strategy too). The customer waits until the transport agent picks it up and, once they arrive at the destination, it stops its execution.

4   Free-Floating Carsharing System

Our proposal is to design and implement a free-floating carsharing system for the simulation software SimFleet. In this section, we describe the system and its agents, and the design of strategies for the carsharing agents in SimFleet.

4.1   System Description

A free-floating carsharing system has a set of *customers* who will use a set of *transports* to travel. Transports will be located anywhere within the borders of a certain city or urban area. Customers know the location of available transports at any time and can issue a booking using, for example, a website or an application. After making a booking request, customers wait for confirmation. If their booking has been accepted, they move to the vehicle (transport) and can access it to travel anywhere. Once the customer has finished using the vehicle, they must leave it properly parked within the mentioned urban area.

There are some problems inherent to free-floating carsharing systems. Since vehicles can be parked anywhere, it may happen that at any given time some users find all the vehicles parked too far away to be able to use them. In an extreme case, the available vehicles may be even further away than their destination and therefore it would not make sense for them to use the service. A system of such flexibility requires careful attention to detect and resolve such situations. In real life, carsharing companies relocate their vehicles from time to time to ensure proper distribution.

Another problem is the recharging or refuelling of the vehicle. Ideally, a customer should find the transport fully charged (refuelled) or at least with enough energy (or fuel) to complete his journey. We understand that carsharing companies are aware of their vehicles charging and periodically charge (or refuel) those who need it, driving them to charging (or petrol) stations or simply charging (or refuelling) them on-site with a generator (or deposit).

To address these issues, the authors in [12] present a "staff" agent who deals with the recharging of vehicles. In addition, customers only book vehicles that will be available within 30 min of the booking being issued. The aim of SimFleet, however, is the development and comparison of the agents' strategies. That is why we decided to give the above-mentioned problems very simple solutions that a future user can improve them to what he or she considers best.

On the one hand, a maximum walking distance for customers has been introduced which indicates how many meters they are willing to walk to get a vehicle. This parameter can be manually defined for each customer in the simulation or not specified to ignore the restriction. A customer with this defined value will not book or use transports that are further away than what the value indicates. This may result in some customers not making their journey if they do not have a vehicle within reach at any time, and hence, the simulation would end with some customers not having reached their destination. In this way, a SimFleet user can also evaluate if his or her initial vehicle distribution is suitable for a certain customer distribution, or even, it could be considered to make relocation of the vehicles in these cases. We must note that the check of the distance between a customer and the available vehicles is done by calculating the straight line distance between them. As it would be expected, the distance that the customer ends up covering is usually greater than this estimate, so, in some cases, customers will exceed maximum walking distance restriction, since the actual route from its origin to the vehicle is usually longer than the straight line distance. We use an optimistic estimate of the distance to avoid calculating the actual route to every available transport, which would overload the routing server in simulations with many agents and increase the simulation time.

On the other hand, to simplify the experiments performed in this work, we assume that the vehicles are recharged on site each time they finish a trip, i.e., it is assumed that the vehicle has its full autonomy for each new customer.

4.2　Agents

Fleet Manager. It acts as the "application" by which customers check available vehicles and their positions. For that, it maintains the updated information about the location and state of every transport in their fleet and sends this information to any customer that requests it.

Transports. These agents act as the vehicles of the system of any type; e-cars by default. Transports are registered to a fleet and managed by their Fleet Manager, whom they will periodically inform about their location and status. Once a booking request of any customer arrives to a transport, it must reply accepting or rejecting it depending on their state. Although ideally customers would only send requests to available transports, in a multi-agent system it may happen that a booking request arrives to the transport when it has already been booked by another customer, or even when it is being used. In such cases, the transport will reject the request. If the transport accepts the request it will change its status to "booked" and wait for its assigned customer to arrive. When the customer arrives, the transport allows him to enter and drives him to its desired location. Upon arriving at the trip destination, the customer finishes using the transport, making it available again.

Customers. Customer agents act as the carsharing system users. Every customer is in its origin position and it has a destination that must reach. For doing so, upon spawning, the customer asks for the available transports to the Fleet Manager. If there are no available transports, the customer would wait for a determined amount of time before asking the Fleet Manager again. Once the customer receives the transports information, it can make a booking request to the vehicle of his choice. If the request gets accepted, the customer will walk to the location of his booked vehicle and, once accessed to it, drive it to his destination. After completing the trip, the customer has achieved his goal and so it stops its execution. As we mentioned earlier, if the customer has a maximum walking distance defined, it will only book transports located within his reach. The simulation will finish once all Customer agents have reached their destination or if the remaining Customer agents can not book any of the available cars, in which case the impossibility of completion of the simulation will be indicated in the output of the execution.

4.3   Design of Intelligent Strategies

SimFleet is coded in Python 3 and, as commented above, makes use of SPADE as a base for the implementation of its agents. A SPADE agent can have one or many *behaviours* that, upon execution, will define the actions of the agent. Besides that, a series of agent-specific methods can be defined for the agent and called by the behaviour execution.

Agents in SimFleet's carsharing system have a *strategy behaviour* that implements their way of interacting with the system and each other during the simulation. These strategies model how agents behave in a carsharing system by means of a finite-state machine that represents the states through which the agent travels and which implement the negotiation and decision-making algorithms to book a vehicle and use it.

The agent strategy behaviour defines its interaction with the system and the other agents. Generally, they are derived from the SPADE class `CyclicBehaviour`, which is a behaviour that keeps executing itself until its goal has been completed. To determine the actions of the agents, an attribute representing their *state* is used. Depending on its state, the agent will pay attention to certain interactions (messages) or simply ignore them. This state is usually changed by the strategy behaviour upon receiving a certain interaction, but it can also be changed internally by the agent itself. Therefore, it is very important to manage the value of the state attribute carefully or the agent behaviour may become unpredictable.

The agents' strategies are implemented as SPADE's `FSMBehaviour`, a behaviour composed of a Finite-State Machine (FSM). In this way, we designed the strategies as a FSM and match every possible value of the agent's state attribute with a different state of its Strategy Behaviour. Also, states for Transport and Customer agents were introduced to reflect their status in the system.

Next, we describe the Strategy Behaviour of the three involved agents, focusing on the actions that take place in each state and the transitions.

Fleet Manager Strategy Behaviour. The Fleet Manager agent is in an end-less state awaiting for messages. It can receive two types of message: a Transport agent informing about its state (available or booked), or a Customer agent asking for the list of available transports. Every time a message from a transport arrives, the Fleet Manager updates its internal list of available transports. When a customer asks for it, the Fleet Manager replies only with the transports whose current state is free.

Transport Strategy Behaviour. A Transport agent is initially waiting for booking requests. When it is booked, it waits for the customer to arrive at the vehicle and finally moves to the destination. A Transport agent can be in one of the following states: (1) waiting to be booked, (2) waiting for the customer to arrive, or (3) driving to the customer's destination (see transitions in Fig. 1).

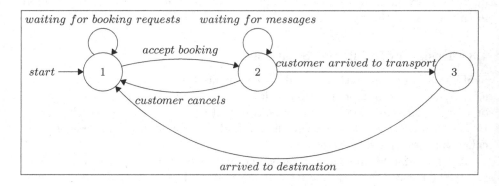

Fig. 1. Transport strategy behaviour as a FSM

Customer Strategy Behaviour. A Customer agent can be in one of the following states: (1) making a booking, (2) waiting for the booking to be accepted, (3) walking to the booked transport's position, (4) inside the transport driving to his destination, or (5) in his destination (see transitions in Fig. 2).

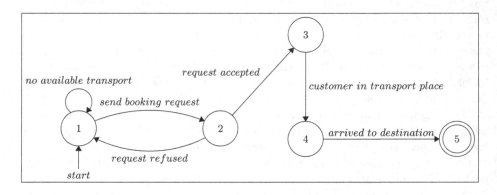

Fig. 2. Customer strategy behaviour as a FSM

Besides the agents and their strategies, some other components of SimFleet were modified such as the addition of Customer agent movement in the web application that acts as a User Interface for the visualisation of the simulation. A SimFleet version with our carsharing implementation is on Github[3].

5   Experimentation

In this section, we show the flexibility of our SimFleet extension by creating and executing different simulations. For that, we make use of the simulation generator, presented in [13], which enables SimFleet users to easily generate scenarios based on real-world data, achieving more realistic simulations. As for the evaluation of the simulations, SimFleet includes many metrics like the time a customer is waiting to be picked up or the delivery time for delivery vehicles. For our system, we use the customer walking distance, in meters; i.e., the distance the customer agent moves from its origin point to their booked vehicle's position.

To run experiments as close to reality as possible, we based the location of the customer agents in a dataset containing origin and destination points of carsharing trips in the city of Turin, Italy, compiled over a period of two months [18]. The origin and destination points of the customer agents matched the ones in the dataset. Then, a certain number of transport agents were placed among the city area following different types of distributions: a *random* distribution, which simply locates vehicles at valid points inside the city area; a *uniform* distribution, which divides the city area as a grid and places a vehicle centred inside each cell; and finally, a *radial* distribution, which places more vehicles in the centre of the city and reduces the number of them in the outer parts. Finally, concerning customer behaviour, we assume that customers always book the closest available transport.

A graphic representation of the city area considered for the simulation can be seen in Fig. 3a. The grey polygon encloses all of the points of the dataset, except the ones corresponding to the airport, which we did not consider as it is outside of the city. The points represent the origin positions of 250 customers (origin of carsharing trips in the dataset). In Fig. 3b, 3c and 3d random, uniform and radial distributions of 125 vehicles are presented, respectively.

We executed simulations with 250 customers, 125 or 250 transports distributed in one of the three aforementioned ways with maximum walking distances of 2000, 1500 and 1000 meters, and compared the mean customer walking distance and its standard deviation. The results can be seen in Table 1. Since the maximum walking distance restriction is checked over an optimistically computed distance, there are cases in which the real distance that the customer walks is considerably higher than the defined maximum. We considered these cases outliers and we do not show them since we understand that real customers would not walk such distances.

These experiments show how our proposal can be of use to free-floating carsharing managers to estimate, for instance, the most appropriate number of

---

[3] https://github.com/jaumejordan/simfleet/tree/feature/car-sharing.

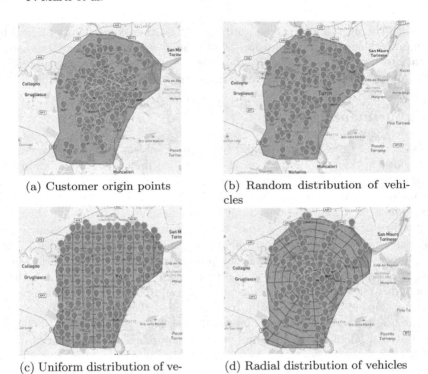

(a) Customer origin points

(b) Random distribution of vehicles

(c) Uniform distribution of vehicles

(d) Radial distribution of vehicles

Fig. 3. Turin city area considered for the simulations

Table 1. Customer walking distance (w.d.) comparison with different simulation configurations

Distribution	# customers	# transports	Max. w.d. (m)	Mean w.d. (m)	σ (m)
random	250	125	2000	1019.7	715.37
uniform				1046.86	727.02
radial				1058.79	752.05
random			1500	953.58	639.12
uniform				928.3	631.5
radial				997.76	636.06
random			1000	799.37	494.27
uniform				830.46	537.28
radial				798	492.99
random		250	2000	967.82	728.07
uniform				944.21	718.36
radial				965.36	754.59
random			1500	886.08	608.45
uniform				886.65	706.79
radial				893.07	662.45
random			1000	759.97	519.4
uniform				769.65	506.39
radial				705.17	549.52

vehicles for their fleets, and how to distribute them through the city to improve customer experience by locating vehicles as close to them as possible. In this particular instance, the alternative vehicle distributions do not present any significant differences in terms of customer walking distance. Of course, system users can define and analyse any other metric they consider relevant.

6   Conclusions

In this practical work, we have designed and implemented a free-floating carsharing system based on SimFleet, a multi-agent urban fleet simulator. With the presented system now integrated, SimFleet offers a large number of configuration options to simulate complex scenarios that reflect real-life fleet operations, giving it more potential to aid in solving urban traffic problems. Although it is not shown in our experimentation, the system can support fleets of many different types in the same simulation. For instance, we could simulate a taxi fleet together with a carsharing one in the same city. Consequently, the simulator is a great tool for evaluating transport and people distributions over actual cities and analysing the effect of different agent strategies.

In future work, we will extend the system implementing other types of carsharing as well as different types of carsharing trips. Besides, we want to explore negotiation and coordination strategies among agents for a better resolution of complex simulation scenarios. Finally, we also consider improving the current system by adding the option of vehicle relocation during the simulation, used when customers have no access to any free vehicle because of their location.

Acknowledgments. This work was partially supported by MINECO/FEDER RTI2018-095390-B-C31 project of the Spanish government. Pasqual Martí and Jaume Jordán are funded by UPV PAID-06-18 project. Jaume Jordán is also funded by grant APOSTD/2018/010 of Generalitat Valenciana - Fondo Social Europeo.

References

1. Ampudia-Renuncio, M., Guirao, B., Molina-Sánchez, R., de Álvarez, C.E.: Understanding the spatial distribution of free-floating carsharing in cities: analysis of the new Madrid experience through a web-based platform. Cities **98**, 102593 (2020)
2. Balac, M., Becker, H., Ciari, F., Axhausen, K.W.: Modeling competing free-floating carsharing operators-a case study for Zurich, Switzerland. Transp. Res. Part C: Emerg. Technol. **98**, 101–117 (2019)
3. Cocca, M., Giordano, D., Mellia, M., Vassio, L.: Free floating electric car sharing design: data driven optimisation. Pervasive Mob. Comput. **55**, 59–75 (2019)
4. Davidsson, P.: Multi agent based simulation: beyond social simulation. In: Moss, S., Davidsson, P. (eds.) MABS 2000. LNCS (LNAI), vol. 1979, pp. 97–107. Springer, Heidelberg (2000). https://doi.org/10.1007/3-540-44561-7_7
5. Deloitte, M.: Car sharing in Europe-business models, national variations, and upcoming disruptions. Dosegljivo (2017). https://www2.deloitte.com/content/dam/Deloitte/de/Documents/consumer-industrial-products/CIP-Automotive-Car-Sharing-in-Europe.pdf

6. Escrivà, M., Palanca, J., Aranda, G.: A jabber-based multi-agent system platform. In: Proceedings of the Fifth International Joint Conference on Autonomous Agents and Multiagent Systems, AAMAS 2006. pp. 1282–1284. Association for Computing Machinery, New York (2006). https://doi.org/10.1145/1160633.1160866
7. Firnkorn, J., Müller, M.: What will be the environmental effects of new free-floating car-sharing systems? The case of car2go in Ulm. Ecol. Econ. **70**(8), 1519–1528 (2011)
8. Gorka, V.C., Helmus, J.R., Lees, M.H.: Simulation of free-floating vehicle charging behaviour at public charging points. Ph.D. thesis, Hogeschool van Amsterdam (2019)
9. Inturri, G., et al.: Multi-agent simulation for planning and designing new shared mobility services. Res. Transp. Econ. **73**, 34–44 (2019)
10. Katzev, R.: Car sharing: a new approach to urban transportation problems. Anal. Soc. Issues Public Policy **3**(1), 65–86 (2003)
11. Laarabi, M.H., Bruno, R.: A generic software framework for carsharing modelling based on a large-scale multi-agent traffic simulation platform. In: Namazi-Rad, M.-R., Padgham, L., Perez, P., Nagel, K., Bazzan, A. (eds.) ABMUS 2016. LNCS (LNAI), vol. 10051, pp. 88–111. Springer, Cham (2017). https://doi.org/10.1007/978-3-319-51957-9_6
12. Niang, N.A., Trépanier, M., Frayret, J.M.: A multi-agent simulation approach to modelling a free-floating carsharing network (2020)
13. Martí, P., Jordán, J., Palanca, J., Julian, V.: Load generators for automatic simulation of urban fleets. In: De La Prieta, F., et al. (eds.) PAAMS 2020. CCIS, vol. 1233, pp. 394–405. Springer, Cham (2020). https://doi.org/10.1007/978-3-030-51999-5_33
14. Palanca, J., Terrasa, A., Carrascosa, C., Julián, V.: SimFleet: a new transport fleet simulator based on MAS. In: De La Prieta, F., et al. (eds.) PAAMS 2019. CCIS, vol. 1047, pp. 257–264. Springer, Cham (2019). https://doi.org/10.1007/978-3-030-24299-2_22
15. Paschke, S., Balać, M., Ciari, F.: Implementation of vehicle relocation for carsharing services in the multi-agent transport simulation matsim. Arbeitsberichte Verkehrs-und Raumplanung **1188** (2016)
16. Picasso, E., Postorino, M.N., Sarné, G.M.: A study to promote car-sharing by adopting a reputation system in a multi-agent context. In: WOA, pp. 13–18 (2017)
17. Saint-Andre, P.: Extensible messaging and presence protocol (XMPP): Core. RFC 6120, RFC Editor, March 2011. http://www.rfc-editor.org/rfc/rfc6120.txt
18. Vassio, L., Giordano, D., Mellia, M., Cocca, M.: Data for: Free Floating Electric Car Sharing Design: Data Driven Optimisation. Anonymized datasaset of 2 months of trips of car sharing users in the city of Turin, Mendeley Data, v1 (2019). https://doi.org/10.17632/drtn5499j2.1

Fatigue Detection in Strength Exercises
for Older People

J. A. Rincon[1][(✉)], A. Costa[2][(✉)], P. Novais[2][(✉)], V. Julian[1][(✉)],
and C. Carrascosa[1][(✉)]

[1] Institut Valencià d'Investigació en Intel·ligència Artificial (VRAIN),
Universitat Politècnica de València, Valencia, Spain
{jrincon,carrasco}@dsic.upv.es, vjulian@upv.es
[2] ALGORITMI Centre, Universidade do Minho, Braga, Portugal
{acosta,pjon}@di.uminho.pt

Abstract. The practice of physical exercise by older people has been proven to have very beneficial effects such as reducing pain and disability and improves quality of life. Today, many older people live alone or do not have the resources to have dedicated coaches or caregivers. Automatic monitoring and control of physical exercise in older people is a key aspect today that can facilitate safe exercise. Fatigue can be a key element in detecting that an exercise may cause problems for a person or, if so, the need to stop or change the exercise. This paper explores this issue by extending a previous exercise monitoring and control system for older people to include fatigue detection.

Keywords: Wearable · Fatigue detection · Elderly · Cognitive Assistants

1 Introduction

Nowadays, the increased life span of humans may lead to a new prospective of how people live their final years [1]. And with growing amount of elder people grows the interest to improving their quality of life [2]. This poses a question: how to monitor their habits (healthy and otherwise) and should be they be checked by an expert?

One path is to have the Human in-the-loop in intelligent systems via *Cognitive Assistants*. This kind of systems can take shape of software, accessible through some common device such as an smartphone, or hardware, using for instance assistant robots [3]. With the help of trained medical personnel these systems can be deployed at an elderly person's house and be able to monitor that person with minimal human intrusion. Furthermore, these systems can actively keep company and entertain the users throughout their daily tasks.

This line of work is presented in a previous project: ME^3CA [4]. ME^3CA is a recommending and monitoring system of physical exercises according to a user physical needs. Additionally, the system adapts to how the users exercises:

© Springer Nature Switzerland AG 2020
C. Analide et al. (Eds.): IDEAL 2020, LNCS 12489, pp. 233–244, 2020.
https://doi.org/10.1007/978-3-030-62362-3_21

if they bore him/her, or if they suppose an excessive physical stress. Then the system tries to look for different exercises that may be addressed to improve the same physical features but in a different way. The present work is an extension of this system, where we try to improve the exercises recommendation. We have improved the system in two ways: being able to assess the amount of effort the user has performing exercises, and evaluate the user's fatigue so to take it into account for the next recommendation.

As in previous versions of these systems, one of our main goals is that this system uses non-invasive sensors if possible. To this we built a wearable knee sleeve that has a force measuring device. This way we can observe (using machine learning methods) the amount of force and stretch angles and detect if the users are getting tired. In this paper, show our current solution and the state of the present prototype.

The rest of the paper is composed of the following sections: Sect. 2 presents the state of the art, displaying a comparison between recent projects and our project; Sect. 3 introduces the problem description and how we have extended the ME^{3}CA to address the observed issues; Sect. 4 describes in depth the architecture and hardware and software used; finally, Sect. 5 presents the conclusions and future work.

2   State of the Art

In this section we present a collection of works that approach the same domain or features, highlighting their contributions to the state of the art.

Skotte et al. [5] shows the validation of a tri-axis accelerometer system for detection of sitting, standing, walking, walking stairs, running, and cycling activities. They have developed a software named Acti4 that detected the physical activity being performed, using threshold values of standard deviation of acceleration and the derived inclination. They claim to have achieved a high accuracy value of over 95% for the previously mentioned activities. They have also compared placement of the accelerometer at a hip and waist level (revealing that hip level is better to detect sitting positions). While the number of tracked samples is high (over 140 h) and the accuracy is great, the type of activities are very distinct and require great body movements (or none at all). Nothing is told about activities that are imminently similar or very discreet, ruling out low physical impact activities, which are the most performed by elderly people.

Hartley et al. [6] present a study using accelerometers attached to the hip and lower leg of 19 persons during 48 h in a hospital scenario. With this study they aimed at detecting if a person is lying, sitting, standing, or moving. They have used two sensors to calculate the leg pitch (bending degree) and if it is moving (horizontal movement and velocity). They report that in this environment the observed people were, over their recorded time, 61.2% lying, 35.6% sitting, 2.1% standing and 1.1% moving. Furthermore, they have conducted an acceptability test through questionnaires that resulted in 100% of satisfaction in wearing the sensors. The authors do not report directly the accuracy of the readings (false positives/false negatives) nor the thresholds used to classify the postural change (although they use quite high degree thresholds to assert the leg pitch). Moreover, the postures detected are very different from each other (apart from moving), thus missing key movements that only slightly differ from the basic postures.

Maman et al. [7] propose a method to detect fatigue of a worker in a specialised environment. To this, they use a Shimmer3 device that incorporates a heart rate sensor and four accelerometers (wrist, hip, torso, ankle). With this data and with the replication of the tasks (Parts Assembly, Supply Pickup and Insertion, Manual Material Handling) they have fusioned the distinct data to extract six features that they believe that are able to describe the tasks being performed and the level of fatigue each worker has. These features are consistent with baseline statistics and are compared to them to detect if there is an increase in fatigue. They have used 6 persons to build their data and create their logistic regression models. They state that their models have 100% accuracy in terms of fatigue prediction. This work was chosen due to its inherent interest and relation to the performance of repetitive, arduous tasks, which are similar to physical exercises. Even more interesting is the relation between fatigue and task rejection (quitting performing a task). Therefore, and due to the similar set of sensors used in our project, fatigue can be used to normalise the emotion detection.

From our research there are not many developments that associate the concept of physical exercise, emotion, and fatigue at an older age. We try to overcome this void with our project, using novel methods to model the levels of fatigue that affect differently elderly people. Reichert et al. [8] have found a significant correlation between physical exercise and emotional state, with an increase valence, energy and clam status, meaning that people that perform physical exercise are more active and more focused. Our work aims to reproduce these findings by engaging elderly people into exercising while measuring their effort and emotional state to keep them interested in the activity but reducing the possibility of injuries.

3   Problem Description

As commented, our proposal is a continuation of a previous research presented in [4]. This new research incorporates a series of devices capable of detecting and classifying the effort made in the movement carried out by users and detecting if fatigue is occurring during the performance of the exercises.

The work presented in [4] presented a system, called **ME³CA**, which was a personal assistant that tried to plan and recommend activities to older people trying to improve their physical activity. To achieve this, the assistant took into account the quality of the movement and also the emotional state of the user. With this information, the assistant tries to suggest new exercises. From a social perspective, the main goal is to maintain elderly people in the comfort of their own homes, avoiding to move them to dedicated facilities, and to closely monitor their health condition forecasting possible critical problems and acting proactively. This is done through the use of non-invasive sensors, such as accelerometers, with which ME³CA attains unequivocal information about the activities and status of users.

However, the problem detected is that the system was not able to determine whether the user was performing the exercise correctly or whether she/he was fatigued during the exercise. This aspect is quite important since it can mean the need to stop the exercise or readjust the exercises to be performed by the user suggested by the assistant.

Bearing this in mind, and as mentioned above, the proposed extension of the system tackles the problem of the analysis of the effort being made by the patient during an exercise. The level of effort allows us to know the possible fatigue of the patient and therefore be able to make a decision regarding the continuation or not of the exercise.

According to [9], elderly people should perform muscle-strengthening exercises of mid to high intensity encompassing all major muscle groups for at least two days a week. Furthermore, activities that focus on leg strength, balance and co-ordination can help elders to avoid falls, as they increase their overall strength and diminish bone and muscle loss. A crucial problem for possible automatic and effective guidance of exercises performed by older people lies in the difficulty of detecting signs of fatigue primarily in strength-related exercises.

Therefore, the proposed improvement of the **ME³CA** assistant focuses on the analysis of strength exercises where it is possible to measure fatigue. For this purpose, a new data capture environment is presented by means of low-cost sensors together with a classification process that allows the assistant to determine the degree of fatigue. This information is taken as input for the assistant in its decision-making process.

In this paper we have focused our interest on a certain type of exercise that will allow us to demonstrate the viability of the proposal. Specifically, we focus on the performance of squats, one of the most typical exercises related to strength in the elderly. In people of this age, it is frequent to have articulation problems, being hips and knees articulations that usually give problems. In the specific case of the knees, processes such as arthrosis produce pain and limitation, so it is necessary to find solutions. Exercise adapted to the capabilities of older people helps reduce pain and disability and improves quality of life. Squatting is a functional exercise, as getting up and down from a chair can be difficult for some older people with limitations.

More specifically, the type of squat to be monitored would have the following sequence of steps (see Fig. 1):

- Stand with the feet shoulder-width apart and the hands down by the sides or stretched out in front for extra balance.
- Lower by bending the knees until they're nearly at a right angle, with the thighs parallel to the floor.
- Keep the back straight and don't let the knees extend over the toes.
- The squat can be done standing in front of a chair with the feet slightly apart

Next section will describe in detail the proposed extension.

Fig. 1. Example of squats which strengthen the entire lower body and core. Source http://www.allwayshomecare.org/best-exercises-older-adults/

4   System Proposal

This section describes the system developed in order to identify fatigue during squatting exercise (see Fig. 2). The proposed system is mainly composed of a set of hardware components and a software module that controls the integrated devices. Regarding the hardware, it was divided into two parts, one of them is a knee pad which is responsible of acquiring the knee movement during the squats. The second part is a wristband which is responsible to measure the arms movement, the heart rate and, also, receive and integrate the data sent by the knee pad.

In the next subsections, these two parts of the hardware component are described in detail.

4.1   Knee Pad

This device has been developed to capture knee movements, taking into account several data obtained by different sensors. Concretely, the device integrates an accelerometer that can measure the knee acceleration in [X,Y,Z] coordinates, a gyroscope that can get the knee orientation in these coordinates [X, Y, Z] and, also, a flex sensor needed to detect the knee flexion and extension. The last component of this device is an Arduino Ble 33[1] (which has a built-in IMU) that is used to get and collect all the information obtained by the different sensors about the knee movement (Fig. 3).

---

[1] https://store.arduino.cc/arduino-nano-33-ble.

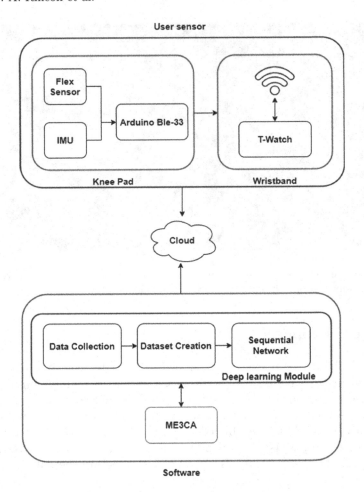

Fig. 2. General view of the proposed extension

Once the information is collected, the Arduino uses a *Low Energy Bluetooth (BLE)* in order to share the acquired data. To do this, different services has been created to be in charge of sending the data to the wristband (but it can be can be sent in a similar way to another devices such as a smartphone).

4.2 Wristband

The wristband has been developed using a T-Wathch (Fig. 4) device. This device incorporates an accelerometer, a low energy bluetooth, a photoplethysmography sensor (PPG) and a WiFi module. The wristband measures the movements of the user's wrist with the accelerometer, it calculates the heart rate and, also controls the reception of data from the knee pad. With the data received from the knee brace and the obtained form the wristband, the system obtains the

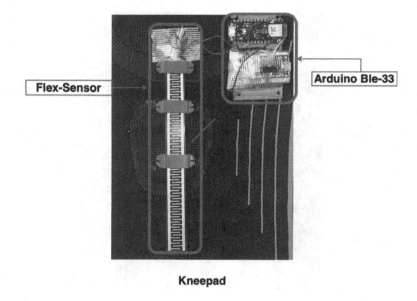

Fig. 3. View of the components of the knee pad including the force flex sensor.

average heart rate and the upper body acceleration. This data is encapsulated and sent to a web service via the WiFi module that is built into the wristband. Nevertheless, this system it is a complement of the **ME³CA**. The information acquired by these systems is used by the **ME³CA**, to improve the activity recommendation. These extra information can used by our system to get the best activity, according by the user.

4.3 Software Description

This section describes the elements that make up the different software tools which were used to classify fatigue during squatting exercises. To do this classification, it was necessary to collect the sensors data. This data was collected throughout a web-service. This web-service receives the data from the T-Watch and the knee pad and stores it to create a data set. The data set is composed by elements formed with the values of each sensor:

- *Flex Sensor value*
- *Knee pad*:
    - Accelerometer: (x, y, z) value
    - Gyroscope: (x, y, z) value
- *T-watch* Accelerometer: (x, y, z) value
- *Heartbeat Average value*

Thirty users between 40 and 50 years of age were used to collect this data. Each user was asked to do 20 squats, having to report the time when the fatigue

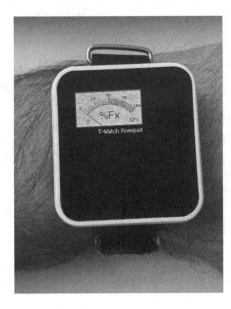

Fig. 4. View of the T-Watch model used in the proposed system.

began. The sampling was made with a frequency of 0.5 Hz, having a sample each 2 s, approximately. As commented above, the information captured at each sample is formed by 11 values (the sensors values).

Upon receipt of the data and created the data set, the next step was to train our deep learning model. To perform the classification and to determine whether the user is fatigued or not, we decided to use a 1D convolutional neural network (1D-CNN) architecture. A 1D CNN is very effective when interesting characteristics are expected from shorter (fixed length) segments of a general data set and when the location of the characteristic within the segment is not of great relevance. This network applies either to the time sequence analysis of sensor data (such as gyro or accelerometer data). It also applies to the analysis of any type of signal data over a fixed length period (such as audio signals).

- Input data: The data has been pre-processed in the web service so that each data log contains 20 time slices, each of 3 min of information from the different sensors. Within each time slice, the eleven values of the accelerometer, gyroscope, flexion sensor, wrist accelerometer and the average of the pulses per minute are stored. This results in a 6000×11 matrix.
- The first layer of 1D CNN: The first layer defines a filter (or also called feature detector) of height 10 (also called core size).
- Second 1D CNN layer: The result of the first CNN layer will be introduced in the second CNN layer. 100 different filters have been redefined to be trained at this level. Using the same structure of the first layer, the output matrix will be size 62×100.

- Max pooling layer: In order to reduce the complexity of the output and avoid overloading the data, the model has a Max pooling layer of three. This means that the size of the output matrix of this layer is only one third of the input matrix.
- Third and fourth 1D CNN layers: The third layer is added in order to learn higher level characteristics. The output matrix obtained after these two layers is a 2×160 matrix.
- Average pooling layer: This layer does not take the maximum value but the average value of two weights within the neural network. Thus the output matrix has a size of 1×160 neurons. For each characteristic detector only one weight remains in the neural network in this layer.
- Dropout layer: In order to reduce the sensitivity of the network to react to small variations in data, 0 weights have been randomly assigned to the neurons in the network. Since we chose a rate of 0.5, 50% of the neurons will receive a zero weight. Using this should further increase our accuracy in unseen data. The output of this layer is still a matrix of 1×160 neurons.
- Fully connected layer with Softmax activation: The last layer will reduce the height vector 160 to a vector of two, since we have two classes we want to predict (Squat, Fatigue). This reduction is done by another multiplication of the matrix. Softmax is used as a trigger function. In this way we force the two outputs of the neural network to be added to one. The value of the output will therefore represent the probability of each of the two classes.

Once the CNN has been performed, the data set was divided in two: 80% was selected to train and the rest (20%) was used to test. Four parameters were taken to generate the classification report: *Precision, Recall, f1-score and Support* (Table 1). The accuracy on test data was of *0.75*.

Table 1. Classification report for test data.

	Precision	Recall	f1-score	Support
Class 1	0.83	0.83	0.83	6
Class 0	0.83	0.83	0.83	6

Precision allows us to determine our classifier's ability to not label a positive instance that is actually negative, that in our case was of *0.83* to *Class 0* and *0.83* to *Calss 1*. It means that for all our instance classified the 0.83 was classified correctly. The Recall represents that for all cases were really positive, it means what percentage was correctly classified. In our case these percentages were 83% for Class 0 and Class 1. The F1 score is a weighted average of accuracy and memory such that the best score is 1.0 and the worst is 0.0. Which in our case was 0.83, for our best generated model. Finally have the Support, this measure is a way to know the number of occurrences of the class in the specified data set. In our case the Support has value of 6 for both classes, it means that our data

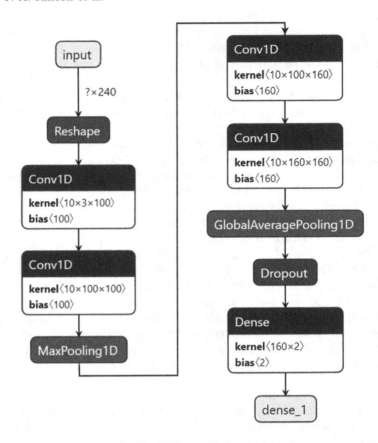

Fig. 5. CNN model structure.

set is a really well-balanced data set. After several training iterations, the best model obtained can be seen in Fig. 5.

To validate the model, we made some experiments using ten users performing squat exercises. The result obtained by this validation process can be seen in the confusion matrix shown in Fig. 6. In this matrix, columns represent the number of predictions for each class, while each row represents the instances in the real class.

It can be observed that we have a success rate around 50% per exercise. We have detected that the main problem to have a greater success rate is due to small movements generated by patients that are not part of the exercise. These movements are such as moving to the sides or moving the arm during the performance of exercises. They also include small spasms or involuntary movements. To try to solve this problem, we are getting more samples and we will try to filter or eliminate those involuntary or voluntary movements of patients.

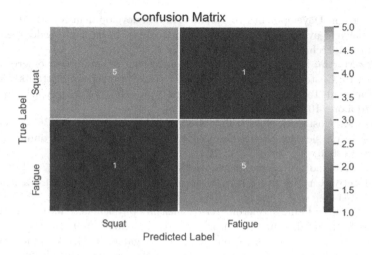

Fig. 6. Confusion matrix.

5   Conclusions

This paper focuses on the problem of analyzing the effort made by a patient, in this case elderly people, during the performance of an exercise. This analysis is an extension of a previous platform called **ME³CA**, which plans and recommends activities to older people. With the approach presented in this paper, the unsuitability of a recommended exercise can be detected by the fatigue produced in the patient. This information is of high interest in order to improve the response of an automated personal assistant.

To do this, the designed system is formed by a set of sensors that capture patient's movements during strengths exercises, and a software module that collects the sensors data and applies a classification model that tries to detect fatigue. The results on the model obtained are quite promising. The system has been tested in isolation for the time being, as it cannot be tested in an integrated way with the **ME³CA** assistant.

As future work the system will be tested with more types of strength exercises, and on the other hand, it will be integrated completely with the **ME³CA** assistant.

Acknowledgements. This work has been supported by FCT - Fundação para a Ciência e a Tecnologia within the R&D Units project scope UIDB/00319/2020 and FCT-Fundação para a Ciência e Tecnología through the Post-Doc scholarship SFRH/BPD/102696/2014 (A. Costa).

References

1. United Nations, Department of Economic and Social Affairs, Population Division. World population ageing 2019. Technical report (2020). (ST/ESA/SER.A/444)

2. United Nations Development Programme. Human development report 2019: Beyond income, beyond averages, beyond today: Inequalities in human development in the 21st century. Technical report (2019)
3. Martinez-Martin, E., del Pobil, A.P.: Personal robot assistants for elderly care: an overview. In: Costa, A., Julian, V., Novais, P. (eds.) Personal Assistants: Emerging Computational Technologies. ISRL, vol. 132, pp. 77–91. Springer, Cham (2018). https://doi.org/10.1007/978-3-319-62530-0_5
4. Rincon, J.A., Costa, A., Novais, P., Julian, V., Carrascosa, C.: ME3CA: a cognitive assistant for physical exercises that monitors emotions and the environment. Sensors **20**(3), 852 (2020)
5. Skotte, J., Korshøj, M., Kristiansen, J., Hanisch, C., Holtermann, A.: Detection of physical activity types using triaxial accelerometers. J. Phys. Act. Health **11**(1), 76–84 (2014)
6. Hartley, P., et al.: Using accelerometers to measure physical activity in older patients admitted to hospital. Curr. Gerontol. Geriatr. Res. **2018**, 1–9 (2018)
7. Maman, Z.S., Yazdi, M.A.A., Cavuoto, L.A., Megahed, F.M.: A data-driven approach to modeling physical fatigue in the workplace using wearable sensors. Appl. Ergon. **65**, 515–529 (2017)
8. Reichert, M.: Within-subject associations between mood dimensions and non-exerciseactivity: an ambulatory assessment approach using repeated real-time and objective data. Front. Psychol. **7** (2016)
9. DC Washington, (ed.): Physical Activity Guidelines for Americans, 2nd edn. U.S. Department of Health and Human Services (2018)

Trading Cryptocurrency with Deep Deterministic Policy Gradients

Evan Tummon[1]($\boxtimes$)(iD), Muhammad Adil Raja[2](iD), and Conor Ryan[2](iD)

[1] Avaya, Mervue, Galway, Ireland
`ettummon@hotmail.com`
[2] Department of Computer Science and Information Systems, University of Limerick, Limerick, Ireland
{`adil.raja,conor.ryan`}`@ul.ie`

Abstract. The volatility incorporated in cryptocurrency prices makes it difficult to earn a profit through day trading. Usually, the best strategy is to buy a cryptocurrency and hold it until the price rises over a long period. This project aims to automate short term trading using Reinforcement Learning (RL), predominantly using the Deep Deterministic Policy Gradient (DDPG) algorithm. The algorithm integrates with the BitMEX cryptocurrency exchange and uses Technical Indicators (TIs) to create an abundance of features. Training on these different features and using diverse environments proved to have mixed results, many of them being exceptionally interesting. The most peculiar model shows that it is possible to create a strategy that can beat a buy and hold strategy relatively effortlessly in terms of profit made.

Keywords: Reinforcement Learning · Cryptocurrency · Trading

1 Introduction

The common consensus is that predicting prices using Machine Learning (ML) is redundant; that is, it barely beats the benchmark of a buy and hold strategy.

Buy and hold strategies are relatively low-risk stock trading strategies. The concept is that one buys stocks and keeps them for an extended period in hopes of profit, regardless of short term fluctuations. It is one of the more straightforward strategies. So an expected outcome of this project was to at least outperform a buy and hold approach. Another concept used in trading stocks or cryptocurrencies is the use of TIs. These indicators are mathematical calculations that indicate price movements based on the stock's data. TIs were used as a feature engineering technique to expand features in this work.

This project aims to research using ML techniques to create an autonomous system that can manage a cryptocurrency portfolio. Though the first subjects that might come to mind for this are time series models or forecasting prices, the approach taken in this study is different. RL will initially be the main focus of this project, with the hopes of integrating other ML approaches to try and

© Springer Nature Switzerland AG 2020
C. Analide et al. (Eds.): IDEAL 2020, LNCS 12489, pp. 245–256, 2020.
https://doi.org/10.1007/978-3-030-62362-3_22

refine the outcome. Before any model can accomplish learning, it must have data. Integration with a cryptocurrency exchange will help retrieve this data. And BitMEX will be the exchange of choice for this integration.

The rest of this paper is organised as follows. In Sect. 2, we discuss the background of the problem at hand. In Sect. 3, we present a review of the related work. Section 4 presents our methodology. Section 5 presents the results. And finally, Sect. 6 presents the concluding remarks.

2   Background

2.1   Cryptocurrency

As mentioned by Milutinovic [8], cryptocurrencies are a digital asset. They are decentralised and extremely secure, which makes them highly admired. Bitcoin established a base for all cryptocurrencies back in 2008. A white paper [9] laid out the foundations for what cryptocurrencies are today. It explained how bitcoin is "a purely peer-to-peer version of electronic cash would allow online payments to be sent directly from one party to another without going through a financial institution" [9]. Figure 1 shows a graphical representation of bitcoins historical prices.

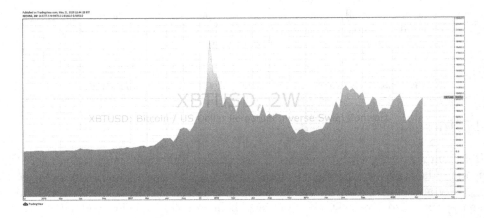

Fig. 1. Bitcoin's prices.

Some interesting aspects to note from Fig. 1 are:

– Bitcoin's price started at close to 0.
– Late 2017 it almost hit a price of 20000 US Dollars.
– Through 2018 it dropped to as low as 3200 US Dollars.
– It currently resides at close to 9000 US Dollars.

It is quite apparent as to what the best strategy would have been for bitcoin back in 2016, that is to buy and hold. A buy and hold strategy is something that the agent should be able to beat. However, it would be difficult to do in this situation where there is an upward trend. There are different techniques to get around this, but a simple one would be to train against a different dataset.

2.2   Exchanges

Cryptocurrency exchanges allow users to buy and sell cryptocurrencies in exchange for other assets, mainly fiat currencies. BitMEX will be the exchange in focus for this project, and it is where the ML model will retrieve its data from. The sheer amount of data that an ML algorithm can use to train on becomes apparent when using a cryptocurrency exchange User Interface (UI). Many cryptocurrencies store their transactions through a technology called Blockchain, which links its transactions using cryptography. Exchanges can retrieve transaction history from Blockchain networks. It is due to technologies like Blockchain that cryptocurrencies are so secure as they are immutable. Transactions can be created and read but never edited. Blockchains are the most popular technology behind cryptocurrencies. However, there are alternatives too [4].

3   Related Work

There is an abundance of literature on using ML to predict the prices of stocks. Papers on cryptocurrency prediction are still few. Consultation of stock papers when considering cryptocurrency problems is possible here. Reason being there is a tremendous correlation between stock and cryptocurrency.

Currently, the most popular way to predict prices is through using time series forecasting algorithms. Auto-Regressive Integrated Moving Average (ARIMA) being the most popular. ARIMA was most recently used in [6].

The authors employed an ARIMA-LSTM hybrid model to successfully outperform extant financial correlation coefficient predictive models using Root Mean Squared Error (RMSE) and Mean Absolute Error (MAE) metrics. These metrics calculate how close a prediction is to real values. Intuitively, the further the prediction is from the actual value, the less profitable the model will be [5]. RMSE and MAE are widely used as indicators of performance when it comes to times series.

Another interesting technique to predict market movement is by doing sentiment analysis in specific forums to evaluate opinions on individual stocks. A recently published paper by Sohangir et al. shows how they achieved successful results using deep learning and sentiment analysis [15]. By optimising the social network for investors and traders, StockTwits, they were able to learn what people were saying on the forums about different markets. With this sentiment analysis, they could then choose to buy stocks that looked most promising. Sohangir et al. initially used a Long Short-Term Memory (LSTM) model to predict sentiment but eventually turned to a Convolutional Neural Network (CNN) as LSTMs were not proving useful. CNNs are widely accepted as the preferred ML algorithm when doing image processing due to their ability to find internal structures of data. This ability is what made them an attractive alternative to LSTMs and ended up outperforming them.

There are also many examples of using RL to predict market prices. A recent example of this was carried out by Xiong et al. [18]. They used the same deep RL model that will be explored in this project, namely DDPG, to optimise a stock trading strategy.

Two different baselines that helped them measure performance are the Dow Jones Industrial (DJI) Average and the traditional min-variance portfolio allocation strategy. The DJI is an index of 30 high profile companies in the USA. On the other hand, the min-variance strategy is a technique to build a portfolio using low volatility investments. Xiong et al. show that it is possible to use DDPG and RL to predict markets by beating the two baselines mentioned above.

One issue with DDPGs and many deep RL algorithms is that they require careful hyper-parameter tuning. This tuning can become tedious most of the times. The results of different hyper-parameters are not apparent until a model completes training, and depending on the model used, training can take some time.

The paper by Lillicrap et al. [7], paved the way for its namesake [1] that deals with the issue of hyper-parameter tuning. This framework uses Bayesian optimisation and Gaussian Processes to determine the optimal hyperparameters, a complicated approach that works exceptionally well.

So far everything mentioned is related to real currency stock markets, which perform very similar to cryptocurrency exchanges. However, the next few papers mentioned will be related to cryptocurrencies. Alessandretti et al. compare three different algorithms for anticipating cryptocurrency prices using ML [3]. The first two algorithms use XGBoost, which is a framework that implements the gradient boosting decision tree algorithm. XGBoost is quite dominant when it comes to tabular or structured data and is used extensively in Kaggle competitions due to its performance and execution speed. The other algorithm used is an LSTM, which was mentioned earlier and is a common theme in stock price prediction. All three algorithms beat the simple moving average baseline. However, it is LSTM that stands out from the rest.

The primary goal of most of the automated trading algorithms is to maximise profit. A bot set up to maximise profit in the shortest time possible could go for the riskiest buys. Maximum profit also has the potential for maximum loss. Shin, Bu, and Cho cover risk management in their work on an automatic bot for low-risk portfolio management [13]. They manage to experiment with risk by tuning the exploration hyperparameter in their deep RL algorithm. The reduction of the exploration parameter reduces the potential for loss, and in the short term, it reduces the potential for profit. However, considering cryptocurrency prices can be volatile, it will benefit the outcome in the long run.

Shin, Bu, and Cho also use CNNs to extract local features from their data. They do this to build up a q-network. As mentioned earlier, image recognition software most commonly uses CNNs. However, here it is shown that they can be useful in other situations as well.

A common obstacle that autonomous trading applications meet is the commission price. On all cryptocurrency exchanges, there are fees associated with buying shares. There are two different fees for different purchasing types. These types are limit-orders and market-orders. As aforementioned, market-orders have a higher commission than limit-orders. This commission discrepancy is because market orders are less complex to place and go through relatively instantly, which is why most trading applications use market orders. However, Schnaubelt has shown a system which optimises the placement of limit orders on cryptocurrency exchanges taking advantage of the lower commission fee [12].

Schnaubelt tests multiple RL algorithms against a virtual exchange to exercise the limit order optimisation. The algorithms used are Deep Double Q-Networks (DDQNs), Backwards-Induction Q-Learning (BIQL) and Proximal Policy Optimisation (PPO). The latter being the most successful. Although fees are minimal, it is still quite detrimental to a model if the commission is not accounted for as the model will behave unexpectedly against a live system.

The most recent examples of using ML in market price anticipation are quite exciting. There are many different strategies that all outperform their chosen benchmarks, which shows that it is possible to make a profit however small it is.

4   Our Methodology

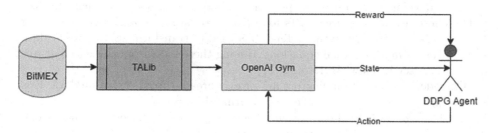

Fig. 2. High level design

As discussed, BitMEX was the source of data. BitMEX has readily available python libraries that allow the user to pull down data related to the exchange. Using the retrieved data, the TA-Lib library will be able to perform a transformation on it to create TIs. OpenAI Gym can then take these TIs as extra features to build up an environment that the DDPG agent can interact with to start training (Fig. 2).

4.1 BitMEX

BitMEX's Python library permits a user to fetch market data, make trades or create third-party clients. BitMEX has two types of exchange, a test exchange and a regular exchange and has user interfaces for both types. When creating a client, the option to choose which exchange to interact with is also available. Testing against the test exchange is convenient when learning how to trade. Everything is carried out in the same manner as the regular exchange. Even depositing imitation bitcoin into a dummy wallet through a mocked bitcoin faucet is possible. Once bitcoin is available, evaluation of trades can take place to understand how BitMEX works.

As aforementioned, BitMEX has an abundance of data. The python libraries permit anyone to interact with all of it. There are options to pull down multiple different cryptocurrencies. Some examples are Bitcoin, Ethereum, Litecoin and Ripple. Each cryptocurrency has a symbol associated with them, which are XBT, ETH, LTC and XRP, respectively.

Adding a fiat currency's symbol is usually done to gauge the cryptocurrency against the said fiat currency. An example is XBTUSD, which is bitcoins price against US Dollars. A user can choose which cryptocurrency data to pull down using these symbols. It is also feasible for a user to pass dates allowing the user to select which time range to view. The default features that BitMEX returns are timestamp, opening price, closing price, high price, low price and volume of trading.

4.2 TA-Lib and Technical Indicators

TA-Lib stands for Technical Analysis Library. It is highly popular among trading software developers requiring to perform technical analysis of financial market data. There are open-source APIs for many programming languages including C/C++, Java, Perl and Python. TA-Lib is used to transform stock market data into TIs using mathematical equations. Having these indicators as extra features on the data-set helps with training [17].

Considering that the indicators are mostly created from the same data, it is highly likely that many are correlated. During data preprocessing, a correlation matrix like Fig. 3a helped determine if there is a correlation between indicators. The matrix stems from twenty seven different TIs of Bitcoin's prices for five consecutive days. The lighter the colours, the more correlation there is. Similarly, the darker the colours, the less correlation there is. Figure 3b is the same matrix but with most correlated features removed.

4.3 Data

Data retrieved from BitMEX is relatively clean. Minimal preprocessing is required, and it is quite possible to train a model directly on the raw data. The only preprocessing to be executed is technical analysis, removal of correlated features and scaling the data to help with performance. Each sample will

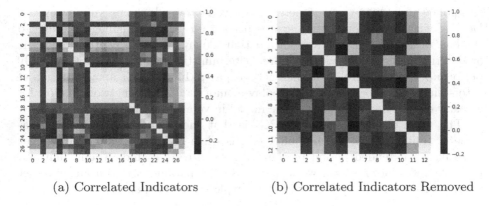

(a) Correlated Indicators   (b) Correlated Indicators Removed

Fig. 3. Correlated vs non-correlated heat matrices.

hold a series of indicators along with opening and closing prices of the share. An investigation into each TI is not required as it is the job of the machine to understand them.

4.4 OpenAI Gym

OpenAI Gym is currently one of the most popular tools to use when developing RL models. It allows users to compare RL algorithms and reproduce results consistently [11].

Gym supplies an interface that is simple to use, and easy to understand. Through implementing the multiple properties of the Gym interface, it unlocks the ability to train a RL model against an environment consistently and cleanly.

4.5 DDPG

The principle reason DDPG is the algorithm of choice is because it can integrate with continuous action spaces without much instability [7]. Many implementations of using RL to trade financial stocks can do so successfully with discrete action spaces [10]. However, the discrete action space leaves the algorithm limited in its choices. Using continuous actions opens a whole new dynamic for the model. The agent can explore an exponentially higher amount of actions, which has only become recently possible due to the aggressive increase in processing power [16].

The reason for DDPGs ability to solve problems with continuous action spaces is due to its deterministic policy gradient nature. In a Deep Q Network, a Q function maps all actions to an expected reward for each state. This Q function can become very difficult to compute as the action space grows. Whereas, policy gradients represent the policy by a probability distribution that is established by the parameters passed into it $\pi\theta(a|s) = P[a|s;\theta]$. The policy distribution selects action a in state s according to parameter vector θ. Policy gradients will sample this policy and adjust the parameters to converge towards the maximum reward. Dealing with continuous action spaces becomes a lot more manageable due to the lack of a map that contains each action. Policy gradients can be stochastic or deterministic. As discussed before, deterministic policy gradients generally perform better in high dimensional action spaces [14]. DDPGs amalgamation of actor-critic framework along with replay memory and target networks, allow for a vastly effective and a highly stable model in volatile environments.

The main disadvantage of DDPGs is that they require a lot of hyperparameter tuning, which can be quite tedious when training large networks that can take an extended period to converge. As already stated, Aboussalah and Lee have introduced a framework that uses hyperparameter optimisation to get around this downfall [1], but will not be explored in this paper.

4.6   Hyperparameter Tuning

The process of hyperparameter tuning can either be manual or else an automated process. For most of the implementation of this framework, hyperparameter tuning has been manual with somewhat successful results.

Optuna was also used to retrieve optimal hyperparameters. The hyperparameters associated with the final results of this paper are based on Optuna. Optuna is a next-generation hyperparameter optimisation software that implements Bayesian methods to optimise hyperparameters [2]. Bayesian optimisation builds a probabilistic model on the function that it is trying to optimise and tries to find the most exciting hyperparameters by minimising a specific error function.

4.7   Neural Network Structure

The actor, critic, actor target and critic target are all made up using neural networks. Naturally, the structure of these networks influences the output of the model. The target networks mimic the value networks, which means they will use the same structure.

Figure 4a shows the actor-network with twenty hidden layers instead of 256 for visualisation purposes. There are also thirteen input nodes to match the thirteen features in our data. Finally, there are two output nodes which refer to the action space of two elements.

Similarly, Fig. 4b shows the critic network with twenty hidden layers instead of 256 for visualisation purposes. The input node for the critic is fifteen as it incorporates the action space and the state space, which is a size of thirteen plus two. There is one output node which denotes the Q value for the action taken.

The actor neural network uses a ReLu activation function for the input and hidden layer, as well as a Softmax activation function for the output.

The critic neural network is similar in that it uses ReLu as the activation function for the input and hidden layer; the difference is that it has no output activation function. The neural network will take the final value returned from the ReLu activation function for the Q value.

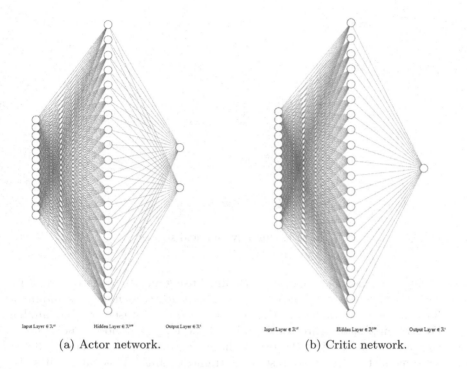

(a) Actor network.   (b) Critic network.

Fig. 4. The actor and critic networks.

4.8   Benchmarks

This project has one benchmark to measure performance against the DDPG algorithm. That is a buy and hold strategy, which has been mentioned previously.

A buy and hold strategy can simply be visualised by the price movement of any stock. The price at the start of the time series will be the purchase price and the price at the end of the time series will be the buy price plus or minus any profits or losses. Therefore depending on stock movement, a buy and hold strategy is not always profitable. A visualisation of five days of Ethereum price movements, seen in Fig. 5, has the same output as a buy and hold strategy.

Through purchasing one Ethereum at the start of the series and holding until the end, the value of that cryptocurrency will be 203.40 US Dollars. By benchmarking an algorithm against this strategy, one can say the algorithm beats the strategy if one stock is worth more than 203.40 US Dollars in the end after buying it for 212.70 US Dollars at the start of the series.

5   Results

The results of this project were quite fascinating. Not only do they show that it can make a profit using an RL algorithm in a cryptocurrency training environment, but it also shows it can do it with multiple different reward systems. The

Fig. 5. Ethereum buy and hold strategy

model that performed the greatest was trained using an Ethereum dataset with a candle size of five minutes. The dataset was built up across nine days and made up 2735 samples. The framework then transformed the dataset into the familiar TIs. The training lasted three days and had the longest period of noise decay used throughout the project. Figure 6a shows the results of the model after it was tested against a separate dataset than trained against. This dataset has also used Ethereum prices. However, it has a candle size of one hour across 20 days of data. The idea to test against a different dataset is to confirm that the training has not over fitted on the initial data set. Figure 6a shows that it can still make a profit. However, Fig. 6b shows the buying and selling pattern, which depicts purchases at low prices, but no sales. It is still a profitable model considering the net worth is on average higher than the buy and hold strategy, but if sales were incorporated, it would leave room for better results.

Table 1. Hyperparameters

REPLAY_BUFFER_SIZE	100000
HIDDEN_SIZE	256
ACTOR_LR	0.0000021
CRITIC_LR	0.0000098
DISCOUNT	0.99
DECAY_PERIOD	2000000
TIMESTEPS	900645
START_TRAINING_AFTER	2000
BATCH_SIZE	128
UPDATE_TARGET_EVERY	1
TAU	0.000005

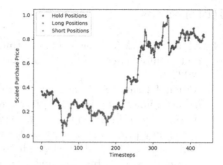

(a) The proposed model vs buy and hold strategy.

(b) Long and short positions.

Fig. 6. Results

The DDPG model returned many different results throughout this project. Many results were exciting, while a few were dismissible. It was quite apparent how much potential this model has in a cryptocurrency trading situation. Running the model in a live environment might output unexpected results as it seems the profit made here is implausible, possibly due to over fitting of the model or else just sheer luck. However, it is undeniable that this method bears a resemblance to a profitable, live stock market, trading method. With more time, it is conceivable that this framework can create a profitable system.

6 Conclusions

There was a constant sense of scepticism when initially researching the idea of training cryptocurrency with an RL model. Many papers conclude that the profit made is only slight in comparison to that of standard baselines [18]. There was initial apprehension in researching this topic over fears that the investigation would be near complete before realising any distinct outcomes were possible. Results shown have instilled confidence that it is, in fact, conceivable to think that a buy and hold strategy is beatable by an RL model.

The use of the DDPG algorithm provided reliable results and helped with the understanding of RL as a whole. Hyperparameter tuning was the primary obstacle, and if the research were to begin from scratch, focusing on automated hyperparameter tuning would be a considerably higher priority early in the implementation.

Another focus that proved timely was the reward system. The leading results were an outcome of using the Omega ratio as the performance metric. Using the direct net profit of a system was also a useful performance metric, but it had its downfalls. The importance of the reward system and the effect it had on results did not become apparent until late in the project, and it is something that could be improved on in the future.

References

1. Aboussalah, A.M., Lee, C.G.: Continuous control with stacked deep dynamic recurrent reinforcement learning for portfolio optimization. Expert Syst. Appl. **140** (2020)
2. Akiba, T., Sano, S., Yanase, T., Ohta, T., Koyama, M.: Optuna: a next-generation hyperparameter optimization framework (2019)
3. Alessandretti, L., ElBahrawy, A., Aiello, L.M., Baronchelli, A.: Anticipating cryptocurrency prices using machine learning. Complexity **2018**, 16 (2018)
4. Baird, L.: The swirlds hashgraph consensus algorithm: fair, fast, byzantine fault tolerance. Swirlds Inc, Technical report SWIRLDS-TR-2016 1 (2016)
5. Bebarta, D.K., Rout, A.K., Biswal, B., Dash, P.K.: Efficient prediction of stock market indices using adaptive neural network. In: Deep, K., Nagar, A., Pant, M., Bansal, J.C. (eds.) Proceedings of the International Conference on Soft Computing for Problem Solving (SocProS 2011) December 20-22, 2011. AISC, vol. 131, pp. 287–294. Springer, New Delhi (2012). https://doi.org/10.1007/978-81-322-0491-6_28
6. Choi, H.K.: Stock price correlation coefficient prediction with ARIMA-LSTM hybrid model, 05 August 2018
7. Lillicrap, T.P., et al.: Continuous control with deep reinforcement learning, 09 September 2015
8. Milutinovic, M.: Cryptocurrency. 2334–9190 (2018). http://ageconsearch.umn.edu/record/290219/
9. Nakamoto, S.: Bitcoin: a peer-to-peer electronic cash system. Manubot (2008)
10. Nevmyvaka, Y., Feng, Y., Kearns, M.: Reinforcement learning for optimized trade execution. In: Proceedings of the 23rd international conference on Machine learning, pp. 673–680 (2006)
11. OpenAI: Gym: A toolkit for developing and comparing reinforcement learning algorithms, 21 October 2019. https://gym.openai.com/
12. Schnaubelt, M.: Deep reinforcement learning for the optimal placement of cryptocurrency limit orders. FAU Discussion Papers in Economics (2020)
13. Shin, W., Bu, S.J., Cho, S.B.: Automatic financial trading agent for low-risk portfolio management using deep reinforcement learning, 07 September 2019
14. Silver, D., Lever, G., Heess, N., Degris, T., Wierstra, D., Riedmiller, M.: Deterministic policy gradient algorithms (2014)
15. Sohangir, S., Wang, D., Pomeranets, A., Khoshgoftaar, T.M.: Big data: deep learning for financial sentiment analysis. J. Big Data **5**(1), 3 (2018)
16. Sutton, R.S.: Reinforcement Learning: An Introduction. Adaptive Computation and Machine Learning, second edn. The MIT Press, Cambridge (2018)
17. TA-Lib: Ta-lib: Technical analysis library - home (2020). https://ta-lib.org/
18. Xiong, Z., Liu, X.Y., Zhong, S., Yang, H., Walid, A.: Practical deep reinforcement learning approach for stock trading, 19 November 2018

TMSA: Participatory Sensing Based on Mobile Phones in Urban Spaces

Luís Rosa[1]([✉]), Fábio Silva[1,2]([✉]), and Cesar Analide[1]([✉])

[1] ALGORITMI Center, Department of Informatics, University of Minho, Braga, Portugal
id8123@alunos.uminho.pt, {fabiosilva,analide}@di.uminho.pt
[2] CIICESI, ESTG, Politécnico do Porto, Felgueiras, Portugal

Abstract. A design for a novel mobile sensing system, called Temperature Measurement System Architecture (TMSA), that uses people as mobile sensing nodes in a network to capture spatiotemporal properties of pedestrians in urban environments is presented in this paper. In this dynamic, microservices approach, real-time data and an open-source IoT platform are combined to provide weather conditions based on information generated by a fleet of mobile sensing platforms. TMSA also offers several advantages over traditional methods using participatory sensing or more recently crowd-sourced data from mobile devices, as it provides a framework in which citizens can bring to light data relevant to urban planning services or learn human behaviour patterns, aiming to change users' attitudes or behaviors through social influence. In this paper, we motivate the need for and demonstrate the potential of such a sensing paradigm, which supports a host of new research and application developments, and illustrate this with a practical urban sensing example.

Keywords: Pedestrian-oriented applications · Participatory sensing · Crowdsourcing · Open-source IoT platform

1 Introduction

The recent development of telecommunication networks is producing an unprecedented wealth of information and, as a consequence, an increasing interest in analyzing such data both from telecoms and from other stakeholders' points of view. In particular, mobile phones with a rich set of embedded sensors, such as an accelerometer, digital compass, gyroscope, Global Positioning System (GPS), microphone, and camera, generate datasets that offer access to insights into urban dynamics and human activities at an unprecedented scale and level of detail.

Collectively, these sensors give rise to a new area of research called mobile phone sensing. They are relevant in a wide variety of domains, such as social networks [1], environmental monitoring [20], healthcare [17], and transportation [21] because of a number of important technological advances. First, the availability

© Springer Nature Switzerland AG 2020
C. Analide et al. (Eds.): IDEAL 2020, LNCS 12489, pp. 257–267, 2020.
https://doi.org/10.1007/978-3-030-62362-3_23

of cheap embedded sensors is changing the landscape of possible applications (e.g. sharing the user's real-time activity with friends on social networks such as Facebook, or keeping track of a person's carbon footprint). Second, to sensing, smartphones come with computing and communication resources that offer a low barrier of entry for third-party programmers (e.g., undergraduates with little phone programming experience are developing applications). In addition, the mobile computing cloud enables developers to offload mobile services to back-end servers, providing unprecedented scale and additional resources for computing on collections of large-scale sensor data and supporting advanced features.

The omnipresence of mobile phones thanks to its massive adoption by users across the globe enables the collection and analysis of data far beyond the scale of what was previously possible. Several observations can now be made thanks to the possibility of crowdsourcing using smartphones. Since 94.01% of the population has mobile network coverage in 138 countries [13] and it is expected to reach more than one billion people by 2022, there is a huge potential for obtaining real-time air temperature maps from smartphones. This will enable a new era of participatory engagement by global population, where the data required is in part willingly provided by citizens via smartphone apps. Although it raises complex societal issues that must be addressed as we face this new era, it also provides an unprecedented opportunity for the citizens of a community to play an active role in the betterment of their community, not only in terms of reporting critical environmental and transportation data, but also in mitigating the very challenges their own data identifies. It will be a cooperative and informed effort, with information technology enabling societal transformation, to address the growing challenges our cities face in the coming decades [19]. Therefore, crowdsourcing opens the door to a new paradigm for monitoring the urban landscape known as participatory sensing.

Participatory sensing applications represent a great opportunity for research and real-world applications. In [12], CarTel is a system that uses mobile phones carried in vehicles to collect information about traffic, the quality of en-route WiFi access points, and potholes on the road. Other applications of participatory sensing include the collection and sharing of information about urban air and noise pollution [9], or consumer pricing information in offline markets [11]. Additionally, they also changed living conditions and influenced lifestyles and behaviours. In indoor activities, energy demand planning [2] will become more challenging, given the projected climate change and its variability. In order to assess the resource implications of policy interventions and to design and operate efficient urban infrastructures, such as energy systems, greater spatial and temporal resolutions are required in the underlying resources that demand data. Besides, they accommodate pedestrian traffic and outdoor activities, and greatly contribute to urban liveability and vitality. Therefore, outdoor spaces that provide a pleasurable thermal comfort experience for pedestrians effectively improve their quality of life [3].

An Internet of Things (IoT) framework can be proposed to support citizens in making decisions about daily activities by a notification system based on a faster access to air temperature data. But data accuracy can be improved by creating an application that promotes participatory sensing. In this tool, the individuals and communities using mobile phones, online social networking and cloud services collect and analyze systematic data. This technology convergence and analytical innovation can have an impact on many aspects of our daily lives and community knowledge can help determine our actions. Since our work is currently focusing on these approaches, this paper will present a blueprint that we have created for air temperature screening that we call Temperature Measurement System Architecture (TMSA).

The rest of this document is structured as follows: Sect. 2 introduces different use cases from the participatory sensing domain which depict the usage of mobile devices from different points of view. Then, based on a state of the art analysis we propose the TMSA that gathers real-time data to allow citizens to access air temperature (and other parameters of weather conditions), and explain how it can influence users' attitudes or behaviors. Section 3 investigates existing IoT building blocks by means of protocols, components, platforms, and ecosystem approaches in the form of a state of the art analysis. Additionally, TMSA infrastructure proposes concrete technologies for the corresponding building blocks of the cloud architecture and their data flow between each. Finally, Sect. 4 concludes this document with a discussion on the resulting architecture specification and its compatibility with the other work packages as well as an outlook on future work within TMSA.

2 Background

To make the best use of currently available literature, we reviewed and herein present an analysis of participatory sensing applications studies with the goal of better understanding the variables that affect daily human activities. Besides, real time monitoring is an emerging technology in both mobile technologies, such as tablets and smartphones, and plays an important role in society. Some studies prove the challenges and limitations to implement non-invasive methods with this kind of sensors. Offering a simple method for mobile sensing could increase the acceptance to monitor and achieve satisfactory results.

2.1 Participatory Sensing via Mobile Devices

With an increasing number of rich embedded sensors, like accelerometer and GPS, a smartphone becomes a pervasive people-centric sensing platform for inferring user's daily activities and social contexts [14]. Additionally, as we mentioned, the number of users with access to mobile technology has increased rapidly around the globe. Besides, data sensing techniques are becoming widely used in various applications including forecasting systems, for example, in [5], the potential of participatory sensing strategies to transform experiences, perceptions, attitudes, and daily routine activities in 15 households equipped with wood-burning stoves in the city of Temuco, Chile. In our previous work [16], crowdsensing data establishes an easy connection between citizens and technology innovation hub to acquire detailed data on human movements. Based or not on empirical observation we can gather air temperature data with the support of people via their mobile phones.

In this work, participatory sensing is aligned with five key principles that spread across areas of design, organizational development, and community action research [18]. It includes:

- Active Citizens: where users work in partnership and contribute to the common good;
- Building Capacities: understanding that sustainable transformation requires a certain amount of trust, which is built through communication and a culture of participation;
- Building Infrastructure and Enabling Platforms: the purposeful devising of structures and platforms support participation, with a focus on sustainability, long-term action and impact;
- Intervention at a Community Scale: harnessing the collective power of the community and using a community-centred approach and community-driven solutions, so that communities become the catalyst of interventions for large-scale and transformative change;
- Enhancing Imagination and Hope: supporting and enhancing the ability for communities to imagine new possibilities, and building a shared vision based on seeing the future in a new light and collectively working towards shared objectives.

Table 1. Limitations and issues in temperature-measuring apps (adapted from [15]).

Context	Restriction	Description
Hardware	Collection of sensor data across mobile platforms and devices	- Participants not charging or turning off their phones - Unavailability of any network connections, hampering data transfer to servers
	Data collection influences data completeness and differ between operating systems	- Collection of sensor data is easier to support on Android than iOS - More apps are available for Android than for iOS - iOS has greater restrictions on accessing system data and background activity
	Battery life in mobile phones	- App collecting sensor data can consume a significant amount of a battery
	Some smartphones are not equipped with a temperature sensor	- Battery temperature sensor can be used to measure temperature
Software	Engagement and retention of users	- User-centered design approach is considered an integral part of any temperature screening app development
	Codesign of temperature screening apps should involve stakeholders	- Increases the likelihood that the app will be perceived as attractive, usable, and helpful by the target population
	Confidential handling and use of data, as well as privacy and anonymity	- Temperature screening data is highly sensitive because of the potential negative implications of unwanted disclosure
	Security and privacy differ by users	- Android and iOS users differ in terms of age, gender and in their awareness of app security and privacy risks

In fact, participatory sensing is a method or an approach to extract any kind of data from the communities which can be used for their own benefit and solve global challenges at a local level. The main sensing paradigm of our study enables monitoring air temperature from mobile phones and, besides, other weather parameters using external sensor networks. Using information collected by pedestrians, the resulting system produces more accurate reports than systems that do not rely on multiple input sources.

2.2 Challenges of Mobile Phone Sensor Technology

Despite the potential of mobile phone sensor technology in air temperature research and the fact that participatory sensing can be useful in developing temperature-measuring apps they pose several key challenges, both at hardware and software level to temperature-measuring apps. If an app collecting sensor data is too resource-intensive, a user's motivation to continue using it decreases [7]. An optimized data collection should therefore be aligned with the expectations of users regarding battery consumption. Other technical and issue restrictions are mentioned in Table 1.

Bigger challenges are confidential handling and use of data, as well as privacy and anonymity (of user data) within apps. Indeed, measuring a person's body temperature is, in theory, to be considered a processing of personal data. This may, for example, be different in cases where the temperature is measured for a few seconds, without there being any direct or indirect link whatsoever to (other) personal data. In that case, the measurement could fall outside the definition of "processing" as defined in the General Data Protection Regulation (GDPR) [8]. Additionally, systems making use of predictive analysis techniques not only collect data but also create information about body temperature status, for example, through identification of risk markers. Therefore, social impact needs to be considered beyond individual privacy concerns.

2.3 Effects of Air Temperature in Daily Human Activities

The air temperature along with other weather condition parameters (rainfall and wind speed), has effects on people's everyday activities [4]. People's daily activities can be inferred, such as place visited, the time this took place, the duration of the visit, based on the GPS location traces of their mobile phones. Based on the collected information, it is possible to estimate, for example, if people are more likely to stay longer at beach, food outlets, and at retail or shopping areas when the weather is very cold or when conditions are calm (non-windy). When compared with people's regular activity information, we can probably research, for instance, how it noticeably affects people's movements and activities at different times of the day. Or we can observe the effect of air temperature, rainfall and wind speed in different geographical areas. Therefore, besides noting how mobile phone data can be used to investigate the influence of environmental factors on daily activities, this work sheds new light on the influence of atmospheric temperature on human behavior.

2.4   Temperature Measurement in Mobile Sensing

Many researchers and engineers have been working on real time and remote monitoring systems to measure temperature. In [10], the author shows an example to illustrate the dependency of air temperature sensor readings on storing positions on the body: front or back trouser pocket, chest pocket, and around the neck (hanging). This work uses the Wet-Bulb Globe Temperature (WBGT) as the benchmark for environmental thermal conditions that reflect the probability of heatstroke, because it consists of three important factors: air temperature, air humidity, and radiant heat [23]. On the other hand, in [22], the authors present a design for a new mobile sensing system (AMSense) that uses vehicles as mobile sensing nodes in a network to capture spatiotemporal properties of pedestrians and cyclists in urban environments.

Given that previous projects indirectly depend on the mobile phone sensing, none of them is able to calculate air temperature. Alternatively, Nguyen Hai Chau, in [6], shows that temperature readings of smartphone batteries can be used to estimate air temperature. In this paper, two statistical models that allow each smartphone to predict the temperature in or out of human clothes' pockets were built. The formulae of models given in Eq. 1 and Eq. 2:

$$T_{air} = T_{battery-out-of-pocket} x 0.903 + 1.074 \qquad (1)$$

and

$$T_{air} = T_{battery-in-pocket} x 1.877 - 35.1, \qquad (2)$$

where T_{air} is the estimated air temperature, $T_{battery-out-of-pocket}$ and $T_{battery-in-pocket}$ the temperatures of a smartphone battery given smartphone context is out of pocket or in pocket, respectively.

Furthermore, smartphones are often carried close to the body, e.g. in pockets of coats, trousers and in people's hands. Therefore, Nguyen Hai Chaur developed a new approach of using two linear regression models to estimate air temperature based on the temperature of an idle smartphone battery given their in-pocket or out-of-pocket positions. Lab test results show that the new approach is better than an existing one in mean absolute error and coefficient of determination metrics. Advantages of the new approach include the simplicity of implementation on smartphones and the ability to create maps of temperature distribution. However, this approach needs to be field-tested on more smartphone models to achieve its robustness.

3   Experimental Study Case

This article surveys the new ideas and techniques related to the use of telecommunication data for urban sensing. We outline the data that can be collected from telecommunication networks as well as their strengths and weaknesses with a particular focus on urban sensing. Moreover, the data contributed from multiple participants can be combined to build a spatio-temporal view of the phenomenon of interest and also to extract important community statistics. Therefore, in the IoT architecture proposed, participatory sensing can achieve a level of coverage in both space and time for observing events of interest in urban spaces.

3.1   Methodology

This chapter presents a state of the art for the building blocks of the IoT system proposed like communication, microservices, data visualization, and device management. For each aspect, we provide a general description on the purpose, requirements, and challenges.

Communication Protocols. The most important protocols for this study are Representational State Transfer (REST) and Message Queue Telemetry Transport (MQTT). REST denotes an architectural style for distributed hypermedia and web service systems. Additionally, it defines a generic interface to resources (referenced by an Uniform Resource Identifier (URI)), consisting of methods like GET for requesting and POST for creating resource representations. Within the World Wide Web (WWW), client applications, like browsers or mobile apps, may invoke these methods, typically via the Hypertext Transfer Protocol (HTTP) where JavaScript Object Notation (JSON) are widely used to encode the transmitted representation.

In its turn, MQTT is a publish-subscribe messaging protocol. The protocol is based on TCP/IP and is designed for small devices with limited network connection. It comprises the publish/subscribe pattern with a broker, which is responsible for receiving messages and sending them to interested parties. The content is identified by a topic. When publishing content, the publisher can choose whether the content should be retained by the server or not. If retained, each subscriber will receive the latest published value directly when subscribing.

Micro-services Integration. A client implementation is provided by the Eclipse Ditto project. It provides a platform to realize the digital twin metaphor and is a framework designed to build IoT applications with an Edge component. The Ditto package is completed with a web front end which allows the developer or administrator to remotely log in and which developers can use to provide a web facing aspect to their own application's configuration needs. It also provides functionality to realize higher level REST API to access devices (i.e. Device as a Service); include notification of state changes and provide metadata-based support to search and select digital twins (i.e. Digital Twin Management).

To integrate between Eclipse Ditto and Firebase a message broker especially dedicated to MQTT is needed. It's the Eclipse Paho MQTT project. Besides, Paho supports clients for various languages, it provides another open-source client implementation: MQTT-Sensors Network (MQTT-SN). In general, developers have to consider security for MQTT themselves. The result is a working system that has been running on 2 microservices, hosted in a Docker Desktop (application for MacOS and Windows machines to build and share containerized applications), via mobile communication or Wi-Fi access points.

Data Processing Components. The increase of device density produces a lot of data and creates new challenges for the Internet infrastructure. In order to solve this problem, we propose a infrastructure called cloud computing services, often called simply "the cloud", which can be categorized into three types–Infrastructure as a Service (IaaS), Platform as a Service (PaaS) and Software as a Service (SaaS)–to relieve this pressure off Big Data. For example, the first service that provides an infrastructure like Servers, Operating Systems, Virtual Machines, Networks, and Storage on a rent basis is used by AmazonWeb Service and Microsoft Azure. Another service is used in developing, testing and main-taining software for Apprenda and Red Hat OpenShift. PaaS is the same as IaaS but also provides additional tools like DBMS and BI service. And lastly, SaaS allows users to connect to applications through the Internet on a subscription basis. Google Applications, Salesforce use it. In this context, the development platform called Firebase can be useful. Because several features are supported, such as encryption, authentication, and cloud data storage, it is ideal for access control, storage and efficient sharing in real-time. Another advantage is it uses the HTTPs communication protocol that supports bidirectional communications encryption between a client and server. Therefore, in our architecture we will use SaaS, a private fragment for storage on smartphone and a public fragment to disperse on Cloud as privacy policy, using Firebase mainly as a Big Data Sharing centre.

3.2   System Design

Figure 1 shows how the IoT Platform architecture works. The design of this architecture includes a secured management and distribution of mobile phone

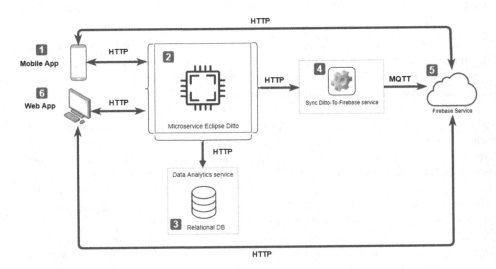

Fig. 1. Architecture proposal for pedestrian-oriented applications.

data as well as the integration of external services and applications. Firstly, (1) mobile phone will detect air temperature using literature's previously described approaches (Eq. 1 and Eq. 2) and get the data. Then the data will be sent to (2) Eclipse Ditto via machine-to-machine (M2M)/"Internet of Things" (IoT) connectivity. Then (3) Eclipse Ditto will save the data to a data analytics service like a relational database. From there, using mobile phone, we can manage client connections to remote systems like social networking services (e.g, Facebook and Twitter) and to exchange Ditto Protocol messages. If the remote system is unable to send messages in the necessary format, there is the option to configure custom payload mapping logic to adapt to almost any message format and encoding. To connect and synchronize Ditto data to the cloud, we added the (4) Sync Ditto-To-Firebase service. (5) Firebase works as a cloud-based machine-to-machine integration platform. The cloud-based philosophy of scalability is at the core of the proposed IoT infrastructure. Additionally, the system records and displays the data. Thereafter (6) the user can get the results from a mobile phone app or web app.

This IoT Platform architecture being based on an IoT Cloud Platform is expected to scale both horizontally to support the large number of heterogeneity devices connected and vertically to address the variety of IoT solutions. However, an IoT architecture comprises core building blocks that are applied across different domains and regardless of the use case or connected Things. Depending on the applicable reference architecture, such building blocks differ in naming and granularity level. Therefore, the IoT schema above shows typically core features for the Eclipse open source software stack for an IoT Cloud Platform with facilities for device management, data storage and management, visualization, analytics, and stream processing.

4 Conclusion and Future Work

Recent advancements in IoT have drawn attention of researchers and developers worldwide. IoT developers and researchers are working together to extend the technology on a large scale and several stakeholders have benefited from the application of smartphone-based sensing. However, multiple gaps remain between this vision and the present state of the art. In particular, additional research is needed to address major issues such as air temperature measurement efficacy, integration of newer analytic approaches including artificial intelligence (AI), privacy issues, and implementation of sensing into actual mobile phones. Moreover, there are a set of options for how device-to-platform connection is made and it depends on the environment and constraints of the device. This is, if the device is outside and moving around we use cellular connectivity; in indoor environments, Ethernet or Wi-Fi connectivity is a better option. And finally, in gateways and edge compute, in some cases the devices can't connect directly to the central cloud platform or other platforms in intermediate layers and instead require the use of an IoT gateway to bridge the gap between local environment and platform. This gateway is required when using Wireless technologies like

Bluetooth and LPWAN, since those don't provide any direct connection to the network or the Cloud. IoT is not only providing services but also generates a huge amount of data. For that reason, the importance of big data analytics is also discussed and can provide accurate decisions that could be used to develop an improved IoT system.

In this survey article, we address the mentioned issues and challenges that must be taken into account to develop applications to measure air temperature based on our IoT architecture. Therefore, due to its reliability and feasibility the proposed layered architecture is useful to many kinds of applications. In future work, we intend to build new model applications based on this proposed IoT architecture. The main goal is to predict our way of living in an era impacted by the viral effects of IoT application (Human-Computer Interaction and Machine-to-Machine connections) and, for example, could be used to foster business process management in the IoT era.

Acknowledgments. This work has been supported by FCT - Fundacao para a Ciencia e Tecnologia within the R&D Units Project Scope: UIDB/00319/2020.

References

1. Abdelraheem, A.Y., Ahmed, A.M.: The impact of using mobile social network applications on students' social-life. Int. J. Instr. **11**(2), 1–14 (2018)
2. Almuhtady, A., Alshwawra, A., Alfaouri, M., Al-Kouz, W., Al-Hinti, I.: Investigation of the trends of electricity demands in Jordan and its susceptibility to the ambient air temperature towards sustainable electricity generation. Energy Sustain. Soc. **9**(1), 1–18 (2019). https://doi.org/10.1186/s13705-019-0224-1
3. Antonini, E., Vodola, V., Gaspari, J., de Giglio, M.: Outdoor wellbeing and quality of life: a scientific literature review on thermal comfort (2020). https://doi.org/10.3390/en13082079
4. Böcker, L., Dijst, M., Prillwitz, J.: Impact of Everyday Weather on Individual Daily Travel Behaviours in Perspective: A Literature Review (2013)
5. Boso, À., Álvarez, B., Oltra, C., Garrido, J., Muñoz, C., Hofflinger, Á.: Out of sight, out of mind: participatory sensing for monitoring indoor air quality. Environ. Monit. Assess. **192**(2) (2020)
6. Chau, N.H.: Estimation of air temperature using smartphones in different contexts. J. Inf. Telecommun. **3**(4), 494–507 (2019)
7. Dennison, L., Morrison, L., Conway, G., Yardley, L.: Opportunities and challenges for smartphone applications in supporting health behavior change: qualitative study. J. Med. Internet Res. **15**(4) (2013). https://doi.org/10.2196/jmir.2583
8. EDPB: Statement on the processing of personal data in the context of the COVID-19 outbreak. (March), 1–3 (2020)
9. Eißfeldt, H.: Sustainable urban air mobility supported with participatory noise sensing. Sustainability (Switzerland) **12**(8) (2020). https://doi.org/10.3390/SU12083320
10. Fujinami, K.: Smartphone-based environmental sensing using device location as metadata. Int. J. Smart Sens. Intell. Syst. **9**(4), 2257–2275 (2016)
11. Gao, G., Sun, Y., Zhang, Y.: Engaging the commons in participatory sensing, pp. 1–14. Association for Computing Machinery (ACM), April 2020

12. Hull, B., Bychkovsky, V., Zhang, Y., Chen, K., Goraczko, M.: CarTel: a distributed mobile sensor computing system. In: Proceedings of the Fourth International Conference on Embedded Networked Sensor Systems, SenSys 2006, pp. 125–138 (2006)
13. International Telecommunication Union: Mobile network coverage by country (2016). https://www.theglobaleconomy.com/rankings/Mobile_network_coverage
14. Predic, B., Yan, Z., Eberle, J., Stojanovic, D., Aberer, K.: ExposureSense: integrating daily activities with air quality using mobile participatory sensing (2013)
15. Ray, P.P.: A survey on Internet of Things architectures (2018)
16. Rosa, L., Silva, L., Analide, C.: Representing human mobility patterns in urban spaces. In: Intelligent Environments 2020, pp. 177–186 (2020)
17. Salehi, H.P.: Smartphone for healthcare communication. J. Healthcare Commun. **03**(03), 34 (2018)
18. Sangiorgi, D.: Transformative services and transformation design. Int. J. Des. **5**(2), 29–40 (2011)
19. Sarwar, M., Soomro, T.R.: Impact of smart phones on society. Eur. J. Sci. Res. **98**(2), 216–226 (2013)
20. Šećerov, I., et al.: Environmental monitoring systems: review and future development. Wirel. Eng. Technol. **10**(01), 1–18 (2019). https://doi.org/10.4236/wet.2019.101001
21. Spyropoulou, I., Linardou, M.: Modelling the effect of mobile phone use on driving behaviour considering different use modes. J. Adv. Transp. **2019** (2019)
22. Vial, A., Daamen, W., Ding, A.Y., van Arem, B., Hoogendoorn, S.: AMSense: how mobile sensing platforms capture pedestrian/cyclist spatiotemporal properties in cities. IEEE Intell. Transp. Syst. Mag. (2020)
23. Yaglou, C.P., Minard, D.: Control of heat casualties at military training centers. A.M.A. Arch. Ind. Health **16**(4), 302–316 (1957)

A Slimmer and Deeper Approach to Network Structures for Image Denoising and Dehazing

Boyan Xu[ID] and Hujun Yin[✉][ID]

The University of Manchester, Manchester M13 9PL, UK
boyan.xu@postgrad.manchester.ac.uk, hujun.yin@manchester.ac.uk

Abstract. Image denoising and dehazing are representatives of low-level vision tasks. The right trade-off between depth and computational complexity of convolutional neural networks (CNNs) is of significant importance to these problems. Wider feature maps and deeper network are beneficial for better performance, but would increase their complexity. In this paper, we explore a new way in network design, to encourage more convolution layers while decrease the width of feature maps. Such slimmer and deeper architectures can enhance the performance while maintain the same level of computational costs. We experimentally evaluate the performances of the proposed approach on denoising and dehazing, and the results demonstrate that it can achieve the state-of-the-art results on both quantitative measures and qualitative performances. Further experiments also indicate that the proposed approach can be adapted for other image restoration tasks such as super-resolution.

Keywords: Image dehazing · Image denoising · Super-resolution

1 Introduction

Image denoising and dehazing aim to restore original images from degradation or to reconstruct enhanced images from given images. They both are highly ill-posed problems and have been perennial topics in image processing and computer vision. The recent developments of deep learning (DL) [14], especially convolutional neural networks (CNNs), have made significant advances in these and many other tasks of computer vision. Thus, learning based approaches have become mainstream solutions to these problems, and further improvement of these networks has become the main goal of current research.

In this paper, we aim to explore useful commonality in denoising and dehazing and to rethink on the structures of CNNs for effectiveness and efficiency. In learning based image restoration, many technical aspects need to be considered. In addition to effectiveness, another important consideration is computational complexity, which is mainly affected by depth of the network and width of its feature maps. The number of channels of the convolution kernels determines the

© Springer Nature Switzerland AG 2020
C. Analide et al. (Eds.): IDEAL 2020, LNCS 12489, pp. 268–279, 2020.
https://doi.org/10.1007/978-3-030-62362-3_24

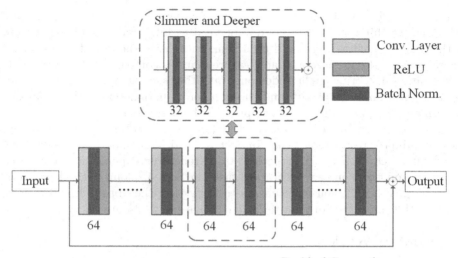

Fig. 1. Conception of our proposed approach. We use a very simple CNN as an example. The network is only consist of convolution, BN and ReLU layers. Two convolution blocks with 64 feature maps are removed and replaced by five blocks with 32 channels.

number of feature maps to be generated, which map features to higher levels progressively. Many studies have verified that networks must have sufficient depths to be able to approximate certain complex features and functions. Otherwise widths of the convolution layers must increase exponentially to have the same effect [19].

Many existing networks have traded off between accuracy and complexity of the networks, which often determine the application performance and computation speed. Both depth and width are important aspects, but their balance is often problem dependent. Inception-Net [27] uses several branches to increase the feature maps and has achieved good performances. However, it performs well mostly in detection and classification tasks, while for low-level vision tasks, different mechanisms are involved. Many tools derived for high-level vision tasks face new challenges when being adapted to image restoration tasks. For example, Lim *et al.* [18] reported that deep super-resolution networks performed better when Batch Normalization (BN) layers were removed.

In addition, simply increasing the network depth and widening the layer width would considerably elevate the computational complexity. So here raises a question, is it possible to achieve better performance without increasing model complexity? We have noticed that in many tasks of inverse problems such as image deblurring, feature maps or number of channels show a tendency of decreasing, from 256 channels [21] to 128 channels [6]. In more recent work such as [32], networks mostly consisted of layers with 32 and 64 channels.

Based on these observations, we propose a novel method; its conception can be seen in Fig. 1. We convert the convolutional layers into new layers with fewer feature maps but with increased number of layers. For image denoising and dehazing, we think that networks do not need to fully understand the content of the images, such as items and their semantics, as in the classification tasks. Therefore they do not need as many feature maps as compared to detection and classification. Experiments have demonstrated that such an approach can reduce the computational complexity and the number of parameters without causing performance degradation, or even improving the performance.

The rest of this paper is organized as follows. In Sect. 2, we briefly review these two image restoration tasks. In Sect. 3, we describe and analyze our proposed approach to network structure. Comprehensive experiments have been conducted on these image restoration tasks in Sect. 4, along with the results. Section 5 gives additional experiments to show that the approach can also be used in other image restoration tasks like image super-resolution.

2 Related Work

Image Denoising. Noise corruption is inevitable during image acquisition and processing and can severely degrade visual quality of acquired images. Image denoising is of extraordinary value to low-level vision from many aspects. First, it is often an essential pre-processing step to remove noise in various computer vision tasks. Second, image denoising is an ideal test bed for evaluating image prior models and optimization methods from the Bayesian perspective [34]. Dabov *et al.* proposed BM3D [3] which achieved enhancement on the sparsity by grouping similar 2-D image fragments (e.g., blocks) into 3-D data arrays. Zhang *et al.* [33] exploited the construction of feed-forward denoising convolutional neural networks (DnCNNs) to embrace the progress in very deep networks and regularization method into image denoising task, and further developed by improve the trade-off between inference speed and performance. In [7], Gharbi *et al.* developed a Monte Carlo denoising method by kernel-splatting architecture, but leading to long processing time.

Image Dehazing. Caused by turbid medium such as mist and dust, the existence of haze leads to poor visibility of photos and adds data-dependent, complicated and nonlinear noise to the images, making haze removal an ill-posed and highly challenging image restoration problem. Many computer vision algorithms can only work well on the scene radiance, which is haze-free. In [10], He *et al.* proposed dark channel prior to remove haze from single image based on the assumption that image patches of outdoor haze free image often have low-intensity values. Li *et al.* [15] proposed an all-in-one method to directly generate clean images through a light-weight CNN, based on a reformulated atmospheric scattering model. Qu *et al.* [23] introduced a GAN for dehazing by using the discriminator to guide the generator to create a pseudo realistic image on a coarse scale, while the enhancer following the generator was required to produce a realistic dehazing image on the fine scale. In [2], Chen *et al.* adopted smoothed dilation technique and leveraged on a gated sub-network to fuse the features from different levels.

3   Proposed Method

The essence of the proposed method is to add cascaded convolution layers with fewer channels compared with the existing layers in a network structure. We choose such a manifestation to split the original convolutional layers (e.g., 2) into more convolutional layers, as the most important reason is to highlight the role of this approach. From another aspect, assuming all parameters in the transferred layers are zero, the structure of the network is actually the same as the original one due to the existence of the residual link from the input to the output. Thus, if we only add these layers in a network, usually the performance would improve. In addition, making the structure slimmer and deeper could be regarded as supplementary to the original convolution layers. However, adding such layers at the first or last layer or connecting too many layers would break the original network structure and hence are not suitable. Thus, when we adopt our approach on an existing network, the goal is to transform its convolution layers without destroying the overall original structure of the network. We then evaluate the effect on image restoration problems with such adaptation.

The proposed method is shown in Fig. 1. We take two ordinary 64-channel convolutional blocks with size of 128 × 128 and kernel size 3 × 3 as an example. The forward between these two convolutional layers is a convolution operation. The number of floating-point operations is given by the following formula

$$FLOPs = H \times W \times (C_{in} \times C_{out} \times K^2 + C_{out}), \qquad (1)$$

where H, W and C_{in} are the height, weight and number of channels of input feature map, C_{out} is the number of output feature maps, and K is the size of convolution kernels. Based on the above assumptions, we calculate the FLOPs of 64-channel convolution is approximately $128 \times 128 \times (64 \times 64 \times 3^2 + 64) = 0.605e^9$ FLOPs or 0.605 GFLOPs. In general, we want to convert the width of convolution layer to half or less of the original. Here, we use 32 channels as an example, resulting five convolution layers lead to four sets of convolution operations, of total $4 \times (128 \times 128 \times (32 \times 32 \times 3^2 + 32)) = 0.606e^9$ FLOPs or 0.606 GFLOPs, similar to the previous case. But the depth has increased greatly from 2 to 5. Splitting into fewer layers is also possible if reduction of computation is required. The number of parameters can be calculated by

$$params = C_{in} \times C_{out} \times K^2 + C_{out}, \qquad (2)$$

which is proportional to FLOPs if the feature maps are not changed. In Sect. 4, we will show that in image restoration tasks, such operations would not cause performance degradation; on the contrary, it often improves the results.

4   Experiments

The proposed method can be applied to noise and haze removal. We choose two state-of-the-art CNNs, which are DnCNN [33], and GCANet [2], respectively. We modify them based on the proposed methodology, while keeping the original structure of networks, including residual links.

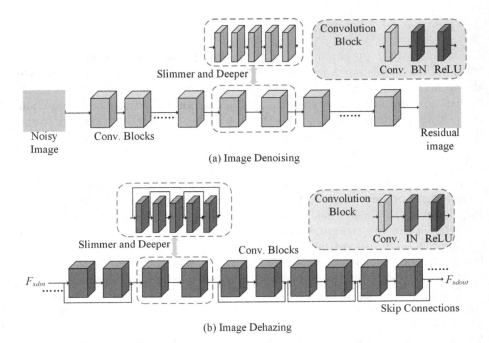

(a) Image Denoising

(b) Image Dehazing

Fig. 2. Network structure of denoising and dehazing: (a) network structure for denoising, modified based on DnCNN [33], (b) network structure for dehazing, modified based on GCANet [24], modified by our proposed slimmer and deeper approach.

4.1   Image Denoising

Network Design. In this paper, we implement our denoising network based on the Denoising Convolutional Neural Network (DnCNN) [33]. DnCNN model has two main features: residual learning formulation and batch normalization (BN), incorporated to speed up training and to boost the denoising performance. Although we have mentioned that BN has bad influence on the performance of low-level vision tasks, we adapt BN here as it is similar to denoising task statistically. By incorporating convolution with ReLU, DnCNN can gradually separate image structure from the noisy observation through the hidden layers. Such a mechanism is similar to the iterative noise removal strategy adopted in methods such as [17] and [28], while DnCNN is trained in an end-to-end fashion. The DnCNN has 17 convolution blocks, we replace the fifth and sixth blocks with our transferred blocks, which have five convolution sub-blocks. Same operation is also done on the 12th and the 13th block. The network structure is shown in Fig. 2 (a), the bottom line. More details of DnCNN can be found in [33].

All our experiments were implemented in PyTorch [22] and evaluated on a single NVIDIA Tesla P100 GPU. To train the network, we randomly cropped images to 40×40 pixel size. Subsequently, the batch size was set to 300 during training. The Adam optimizer was used to train our models. The initial learning rate was set to $1e-3$.

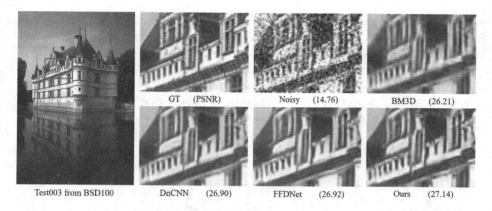

Fig. 3. Visual comparison of our image denoising method with the state-of-the-art methods, with the noise level of 50.

Table 1. Quantitative comparisons of image denoising on the BSD68 [25] dataset. The comparisons are based on PSNR.

Methods	BM3D [3]	WNNM [17]	EPLL [28]	TNRD [8]	DnCNN [33]	Ours
$\sigma = 15$	31.07	31.37	31.21	31.42	31.73	**31.96**
$\sigma = 25$	28.57	28.83	28.68	28.92	29.23	**29.54**
$\sigma = 50$	25.62	25.87	25.67	25.97	26.23	**26.77**

Result. We tested the proposed network on the task of removing additive Gaussian noise of three levels (15, 25, 50) from a gray-scale noisy image. Following the same experimental protocols used by the previous studies, we used 800 training images from DIV2K dataset [29] as training set and tested the results of the proposed denoising network using the BSD68 [26] dataset. Quantitative results are given in Table 1 and qualitative results with $\sigma = 50$ are shown in Fig. 3. It is observed from Table 1 that the proposed network outperforms the previous methods for all three noise levels.

4.2 Image Dehazing

Network Design. In this paper, we implement our dehazing network based on GCANet [2], shown in Fig. 2 (b) - bootom line. Note that Fig. 2 (b) only illustrates the smoothed dilated convolution period of GCANet, F_{sdin} denotes the input of the first smoothed dilated, and F_{sdout} is the input of gated fusion. The convolution blocks denote the smoothed dilated Resblock in original paper [2]. Given a hazy input image, GCANet first processes it into feature maps by the encoder part, then enhances them by aggregating more context information and fusing the features of different levels without down-sampling. Specifically, the smoothed dilated convolutions leverage an extra gate sub-network. The enhanced feature maps are finally decoded back to the original image space to obtain the target haze residue. By adding it onto the input hazy image, GCANet obtains the final dehazed image. We replaced the second and fifth Smooth Dilated Residual Block with our transferred blocks.

All our experiments were implemented in PyTorch [22] and evaluated on a single NVIDIA Tesla P100 GPU. To train the network, we randomly cropped images to 256×256 pixel size. Subsequently, the batch size was set to 21 during training. The Adam optimizer was used to train our models. The initial learning rate was set to 0.01, and decayed by 0.1 for every 40 epochs. By default, it was trained for 150 epochs.

Fig. 4. Qualitative comparisons with different dehazing methods: DCP [10], AOD-Net [15] and GCANet [2].

Result. To evaluate the proposed dehazing network, we trained and tested it on the RESIDE [16] dataset. RESIDE dataset contains a training subset of 13,990 samples of indoor scenes and a few test subsets. We used a subset SOTS (Synthetic Objective Testing Set), which contains 500 indoor scene samples for evaluation. Table 2 shows the results on RESIDE-SOTS datasets, respectively. Figure 4 depicts examples of the results obtained by the proposed network and others on the same input images. It can be noted that our results match the original images better in colour.

5   Additional Experiments on Image Super-Resolution

In this section, we test our approach for super-resolution tasks, to show the universal value of slimmer and deeper structure on image restoration tasks. Image Super-resolution (SR) is to estimate a high-resolution (HR) image from its low-resolution (LR) input. SR have attracted extensive attention in the computer vision community due to its wide range of applications. Dong *et al.* proposed SRCNN [4] to adopt deep convolutional network into solving image super-resolution. Lim *et al.* [18] proposed improved the performance by removing

Table 2. Quantitative comparisons of image dehazing on the SOTS indoor dataset from RESIDE.

	DCP [10]	GAP [38]	AOD-Net [15]	GFN [24]	GCANet [2]	Ours
PSNR	16.62	19.05	19.06	22.30	30.23	**31.15**
SSIM	0.82	0.84	0.85	0.88	0.98	**0.9814**

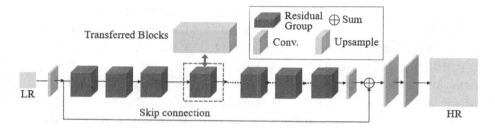

Fig. 5. Super-resolution network modified from RCAN [36] with proposed approach.

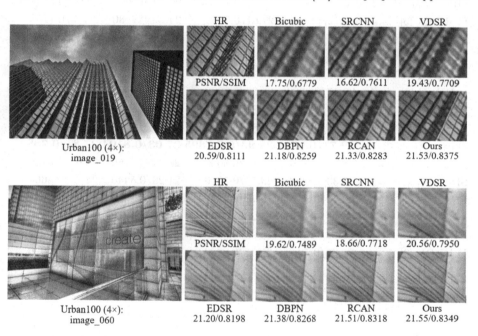

Fig. 6. Visual comparison for 4× SR with BI model on Urban100 with the state-of-the-art methods: SRCNN [4], FSRCNN [5], VDSR [12], EDSR [18], DBPN [9], and RCAN [36], respectively.

unnecessary modules in conventional residual networks. Based on [18], Zhang *et al.* [36] adopted residual in residual (RIR) and residual channel attention networks (RCAN) [36] to further improve the network.

Table 3. Quantitative results with BI degradation model. Best and second best results are highlighted and underlined, respectively.

Methods	Scale	Set5	Set14	Urban100	B100
Bicubic	×2	33.66/0.9299	30.24/0.8688	26.88/0.8403	29.56/0.8431
SRCNN [4]	×2	36.66/0.9542	32.45/0.9067	29.50/0.8946	31.36/0.8879
FSRCNN [5]	×2	37.05/0.9560	32.66/0.9090	29.88/0.9020	31.53/0.8920
VDSR [12]	×2	37.53/0.9590	33.05/0.9130	30.77/0.9140	31.90/0.8960
EDSR [18]	×2	38.11/0.9602	33.92/0.9195	32.93/0.9351	32.32/0.9013
SRMDNF [35]	×2	37.79/0.9601	33.32/0.9159	31.33/0.9204	32.05/0.8985
D-DBPN [9]	×2	38.09/0.9600	33.85/0.9190	32.55/0.9324	32.27/0.9000
RDN [37]	×2	38.24/0.9614	34.01/0.9212	32.89/0.9353	32.34/0.9017
RCAN [36]	×2	<u>38.27</u>/<u>0.9614</u>	<u>34.12</u>/<u>0.9216</u>	<u>33.34</u>/<u>0.9384</u>	<u>32.41</u>/<u>0.9027</u>
Ours	×2	**38.29/0.9618**	**34.15/0.9217**	**33.42/0.9396**	**32.47/0.9033**
Bicubic	×4	28.42/0.8104	26.00/0.7027	23.14/0.6517	25.96/0.6675
SRCNN [4]	×4	30.48/0.8628	27.50/0.7513	24.52/0.7221	26.90/0.7101
FSRCNN [5]	×4	30.72/0.8660	27.61/0.7550	24.62/0.7280	26.98/0.7150
VDSR [12]	×4	31.35/0.8830	28.02/0.7680	25.18/0.7540	27.29/0.7260
EDSR [18]	×4	32.46/0.8968	28.80/0.7876	26.64/0.8033	27.71/0.7420
SRMDNF [35]	×4	31.96/0.8925	28.35/0.7787	25.68/0.7731	27.49/0.7337
D-DBPN [9]	×4	32.47/0.8980	28.82/0.7860	26.38/0.7946	27.72/0.7400
RDN [37]	×4	32.47/0.8990	28.81/0.7871	26.21/0.8028	27.72/0.7419
RCAN [36]	×4	<u>32.63</u>/<u>0.9002</u>	<u>28.87</u>/<u>0.7889</u>	<u>26.82</u>/<u>0.8087</u>	<u>27.77</u>/<u>0.7436</u>
Ours	×4	**32.71/0.9011**	**28.93/0.0.7895**	**27.03/0.8104**	**27.81/0.7899**
Bicubic	×8	24.40/0.6580	23.10/0.5660	20.74/0.5160	23.67/0.5480
SRCNN [4]	×8	25.33/0.6900	23.76/0.5910	21.29/0.5440	24.13/0.5660
FSRCNN [5]	×8	20.13/0.5520	19.75/0.4820	21.32/0.5380	24.21/0.5680
VDSR [12]	×8	25.93/0.7240	24.26/0.6140	21.70/0.5710	24.49/0.5830
EDSR [18]	×8	26.96/0.7762	24.91/0.6420	22.51/0.6221	24.81/0.5985
D-DBPN [9]	×8	27.21/0.7840	25.13/0.6480	22.73/0.6312	24.88/0.6010
RCAN [36]	×8	<u>27.31</u>/<u>0.7878</u>	<u>25.23</u>/<u>0.6511</u>	<u>23.00</u>/<u>0.6452</u>	<u>24.98</u>/<u>0.6058</u>
Ours	×8	**27.39/0.7891**	**23.28/0.6523**	**23.14/0.6470**	**25.02/0.6527**

Network Design. Our super-resolution network is based on Residual Channel Attention Networks (RCAN) [36]. RCAN mainly consists of four parts: shallow feature extraction, residual in residual (RIR) deep feature extraction, upscale module and reconstruction part. Here, we chose two of the 10 residual groups, and reduced their channel number to half of the original and set the number of residual channel attention blocks to twice as the original. The network structure can be seen in Fig. 5, we transferred the fourth and the seventh residual group (counting from the first on the left). Other details were the same as in the original RCAN, thus more details could be found in [36].

Our model was trained by the ADAM optimizer [13] with $\beta_1 = 0.9$, $\beta_2 = 0.999$, and $\epsilon = 10\text{-}8$. All the experiments were implemented in PyTorch [22] and evaluated on a single NVIDIA Tesla P100 GPU. Subsequently, the batch size was set to 32 during training. The initial learning rate was set to 1e-4, and decreased by 0.5 every 200 epoch. We also adapted data augmentation on training images, which were randomly rotated by 270°, 180°, and 90°.

Result. Following [18], [36], we used 800 training images from DIV2K dataset [29] as training set. We used four standard benchmark datasets for testing: Set5 [1], Set14 [31], B100 [20], and Urban100 [11]. We conducted the experiments with Bicubic (BI) . The SR results were evaluated with PSNR and SSIM [30] of transformed YCbCr space. We demonstrate the quantitative results in Table 3 and qualitative results in Fig. 6. One can observe from Table 3 that the proposed network outperforms the previous methods for 4× super-resolution levels.

6   Conclusions

In this paper, we gave a rethink the balance between width of feature maps and depth of the networks for image denoising and dehazing. We explored deeper and slimmer structures and such network components as substitutes in any original networks. Experimental results show that such approach can reduce computational complexity without losing or even improving the performance. This can be explained by that in low-level vision tasks, decreasing the width of feature maps appropriately would not lead to performance deterioration but can significantly reduce the computational costs. With increased depth, our transferred network often outperforms the existing network with similar complexity and number of parameters. Additional experiments indicate that such structure can also benefit super-resolution tasks.

References

1. Bevilacqua, M., Roumy, A., Guillemot, C., Alberi-Morel, M.L.: Low-complexity single-image super-resolution based on nonnegative neighbor embedding. In: Proceedings of the British Machine Vision Conference (BMVC). BMVA Press (2012)
2. Chen, D., et al.: Gated context aggregation network for image dehazing and deraining. In: IEEE Winter Conference on Applications of Computer Vision (WACV), pp. 1375–1383. IEEE (2019)
3. Dabov, K., Foi, A., Katkovnik, V., Egiazarian, K.: Image denoising by sparse 3-D transform-domain collaborative filtering. IEEE Trans. Image Process. **16**(8), 2080–2095 (2007)
4. Dong, C., Loy, C.C., He, K., Tang, X.: Image super-resolution using deep convolutional networks. IEEE Trans. Pattern Anal. Mach. Intell. **38**(2), 295–307 (2015)

5. Dong, C., Loy, C.C., Tang, X.: Accelerating the super-resolution convolutional neural network. In: Leibe, B., Matas, J., Sebe, N., Welling, M. (eds.) ECCV 2016. LNCS, vol. 9906, pp. 391–407. Springer, Cham (2016). https://doi.org/10.1007/978-3-319-46475-6_25

6. Gao, H., Tao, X., Shen, X., Jia, J.: Dynamic scene deblurring with parameter selective sharing and nested skip connections. In: Proceedings of the IEEE Conference on Computer Vision and Pattern Recognition (CVPR), pp. 3848–3856 (2019)

7. Gharbi, M., Li, T.M., Aittala, M., Lehtinen, J., Durand, F.: Sample-based Monte Carlo denoising using a kernel-splatting network. ACM Trans. Graph. (TOG) **38**(4), 1–12 (2019)

8. Gu, S., Xie, Q., Meng, D., Zuo, W., Feng, X., Zhang, L.: Weighted nuclear norm minimization and its applications to low level vision. Int. J. Comput. Vision **121**(2), 183–208 (2017)

9. Haris, M., Shakhnarovich, G., Ukita, N.: Deep back-projection networks for super-resolution. In: Proceedings of the IEEE Conference on Computer Vision and Pattern Recognition (CVPR), pp. 1664–1673 (2018)

10. He, K., Sun, J., Tang, X.: Single image haze removal using dark channel prior. IEEE Trans. Pattern Anal. Mach. Intell. **33**(12), 2341–2353 (2010)

11. Huang, J.B., Singh, A., Ahuja, N.: Single image super-resolution from transformed self-exemplars. In: Proceedings of the IEEE Conference on Computer Vision and Pattern Recognition (CVPR), pp. 5197–5206 (2015)

12. Kim, J., Kwon Lee, J., Mu Lee, K.: Accurate image super-resolution using very deep convolutional networks. In: Proceedings of the IEEE Conference on Computer Vision and Pattern Recognition (CVPR), pp. 1646–1654 (2016)

13. Kingma, D.P., Ba, J.: Adam: a method for stochastic optimization. arXiv preprint arXiv:1412.6980 (2014)

14. LeCun, Y., Bengio, Y., Hinton, G.: Deep learning. Nature **521**(7553), 436 (2015)

15. Li, B., Peng, X., Wang, Z., Xu, J., Feng, D.: An all-in-one network for dehazing and beyond. arXiv preprint arXiv:1707.06543 (2017)

16. Li, B., Ren, W., Fu, D., Tao, D., Feng, D., Zeng, W., Wang, Z.: Benchmarking single-image dehazing and beyond. IEEE Trans. Image Process. **28**(1), 492–505 (2018)

17. Li, S.Z.: Markov Random Field Modeling in Image Analysis. Springer, Heidelberg (2009). https://doi.org/10.1007/978-1-84800-279-1

18. Lim, B., Son, S., Kim, H., Nah, S., Mu Lee, K.: Enhanced deep residual networks for single image super-resolution. In: Proceedings of the IEEE Conference on Computer Vision and Pattern Recognition Workshops (CVPRW), pp. 136–144 (2017)

19. Lu, Z., Pu, H., Wang, F., Hu, Z., Wang, L.: The expressive power of neural networks: a view from the width. In: Advances in Neural Information Processing Systems (NIPS), pp. 6231–6239 (2017)

20. Martin, D., Fowlkes, C., Tal, D., Malik, J.: A database of human segmented natural images and its application to evaluating segmentation algorithms and measuring ecological statistics. In: Proceedings IEEE International Conference on Computer Vision (ICCV), vol. 2, pp. 416–423. IEEE (2001)

21. Nah, S., Hyun Kim, T., Mu Lee, K.: Deep multi-scale convolutional neural network for dynamic scene deblurring. In: Proceedings of the IEEE Conference on Computer Vision and Pattern Recognition (CVPR), pp. 3883–3891 (2017)

22. Paszke, A., et al.: Pytorch: an imperative style, high-performance deep learning library. In: Advances in Neural Information Processing Systems (NIPS), pp. 8024–8035 (2019)

23. Qu, Y., Chen, Y., Huang, J., Xie, Y.: Enhanced pix2pix dehazing network. In: Proceedings of the IEEE Conference on Computer Vision and Pattern Recognition (CVPR), pp. 8160–8168 (2019)
24. Ren, W., et al.: Gated fusion network for single image dehazing. In: Proceedings of the IEEE Conference on Computer Vision and Pattern Recognition (CVPR), pp. 3253–3261 (2018)
25. Roth, S., Black, M.J.: Fields of experts: a framework for learning image priors. In: Proceedings of the IEEE Conference on Computer Vision and Pattern Recognition (CVPR), vol. 2, pp. 860–867. IEEE (2005)
26. Roth, S., Black, M.J.: Fields of experts. Int. J. Comput. Vision **82**(2), 205 (2009)
27. Szegedy, C., Ioffe, S., Vanhoucke, V., Alemi, A.A.: Inception-v4, inception-resnet and the impact of residual connections on learning. In: Thirty-First AAAI Conference on Artificial Intelligence (AAAI) (2017)
28. Szegedy, C., et al.: Going deeper with convolutions. In: Proceedings of the IEEE Conference on Computer Vision and Pattern Recognition (CVPR), pp. 1–9 (2015)
29. Timofte, R., Agustsson, E., Van Gool, L., Yang, M.H., Zhang, L.: Ntire 2017 challenge on single image super-resolution: methods and results. In: Proceedings of the IEEE Conference on Computer Vision and Pattern Recognition workshops (CVPRW), pp. 114–125 (2017)
30. Wang, Z., Bovik, A.C., Sheikh, H.R., Simoncelli, E.P.: Image quality assessment: from error visibility to structural similarity. IEEE Trans. Image Process. **13**(4), 600–612 (2004)
31. Zeyde, R., Elad, M., Protter, M.: On single image scale-up using sparse-representations. In: Boissonnat, J.-D., et al. (eds.) Curves and Surfaces 2010. LNCS, vol. 6920, pp. 711–730. Springer, Heidelberg (2012). https://doi.org/10.1007/978-3-642-27413-8_47
32. Zhang, H., Dai, Y., Li, H., Koniusz, P.: Deep stacked hierarchical multi-patch network for image deblurring. In: Proceedings of the IEEE Conference on Computer Vision and Pattern Recognition (CVPR), pp. 5978–5986 (2019)
33. Zhang, K., Zuo, W., Chen, Y., Meng, D., Zhang, L.: Beyond a Gaussian denoiser: residual learning of deep CNN for image denoising. IEEE Trans. Image Process. **26**(7), 3142–3155 (2017)
34. Zhang, K., Zuo, W., Zhang, L.: FFDNet: toward a fast and flexible solution for CNN-based image denoising. IEEE Trans. Image Process. **27**(9), 4608–4622 (2018)
35. Zhang, K., Zuo, W., Zhang, L.: Learning a single convolutional super-resolution network for multiple degradations. In: Proceedings of the IEEE Conference on Computer Vision and Pattern Recognition (CVPR), pp. 3262–3271 (2018)
36. Zhang, Y., Li, K., Li, K., Wang, L., Zhong, B., Fu, Y.: Image super-resolution using very deep residual channel attention networks. In: Ferrari, V., Hebert, M., Sminchisescu, C., Weiss, Y. (eds.) ECCV 2018. LNCS, vol. 11211, pp. 294–310. Springer, Cham (2018). https://doi.org/10.1007/978-3-030-01234-2_18
37. Zhang, Y., Tian, Y., Kong, Y., Zhong, B., Fu, Y.: Residual dense network for image super-resolution. In: Proceedings of the IEEE Conference on Computer Vision and Pattern Recognition (CVPR), pp. 2472–2481 (2018)
38. Zhu, Q., Mai, J., Shao, L.: A fast single image haze removal algorithm using color attenuation prior. IEEE Trans. Image Process. **24**(11), 3522–3533 (2015)

PC-OPT: A SfM Point Cloud Denoising Algorithm

Yushi Li and George Baciu$^{(\boxtimes)}$

The Hong Kong Polytechnic University, Kowloon, Hong Kong
{csysli,csgeorge}@comp.polyu.edu.hk

Abstract. Different from 3D models created by digital scanning devices, *Structure From Motion* (SfM) models are represented by point clouds with much sparser distributions. Noisy points in these representations are often unavoidable in practical applications, specifically when the accurate reconstruction of 3D surfaces is required, or when object registration and classification is performed in deep convolutional neural networks. Outliers and deformed geometric structures caused by computational errors in the SfM algorithms have a significant negative impact on the post-processing of 3D point clouds in object and scene learning algorithms, indoor localization and automatic vehicle navigation, medical imaging, and many other applications. In this paper, we introduce several new methods to classify the points generated by the SfM process. We present a novel approach, *Point-Cloud Optimization* (PC-OPT), that integrates density-based filtering and surface smoothing for handling noisy points, and maintains the geometric integrity. Furthermore, an improved *moving least squares* (MLS) is constructed to smooth out the SfM geometry with varying scales.

Keywords: 3D noise reduction · Outlier removing · SfM · Density-based clustering · Surface smoothing

1 Introduction

Three-dimensional geometry reconstruction from structured lighting or RGBD cameras is generally characterized by three main steps: (a) generate point samples, (b) for each point estimate the point depth, (c) match known object features. In this paper, we address in depth the following two questions: (1) how noisy are these point clouds? and (2) how can we reduce the noise and by how much?

The rapid development of automatic reconstruction techniques has made the acquisition of implicit 3D models a trivial process. But, this process has a critical bottleneck. The raw point cloud data is very noisy leading to severe flaws in the reconstructed geometry. Currently, there are a variety of methods to attain point clouds mainly from structured lighting such as laser-based scanning, light scanning, LiDAR scanning, and multi-view stereo [1]. Since the scattered point

© Springer Nature Switzerland AG 2020
C. Analide et al. (Eds.): IDEAL 2020, LNCS 12489, pp. 280–291, 2020.
https://doi.org/10.1007/978-3-030-62362-3_25

clouds maintain most of the structural information of the reconstructed 3D scene, they are normally adopted in further processing such as dense reconstruction, or grid generation. However, problems arise when *Structure From Notion* (SfM) is employed [2]. This is because SfM-based reconstruction algorithms create 3D point from triangulating matched image features rather than the depth information obtained by scanners.

From the perspective of model visualization, the central drawback of SfM is that the constructed representations generally have lower-resolution and more noise in comparison with the point set created by scanning devices. This makes further processing of SfM reconstruction results susceptible to outliers, noise and highly detailed structure. This disadvantage often leads to a negative impact on the complementary refinement of point clouds such as surface reconstruction, rendering, and object recognition. Hence, the major factors that affect the use of SfM point clouds are:

1. **Outliers**: SfM reconstruction is achieved by image feature matching which makes it possible to generate a large number of errors in feature correspondence. Complex geometric structure, transparent surface and varying reflection also play essential roles in the generation of outliers. Therefore, redundant outliers cannot be avoided in SfM results.

2. **Noise**: There are always some points, not too far from the main cloud structure, that do not belong to the locally fitted structural segments. These points lead to the deformation of the 3D model, which is a problem in an SfM point cloud.

3. **Density Variance**: SfM cannot provide details as the scanning techniques and the point clouds rebuilt by this method are normally sparse and not consistent in density. Varied levels of sparsity makes removing outliers and filtering noise much more difficult than operating on a model with uniform density.

In spite of these problems, SfM algorithms have their unique merits. The SfM approach does not require high-cost scanning devices, such as laser scanners or LiDAR, and it is more adaptive to large-scale reconstruction by comparison with scanning methods. Different from previous point set improvement methods that handle outliers and noise independently, our main contributions can be summarized as follows:

1. **Point Classification**: we introduce a simple but novel integrated strategy to distinguish outliers, noise, and structural points of SfM point clouds that originate from scanning arbitrary geometric shapes;

2. **Point Preservation**: we present an adaptive approach to remove points that deform the underlying object geometry in order to preserve the essential geometric information of the SfM point clouds; furthermore, our method works directly on multi-scale point clouds;

3. **Improved MLS**: we propose an improved *moving least square* (MLS) to achieve an efficient and effective 3D surface smoothing and integrate it into our sparse point cloud denoising algorithm.

2   Related Work

The denoising approaches of point cloud can be grouped into two main categories, outlier filtering and surface smoothing. Specifically, outliers are the points that are distant from the actual underlying surface. They are commonly a result of geometric distortion. Adaptively handling the outliers is not trivial on account that they are randomly distributed in space. Occasionally, the distribution of outliers is dense rather than sparse. This increases the difficulty of distinguishing outliers from structural points. Removing 3D outliers involves statistical outlier identification techniques, density evaluations, and geometric relationships. Schall et al. [2] proposed a likelihood function with non-parametric kernel density estimation to automatically identify and remove scattered noise in dense point clouds. For removing the outliers caused by measurement errors in laser scanners, Rusu et al. [3] introduced *Statistical Outlier Removal* (SOR) filter. This is currently one of the most popular tool used in point cloud cleaning. The basic idea of this approach is defining an interval by global mean distance and standard deviation to threshold the outliers. A similar 3D noise filter was proposed by Daniel Girardeau-Montaut in CloudCompare [4]. Different from SOR, the noise filter accessible in this paper replaces the distance to point neighbors by the distance to the locally fitted plane. Steinke et al. [5] combined the support vector machine (SVM) with a linear loss function to deal with outlying noise.

Many new approaches have been proposed to smooth out the model surface. According to the summary provided by [1], surface smoothing can be categorized as: local smoothing, global smoothing, and piece-wise smoothing. Local methods are the most general. The foremost characteristic of this class is that smoothing is only performed on the proximate region of the focusing data. As a consequence, local smoothing has the natural advantage in dealing with noise. The most representative branch of local smoothing, *moving least squares* (MLS), plays an essential role in modeling surface reconstruction. The core idea of MLS is to approximate the scattered points in the same neighborhood as a low-degree polynomial. The original prototype of MLS was proposed by [6]. Originally, this method was used extensively in approximating data to polynomials in 2D space. Because MLS is good at thinning the curve-like point clusters, an improved algorithm using Euclidean minimum spanning tree was proposed to reconstruct the curves without self-intersections from unorganized points by [7]. Inspired by the work of [8], Alexa et al. [9] presented a classic MLS formation which treats the surface smoothing as a point projection, and depicted the mathematical models of this procedure. This projection procedure is widely applied in the research literature on surface smoothing. [10] associated implicit MLS-based surface definition with unstructured point sets to provide a robust modeling technique. Later, [11] combined a weighting function to extend the point set surfaces to irregular settings. For improving the stability of MLS surfaces, [12] put forward a new point set surface definition that substitutes local plane by higher order algebraic sphere. This approach is more suitable to objects with relatively smooth rounded shapes. On such objects, this approach performs better than the MLS methods based on planar fitting. In addition, [13] introduced the algebraic sphere can be fitted robustly by an adaptive MLS formalism.

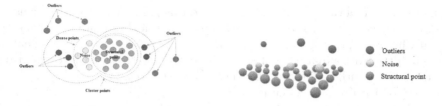

(a) 2D schematic diagram of various points defined in section 3

(b) 3D schematic diagram of the relations among outliers, noise and structural points

Fig. 1. Point definitions. In sub-figure (a), we reduce the 3D structure of a point cloud to a 2D plane and propose visual illustration of various points defined in Sect. 3. This figure demonstrates the inclusion and intersection relations among the point sets defined in this section. In sub-figure (b), 3D structural, outliers and noise points are marked with blue, yellow and red respectively. The yellow points (noise) are close to the structural points but not close enough to the local plane fitted by structural points, while the clustering outliers are much more distant. (Color figure online)

Another two subclasses of local smoothing are *hierarchical partitioning* (HP) and *locally optimal projection* (LOP). In HP methods, [14] constructed the tree structure on the basis of the residual error resulted from local quadratic polynomial fitting. Different from MLS and HP approaches, LOP methods do not need to make use of least squares fit. [15] fitted the neighborhood point sets onto a multi-variate median.

3 Point Definitions

The potential outliers and noise points always drift away from the primary point cluster. Even when some of these points are surrounded by clusters representing regional borders or surfaces, they can still be registered as visual noise. On the basis of this observation, we attempt to formalize an effective strategy to identify the points required for representing constitutional models and discard the rest. This strategy contains five interrelated classifications to determine whether a point belongs to a structure or needs to be removed. In Fig. 1 (a), we can easily see the inclusion and intersection relations among the points defined in this section. Let $\mathbf{P} = \{\mathbf{p_i} = (x_i, y_i, z_i) \mid i = 1, 2, ..., n\}$ be a point set reconstructed by SfM in $\mathbf{R}^3$ that consists of the information approximating the geometric structure of the object, and let $\mathbf{d}(\mathbf{p_a}, \mathbf{p_b})$ be the metric distance between two points $\mathbf{p_a}$ and $\mathbf{p_b}$. Then, the following five definitions represent the classes of points.

Cluster point – A point $\mathbf{p_c}$ is called cluster point if: (1) it is defined as the core point; or (2) it is defined as the border point. The definitions of core and border points are discussed in DBSCAN clustering section. The point cluster is denoted as $\mathbf{P_{cluster}}$.

Dense Point – In assigned clusters, a point $\mathbf{p_c}$ is recognized as dense point $\mathbf{p_d}$ if its neighborhood region is dense enough. A measure of density can be formulated based on the kNN neighbors around this point. We propose a density function to evaluate the dense level of each point and a thresholding method to confirm whether a point is dense or not in the local point cluster it is affiliated with. The detailed computation of this process will be discussed in the section of density-based outliers filtering. A dense point set is defined as $\mathbf{P_{dense}}$.

Structural point – In a dense point set, a point $\mathbf{p_d}$ is defined as structural point $\mathbf{p_s}$ if its vertical distance to a locally fitting plane (estimated from a local neighborhood of the corresponding dense point) is small enough. That is, it can be considered as a part of the principal geometric structure represented by a point cloud. The set of structural point is represented as $\mathbf{P_{struct}}$.

Clustering Outlier – There are two kinds of points defined as outliers $\mathbf{p_o}$: (1) the points that are not assigned to any cluster; and (2) the points that are assigned to clusters but not dense.

Noise – The point that belongs to a dense point set, but it is not close enough to its locally fitted plane, is defined as noise $\mathbf{p_n}$. In light of our analysis and observations, a noisy point is the dominant cause of structural deformation in 3D point clouds. In Fig. 1 only points that comply with the definition of a structural point are retained after the outliers and noise are removed.

4   Adaptive Density-Based Outlier Removal

The outlier removal method presented in this paper is inspired by the properties of DBSCAN clustering and the Gaussian filter. In this section, a practical methodology is proposed to estimate the original radius parameter needed by DBSCAN algorithm. We then discuss how to embed our density function within the GMM for removing the outlier points which are normally sparser than others from the common point set clustered by the DBSCAN method. In DBSCAN, most distant outliers are not classified into any cluster. In addition, if ε is too large, the resulting point cloud tends to have only a few clusters with a large number of points. This makes a big impact on the local density variation in our adaptive density-based filtering and more noisy objects would result after clustering. On the other hand, if this parameter is quite small, many clusters with small size would be found. Meanwhile some useful border or core points would be removed during the process of clustering, which would diminish the integrity of the reconstruction models. Aside from modifying the definitions of DBSCAN method that meet our requirements, a simple function is set forth to compute ε based on the global average kNN distance of the SfM point cloud:

$$d_{avg} = \frac{1}{nk} \sum_{i=1}^{n} \sum_{j=1}^{k} d\left(\mathbf{p_i}, \mathbf{p_j}\right); \qquad \text{then, } \varepsilon = \alpha \cdot d_{avg} \left(\frac{1}{e}\right)^{d_{avg}} \qquad (1)$$

where α is a constant adjustment parameter, n is the size of the point set, and k is the kNN neighboring parameter. The parameter α is adopted to associate with d_{avg} for adjusting the neighborhood scope of each point in the clustering process of DBSCAN. Without α, the ε computed directly from d_{avg} often leads to assigning unwanted outliers into clusters, which will cause negative effect in the performance of outlier removing. If α is set small (e.g. 0.1–0.3 for the large-scale models applied in our experiments), a large number of potential structural points would be removed as outliers during the clustering stage. Large α setting would preserve more outliers in the resulting point cloud.

The parameter ε required in DBSCAN clustering can be considered as the radius applied to evaluate the ε-reachable and density connected state of each independent object in the point set. But this parameter is a global parameter and it does not adapt easily to non-uniform local data. Although we choose the average kNN distance of a point set to be ε for complying with SfM point clouds with different scales, the compactness of these models is still susceptible to outliers that are far from the structural point clusters.

The key idea in our adaptive density-based outlier removal is to construct a function to estimate the local density of each point appropriately and eliminate the points with the lowest local densities via a threshold operation determined by a larger local region rather than the global set. More specifically, the k nearest neighbors of each point that remained after clustering are found by the kd-tree algorithm. Then, the Euclidean distance $(d_1, ..., d_k)$ from each neighbor to the centering point is calculated. The density value of each point is assessed by:

$$\rho_p = \frac{1}{(c + e^{d_p})} \quad \text{where} \quad d_p = \frac{1}{k} \sum_{i=1}^{k} d(\mathbf{p}, \mathbf{p_i}), \ \ (i = 1...k) \tag{2}$$

Note that d_p is the average distance from k nearest neighbors to the center point, and c is a constant parameter. We set c to 3 in all experiments. This controls the shape of the function described by Eq. 2 for assessing an appropriate density value of each point. Because the density ρ_p is estimated depending on each point and its k nearest neighbors, this density estimation function overcomes the disadvantage of the global method, and it gives the dense status of each point locally.

Outliers surrounding a 3D point cloud can be filtered out by our proposed adaptive method based on the Gaussian mixture model to handle more complex situations. The multivariate parameters of the mixed model such as μ and σ are estimated by *expectation-maximization* (EM) algorithm. In general, it is difficult to choose the thresholding parameter β used in our method as it is data dependent. Fortunately, a mixture of Gaussian curves improves the fit of the distribution explicitly. Since estimating β is based on the density distribution of each cluster, using β is more adaptive than a globally fixed parameter, and a local estimation function can be employed to evaluate β:

$$\beta = \left(\gamma^{\rho_{\max} - \rho_{\min}} - \epsilon \right) \tag{3}$$

where $\rho_{\max}$ and $\rho_{\min}$ are the maximum and minimum density values of the point cluster. Here, γ and ϵ are constants set to 1.01 and 0.3–0.6, respectively. To adapt the Gaussian mixture model, the new $\mathbf{P_{dense}}$ is defined as follows:

$$\mathbf{P_{dense}} = \{\mathbf{p} \in \mathbf{P_{cluster}} \mid \rho_p > \mu_l - \beta \cdot \sigma_l\} \tag{4}$$

Notably, μ_l and σ_l are the parameters applied in the mixture model.

5   MLS-Based Smoothness

The MLS method builds a polynomial to approximate a continuous curve or surface over a point set. The basic procedure of this approach starts with an arbitrary point $\mathbf{p}_*$ and its N neighbor points in the data set. It then estimates the local polynomial approximation that best fits the data with the formulation of weighted least squares given by Eq. 5. Then, it projects $\mathbf{p}_*$ to the surface represented by this polynomial. This procedure repeats over the entire domain of a point set, and the locally fitted functions are blended to create a global fit.

$$\min_f \sum_i^N w\,(\mathbf{p}_i - \mathbf{p}_*)\,\|f(\mathbf{p}_i) - f_i\|^2 \tag{5}$$

where f is the polynomial that describes the fitted plane and $w\,(\mathbf{p}_i - \mathbf{p}_*)$ is the weighting function that indicates the effect of the points adjacent to $\mathbf{p}_*$. We use the minimization model proposed in [9] to implement the point projection as follows:

$$\min_n \sum_i^N w\,(\mathbf{p}_i - \mathbf{p}_*)\,(\mathbf{n} \cdot (\mathbf{p_i} - \mathbf{p}_*))^2 \tag{6}$$

where $\mathbf{n}$ is the normal vector orthogonal to the local surface passing through $\mathbf{p}_*$. Here, we propose a minimization method to approximately estimate the initial $\mathbf{n}$. The partial derivative of the above equation can be rewritten as:

$$(\mathbf{W} \odot (\mathbf{A} \cdot \mathbf{n}))^T \cdot \mathbf{A} \tag{7}$$

for $\mathbf{W}$ is a $N \times 1$ vector containing the weights of all N neighbors of $\mathbf{p}_*$, and $\mathbf{A}$ is the $N \times 3$ matrix composed of $\mathbf{p}_i - \mathbf{p}_*$, $i = 1, ..., N$. Note that $\odot$ means element-wise multiplication. We integrate this function by adding a penalty to the constraint $\|\mathbf{n}\| = 1$. We propose a minimization function to estimate the initial $\mathbf{n}$:

$$\min_n \{\|\mathbf{M} \cdot \mathbf{n}\|^2 + \lambda(1 - \|\mathbf{n}\|^2)\} \tag{8}$$

In the above equation, λ is the penalty parameter used to tune the effect of the penalty constraint $\|\mathbf{n}\| = 1$ on this minimization problem. In our experiments we set λ to 10^7.

6   PC-OPT Algorithm

The PC-OPT algorithm integrates and significantly improves upon four major developments in the current research: **DBSCAN**, **GMMfilter**, **MLS**, and **FND** (further noise filter). In particular, the **KDtreeneighbor** algorithm is

```
Input: point set P, k, ε₁, ε₂
Output: optimized model optm
  ε = Avgdist(P, k);
clusters = DBSCAN(P, ε);
for cluster in clusters do
    if len(cluster) ≥ k then
        if len(cluster) ≥ m then
            smoothpts ← array used to store smoothed points;
            filteredpts=GMMfilter(cluster, k, ε₁);
            for i = 0; i < len(filteredpts) do
                if len(filteredpts) ≥ m then
                    nearestpts=
                    KDtreeneighbor(filteredpts, m);
                else
                    nearestpts=
                    KDtreeneighbor(filteredpts, len(filteredpts)
                end
                smoothpt=MLS(nearestpts, filteredpts[i]);
                insert smoothpt into smoothpts;
            end
            structuralpts=FND(smoothpts, k);
            insert structuralpts into optm;
        else
            filteredpts=Densfilter(cluster, k, ε₂);
            insert filteredpts into optm
        end
    end
end
return optm
```

the kdtree-based procedure used to determine k-nearest neighbors of each point that passes the thresholding of **GMMfilter**. The function **Densfilter** is different from **GMMfilter**; it is built on the first definition of $\mathbf{P_{dense}}$ (Eq. 4). This is because the Gaussian mixture is not suitable in removing outliers if the size of input point cluster is small. We set the parameter m to 100 in our experiments for deciding to use **GMMfilter** or **Densfilter**. In this paper, we employ RANSAC as the **FND**. Finally, the core part of this algorithm put forward in this paper can be summarized as follows.

7   Experiments and Evaluation

We compare the performance of the PC-OPT with two other widely applied outlier and noise removing approaches [3,4]. Our experimental results demonstrate that our approach cleans and smooths out the SfM reconstruction models with distinct scales and various densities robustly. We choose typical datasets that are popular in 3D reconstruction: Template [16], Bunny [17], Aachen [18], Notre Dame Arco Valentino, Palazzo Carignano, and Villa La Tesoriera.

7.1   Quantitative Analysis

In this paper, we apply the confusion matrix to evaluate the performances of different methods on synthetic data and design a simple *Nearest Neighbor Distance* (NND) to analyze the processed results of real SfM point clouds. Four metrics (precision, accuracy, recall and F-measure) estimated from the confusion matrix are employed in our evaluation of synthetic data results as follows:

1. **Precision**: expresses the ability of an algorithm to identify structural points;
2. **Accuracy**: reveals the ratio of structural and non-structural (outliers and noise) points that a method can accurately identify in a point cloud;
3. **Recall**: justifies the ability of an approach to maintain structural points;
4. **F-1 score**: provides an optimal blend of precision and recall to comprehensively assess the abilities of an algorithm in the identification of structural points and their maintenance.

In the synthetic experiments, the structural and non-structural points are labeled before the computation process of the confusion matrix. The details of the confusion matrix analysis of synthetic data is discussed in Sect. 7.2.

7.2   Experiments on Synthetic Data

Before evaluating the performances of distinctive methods on arbitrarily reconstructed scene models, the robustness of these measurements are examined on synthetic data. In this subsection, our experiments are ran on two point sets representing as spheres, while they have 2000 and 5000 structural points respectively. We add 400−1600 unexpected outliers and noise points to the synthetic spheres. The structural points are randomly diffused on the surface of a sphere. In addition, for testing the performances of SOR and CloudCompare noise filters, we use the optimal parameters of these two methods.

In contrast to the point spheres processed by CloudCompare, the output of SOR is much better. More specifically, points of synthetic sphere are maintained well and most deviated points are removed. Nevertheless, there are still many noisy points left close to the sphere surface. This means that SOR is not robust enough to handle the noisy points surrounding the main geometric structure.

Not only our method is highly robust in the outlier removal process, but it is also locally adaptive in reducing the spatial local noise. The kNN parameter in all these approaches is set to 6, and the thresholding parameter ϵ of Eq. 3 is determined as 0.1. Notably, the density value shown in this figure is estimated using Eq. 2.

Table 1 contains the confusion matrix of the experimental results generated by different algorithms. CloudCompare fairs poorly at maintaining the geometric structure of a point cloud due to its low recall values. Meanwhile, the precision results indicate that the SOR filter is deficient in identifying structural points. In comparison with these two methods, PC-OPT provides the optimal results in most estimation metrics (precision, accuracy, F-1 score), except recall. This means that our algorithm is good at both of filtering unwanted points and maintaining essential geometry.

Table 1. Confusion matrix analysis of synthetic spheres 2 (with 5000 structural points)

Dataset	Sphere 2 (400)			Sphere 2 (800)			Sphere 2 (1200)			Sphere 2 (1600)		
Metric	Our	SOR	CC	Our	SOR	CC	Our	SOR	CC	Our	SOR	CC
Precision	99.8%	98.5%	80.5%	99.6%	96.1%	66%	99.1%	92.6%	73.5%	98.6%	87.4%	69.9%
Accuracy	99.2%	98.6%	14.1%	99.2%	96.5%	18.8%	98.8%	93.5%	37.3%	98.4%	89.1%	38.2%
Recall	99.2%	100%	9.5%	99.5%	100%	11.9%	99.4%	100%	26.9%	99.3%	100%	32.4%
F-1 score	99.6	99.2	17	99.5	98	20.2	99.3	96.1	39.4	99	93.3	44.3

7.3 Experiments on SfM Point Clouds

We rebuilt 3D point clouds from real shot photographs in order to evaluate the performance of the algorithm presented in this paper in practical scenarios. Different from synthetic data, the point set generated from SfM reconstruction generally has a more complex structure and a larger variance in the point density distribution. Moreover, a glaring trait of SfM point cloud is that the structure is sometimes fragmented. In other words, the 3D point set created by SfM is not as densely connected and compact as the results scanned by laser devices. This makes the cleaning approaches designed for laser-based reconstruction ineffective. Namely, removing the outliers and noise from real data generated by SfM is more challenging than processing a synthetic set or a laser reconstruction result. At the same time, we want to retain the integrity of the geometric structure. In this subsection, we discuss the experimental consequences derived from Cloud-Compare noise filter, SOR and our algorithm. Specifically, in our experiments, the point number used for the mean distance estimation in SOR is tested in a large range from 3 to 30, while the multiplier threshold is set from 1 to 10. In CloudCompare noise filter, the kNN parameter is tested from 3 to 30.

Experiments on Small-Scale Objects

A point cloud resulted from SfM usually represents the shape of an object, typically its surface. Under normal conditions, the interaction between the external structures of small-scale objects are easy to present by point clouds. However, superficial densities of small-scale reconstruction models are more uniform than large-scale models. In addition, the small-scale models usually have better continuity comparing with large-scale ones. This implies that the outliers of small-scale objects are often less problamatic than their counterparts of the large-scale objects, whereas small-scale model is more susceptible to the noise close to its surface structure. We choose two reconstruction models with varying surface densities, Fig. 2.

Experiments on Large-Scale Scenes

Our comparisons were performed on five distinctive large-scale point clouds. In our practical experiment, kNN parameters k is set to 8. Similar to the results

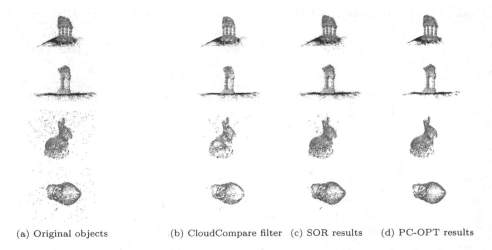

(a) Original objects   (b) CloudCompare filter (c) SOR results   (d) PC-OPT results

Fig. 2. Experimental results of small-scale models. In (a), the subfigures show the original small-scale point clouds of template and bunny.

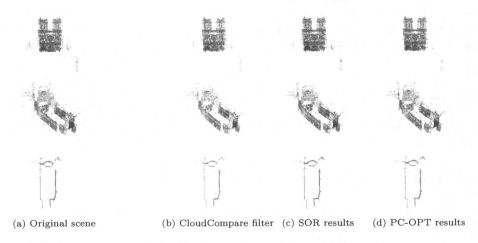

(a) Original scene   (b) CloudCompare filter (c) SOR results   (d) PC-OPT results

Fig. 3. Experimental results of large-scale outdoor models that present parts of archi-tectures. PC-OPT outperforms both of CloudCompare filter and SOR.

obtained from synthetic data and small-scale models, CloudCompare noise filter removes too many points, while the SOR method reserves numerous objects in experimental point sets.

8 Conclusions

In this paper, we propose new types of feature points in point cloud data sets. Based on these features, a novel approach, called PC-OPT, to optimize

SfM reconstruction is proposed. Our new PC-OPT algorithm integrates outlier removal and 3D surface smoothing and performs better on both synthetic data and real models.

References

1. Berger, M., et al.: A survey of surface reconstruction from point clouds. In: Computer Graphics Forum, vol. 36, no. 1 (2017)
2. Schall, O., Belyaev, A., Seidel, H.-P.: Robust filtering of noisy scattered point data. In: Proceedings Eurographics/IEEE VGTC Symposium Point-Based Graphics. IEEE (2005)
3. Rusu, R.B., Cousins, S.: 3D is here: point cloud library (PCL). In: 2011 IEEE International Conference on Robotics and Automation. IEEE (2011)
4. Girardeau-Montaut, D.: CloudCompare (2016)
5. Steinke, F., Schölkopf, B., Blanz, V.: Support vector machines for 3D shape processing. In: Computer Graphics Forum, Amsterdam, vol. 24, no. 3. (1982, 2005)
6. Lancaster, P., Salkauskas, K.: Surfaces generated by moving least squares methods. Math. Comput. **37**(155), 141–158 (1981)
7. Lee, I.-K.: Curve reconstruction from unorganized points. Comput. Aided Geom. Design **17**(2), 161–177 (2000)
8. Levin, D.: Mesh-independent surface interpolation. In: Brunnett, G., Hamann, B., Müller, H., Linsen, L. (eds.) Geometric Modeling for Scientific Visualization. Mathematics and Visualization, pp. 37–49. Springer, Heidelberg (2004). https://doi.org/10.1007/978-3-662-07443-5_3
9. Alexa, M., et al.: Computing and rendering point set surfaces. IEEE Trans. Vis. Comput. Graph. **9**(1), 3–15 (2003)
10. Pauly, M., Keiser, R., Kobbelt, L.P., Gross, M.: Shape modeling with point-sampled geometry. ACM Trans. Graph. (TOG) **22**(3), 641–650 (2003)
11. Adamson, A., Alexa, M.: Anisotropic point set surfaces. In: Proceedings of the 4th International Conference on Computer Graphics, Virtual Reality, Visualisation and Interaction in Africa (2006)
12. Guennebaud, G., Gross, M.: Algebraic point set surfaces. In: ACM SIGGRAPH 2007 papers, p. 23-es (2007)
13. Chen, J., et al.: Non-oriented MLS gradient fields. In: Computer Graphics Forum, vol. 32, no. 8 (2013)
14. Ohtake, Y., et al.: Multi-level partition of unity implicits. In: ACM SIGGRAPH 2005 Courses, p. 173-es (2005)
15. Lipman, Y., et al.: Parameterization-free projection for geometry reconstruction. ACM Trans. Graph. (TOG) **26**(3), 22-es (2007)
16. Seitz, S.M., et al.: A comparison and evaluation of multi-view stereo reconstruction algorithms. In: 2006 IEEE Computer Society Conference on Computer Vision and Pattern Recognition (CVPR 2006), vol. 1. IEEE (2006)
17. Kolev, K., et al.: Continuous global optimization in multiview 3D reconstruction. Int. J. Comput. Vision **84**(1), 80–96 (2009). https://doi.org/10.1007/s11263-009-0233-1
18. Sattler, T., Leibe, B., Kobbelt, L.: Improving image-based localization by active correspondence search. In: Fitzgibbon, A., Lazebnik, S., Perona, P., Sato, Y., Schmid, C. (eds.) ECCV 2012. LNCS, vol. 7572, pp. 752–765. Springer, Heidelberg (2012). https://doi.org/10.1007/978-3-642-33718-5_54

GELAB and Hybrid Optimization Using Grammatical Evolution

Muhammad Adil Raja[1,2,3](✉) ⓘ, Aidan Murphy[1,2,3] ⓘ, and Conor Ryan[1,2,3] ⓘ

[1] Bio-computing and Developmental Systems (BDS) Research Group,
University of Limerick, Limerick, Ireland
[2] The Irish Software Research Centre, Lero, University of Limerick, Limerick, Ireland
[3] Department of Computer Science and Information Systems,
University of Limerick, Limerick, Ireland
{adil.raja,aidan.murphy,conor.ryan}@ul.ie
http://bds.ul.ie/

Abstract. Grammatical Evolution (GE) is a well known technique for program synthesis and evolution. Much has been written in the past about its research and applications. This paper presents a novel approach to performing hybrid optimization using GE. GE is used for structural search in the program space while other meta-heuristic algorithms are used for numerical optimization of the searched programs. The hybridised GE system was implemented in GELAB, a Matlab toolbox for GE.

Keywords: Grammatical evolution · Genetic algorithms · Simulated annealing · Swarm optimization · Hybrid optimization · GELAB

Much has been written about the theory and applications of GE in the past. GE is a technique that takes inspiration from Darwinian evolution and genetics. From an evolutionary point of view, it embraces the Darwinian worldview in which the survival of the fittest is the guiding principle in the evolution of living systems. From a genetics standpoint, it is a proponent of the fact that a certain genotypic sequence gives rise to a certain phenotypic structure. The phenotype, physical traits, or properties of a species, are the direct result of its genotype. Any evolutionary changes at the genotypic level manifest themselves as phenotypic changes in the individual during the course of evolution.

GE is a mathematical modeling or a program search technique. It has been quite successful in deriving human competitive solutions to problems requiring mathematical treatment [9]. However, little research has been reported about utilizing GE in hybrid optimization. In this paper, a novel scheme for performing hybrid optimization using GE and a handful of other meta-heuristic algorithms is proposed. We have run our scheme on numerous benchmark problems for Symbolic Regression (SR). The results are promising and are better than the state of the art results. The proposed scheme is also computationally efficient. The system is integrated with GELAB, a Matlab toolbox for GE.

The rest of the paper is organized as follows: In Sect. 1, the GE algorithm is described. In Sect. 2 we talk about hybrid optimization from a retrospective

C. Analide et al. (Eds.): IDEAL 2020, LNCS 12489, pp. 292–303, 2020.
https://doi.org/10.1007/978-3-030-62362-3_26

perspective. Section 3 discusses how the capability of hybrid optimization is achieved using GELAB. Section 4 discusses the hybrid optimization algorithm and some of its technical underpinnings. Section 5 presents details of the benchmark experiments and Sect. 6 contains the results of these experiments. Finally, Sect. 7 concludes the paper.

1 Grammatical Evolution

GE is a variant of Symbolic Regression (SR). However, GE differs from GP in certain ways. GE was first proposed by the Bio-computing and Developmental Systems (BDS) Research Group, Department of Computer Science and Information Systems (CSIS), University of Limerick, Ireland[1] [9]. As mentioned earlier, GE is inspired by Darwinian evolution. Given a user-specified problem it creates a large population of randomly generated integer arrays. These arrays are called genotypes. Each of the integer arrays is mapped to a corresponding program. This mapping process is central to GE. Mapping is performed using the production rules of a grammar specified in Backus Naur Form (BNF) form. Although various types of grammars can be used, a context-free grammar is most commonly employed. Figure 1b shows the conceptual diagram of GE's mapping process.

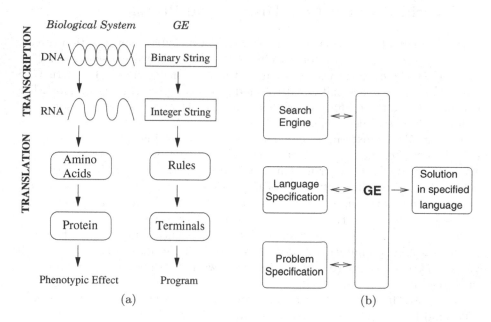

Fig. 1. (a) Genotype to phenotype mapping in biological systems and in GE. (b) Conceptual diagram of GE's mapping process.

---

[1] Web: http://bds.ul.ie/libGE/.

GE eventually evaluates each of the programs by using the problem data and assigns it a fitness score. After that, reproduction of programs is carried out over a certain number of generations. Genetic operators of selection, crossover, and mutation are applied to produce an offspring population. Subsequently, fitness evaluation of the child population is done. Finally, replacement is applied as a final step to remove the undesirable solutions and to retain better candidates for the next iteration of the evolutionary process. Evolution continues until the desired solution is found or the stopping criterion is met.

To this end, the search space in GE is the set of all the possible computer programs specified by the grammar.

To yield computer programs, GE requires a source that can generate a large number of genomes. To accomplish this the GE mapper is normally augmented with an integer-coded Genetic Algorithm (GA). The GA creates a large number of integer-coded genomes at each iteration. The genomes are then mapped to the corresponding genotype using the mapper. Pseudo-code of GA is given in Algorithm 1.

GELAB was introduced recently in [7]. It is a Matlab toolbox for GE. Its mapper is written in Java. GELAB is user-friendly and interactive. In this paper, we are reporting our capability to hybridize GE with other meta-heuristic algorithms using GELAB.

2   Hybrid Optimization: History and Practice

In the context of Evolutionary Algorithms (EAs), hybrid optimization means to integrate two algorithms for learning. One of the algorithms searches for an appropriate structure of the solution to the problem at hand. The structure here refers to the exact mathematical model or a computer program. The other algorithm tunes the coefficients of this structure. There are several reasons why such a scheme is so desirable.

GP or GE are extremely popular schemes for solving user-specified problems. Before their advent, a practitioner would take a problem, contemplate on the mathematical model for its solution, and fix it. Eventually the practitioner would try to tune the coefficients of the presupposed model with a numerical optimization algorithm. Traditionally various line search and trust region algorithms were in vogue. Examples of these include gradient descent, Levenberg Marquardt (LVM), Quasi Newton etc. Later, advancement in meta-heuristic algorithms and computational power gave users the options of algorithms such as Simulated Annealing (SA), GAs, Particle Swarm Optimization (PSO) etc. for numerical optimization.

If the user failed to optimize the model to a desirable level, the only choice left would be to contemplate on the model again and come up with a different structure. Tuning of coefficients would be the left out chore that would again be done with the help of one of the aforementioned algorithms.

There are numerous problems with this approach. The foremost of these is the human bias involved in choosing a structure of a solution. The second is the time and effort it takes by an engineer in setting and resetting the structure in case of repeated lack of luck in finding the right solution.

In the meantime GP was invented. Then gradually its variants, including GE, came along to stand by its side, and often even to compete with it. GP and GE had this capability to search whole program spaces to find the optimal solution. The solution here refers to both the right structure for the mathematical model (or computer program) as well as the right values of the coefficients. As a matter of fact, GP style algorithms specialize in finding the right solution in terms of the structure. As far as the values of the coefficients are concerned, they are normally chosen randomly and no specialized scheme is employed to tune them. To be more precise, most GP systems employ ephemeral random constants as values for various coefficients of a model. To this end, coefficient optimization in GP (or GE) is rather whimsical.

Afterwards, schemes for hybrid optimization were introduced. The idea was normally to hybridize a GP system with a numerical optimization algorithm. In such a scheme, the GP system would search for an appropriate structure of the solution for a problem at hand. At the same time, the numerical optimization algorithm would search for a set of values for the coefficients of the solution model. For instance, in [3] Howard and Donna proposed a hybrid GA-P algorithm. GP was used to find optimal expressions for problem solving and a GA was used to tune the coefficients of the GP trees/expressions during evolution. Similarly in [11] Topchy and Punch have used the gradient descent algorithm for the local search of leaf coefficients of GP trees. Moreover, quasi-Newton method has been used to achieve the same objective in [5].

3   Hybrid Optimization Using GELAB

As described in the previous sections, hybrid optimization employs two levels of optimization. At the first level, a program synthesis algorithm is used to derive the structure for the target mathematical model. At the second level, a numerical optimization algorithm is employed to tune the coefficients of the derived model. GELAB was employed for program synthesis. For numerical optimization, many options from Matlab's numerical and global optimization toolboxes were available. The numerous algorithms that were integrated with GELAB are SA, GA, PSO, the LVM algorithm and the Quasi Newton method. Integration with other algorithms is also possible. However, the work that is reported in this paper is confined to the use of SA, GA and PSO only. Below, an example of how hybrid optimization actually works and how it is accomplished in GELAB.

Consider the following target expression:

$$y = -2.3 + (0.13 * sin(x)) \tag{1}$$

In this, y is a function of an independent variable, x. The function has two constants. The expression also has a non-linear function, sin. In deriving such an expression with hybrid optimization, GE is leveraged to derive the structure of the model including any functions such as *sin*, as in Eq. 1. The optimum values for the constants of the target expression are derived by further tuning the derived models either with SA, GA or PSO.

In GELAB this is accomplished by allowing the grammar to have constants in its specification. This is shown in Listing 1.1. The noticeable thing about this is that constants are specified in the form of w(1), w(2) etc. Moreover, variables are specified in the form of X(:, 1), X(:, 2) etc. The reason for this specification is that it is suitable for Matlab's *eval* function to solve expression that have data variables as well as any coefficients specified in this manner. So, as it stands, Matlab inteprets X(:, 1) as all instances of the first input variable (a vector), and so on. Similarly, w(1) is interpreted as the first tunable coefficient (a scalar), and so on. Specifying variables and constants in this way allows GELAB to construct programs that will have these building blocks as constituents. Evolution is then guided by following the principle of survival of the fittest and a function that is behaviorally closer to the target expression (Eq. 1) is searched.

Listing 1.1. A context free grammar used for hybrid optimization in GELAB.

```
<expr>    ::=     (<expr> <op> <expr>)
                | <u−pre−op>(<expr>)
                | <b−pre−op>(<expr>,<expr>)
                | <var>
                | <const>
<op>      ::=    + | − | .*
<u−pre−op>       ::=     ge_square | ge_cube |
                        ge_sin | ge_cos |
                        ge_log | ge_exp |
                        ge_sqrt
<b−pre−op>       ::=     ge_divide
<var>     ::=    X(:,1) | X(:,2) | X(:,3) |
                 X(:,4) | X(:,5)
<const> ::=      w(1) | w(2) | w(3) | w(4) |
                 w(5) | w(6)
```

As an individual is created, it receives randomly chosen terminals (variables and constants) and non-terminals (functions). While the values for variables (X(:, 1), X(:, 2), $\cdots$, X(:, N)) are assigned by the input data, the values for constants (w(1), w(2), $\cdots$, w(6)) are randomly assigned initially. Initially the individual is evaluated using MSE only. During this evaluation, whatever the random values for constants the individual has, are used. In subsequent generations, where the individual has to be optimized with hybrid optimization, the hybrid algorithm also treats variables (X(:, 1), X(:, 2), $\cdots$, X(:, N)) as (constant) input data. Moreover, it also treats constants (w(1), w(2), $\cdots$, w(6)) as tunable coefficients.

4   Hybrid Algorithm

As discussed earlier, hybrid optimization employs an algorithm for searching for a program (or a mathematical model) with the right structure from the program space. Additionally, it employs a numerical optimization algorithm to find the appropriate values for the coefficients for the programs being found. GE has been employed for the former and a number of meta-heuristic algorithms for the latter.

In an ideal case, hybrid optimization should be able to tune the coefficients of all the programs (mathematical models) being found by the program search algorithms. However, this requirement can make hybrid optimization quite compute intensive [11]. To address this issue, numerous compromises were made to retain the computational efficiency as well as to be able to derive more accurate results. The compromises, as well as particular details of the algorithm, are listed as follows.

4.1   The Basic Algorithm

A scheme in which multiple fitness functions are used during the course of evolution was employed. As an individual is born (in the initial generation or subsequent generations due to reproduction), it is simply subjected to Mean Squared Error (MSE) based fitness evaluation. As an individual acquires an age of two (by surviving two generations of evolution), it is subjected to hybrid optimization and fitness is evaluated based on that. After a certain number of generations, when the age of the individual has reached a certain higher level, evaluation is done based on scaled-MSE (MSE_s) [4]. Although this mature age is configurable, it was set to two generations in the experiments.

Moreover, not all individuals undergo hybrid optimization due to the high computational costs associated with the process. An individual is chosen for this with a certain probability. In the experiments, this configurable probability is set to 0.25.

MSE_s is given in Eq. 2.

$$MSE_s(y, t) = 1/n \sum_i^n (t_i - (a + by_i))^2 \tag{2}$$

where y is a function of the input parameters (a mathematical expression), y_i represents the value produced by a GE individual and t_i represents the target value. a and b adjust the slope and y-intercept of the evolved expression to minimize the squared error. They are computed according to Eq. (3).

$$a = \bar{t} - b\bar{y}, b = \frac{cov(t, y)}{var(y)} \tag{3}$$

Algorithm 1. Basic GE Algorithm (calls Algorithm 2 on line number) 10.

```
parentPop=initPop ;
parentPop=genotype2phenotypeMapping ( parentPop );
parentPop=evalPop ( parentPop ).

for ( i =1:numGens)
        childPop=selection ( parentPop );
        childPop=crossover ( parentPop );
        childPop=mutation ( parentPop );
        childPop=genotype2phenotypeMapping ( childPop );
        childPop=evalPop ( childPop );
        parentPop=replacement ( parentPop , childPop );
end
```

4.2 Of Age and Eligibility

Computational intensiveness of hybrid algorithms has led to such compromises in the past, too, where only a fraction of the whole GP population is further tuned with a numerical optimization algorithm [11]. This work introduces an age-based eligibility scheme for hybrid optimization. To this end, as a program is created it is evaluated using MSE. At this point, it is assigned an age of zero. Moreover, its fitness function remains MSE till the time it is one generation old. The age of an individual is incremented by one at the point of fitness evaluation at every generation.

As the age of an individual reaches two generations, it becomes eligible for hybrid optimization. At this stage it is treated with the meta-heuristic with a certain probability as discussed in Sect. 4.1. The individual remains eligible for hybrid optimization till the time it is five generations old.

After this, the fitness function is changed to linear scaling (MSE_s) for all the subsequent generations. All of these steps are shown in Algorithm 2.

4.3 Of Improvability

As mentioned earlier, we have used three meta-heuristic algorithms for hybrid optimization. These are SA, a GA and PSO.

Since meta-heuristic algorithms are stochastic search algorithms, it is a customary practice in Machine Learning (ML) to run the algorithm on the same instances of the problem multiple times before the results can be trusted. The reason is to achieve more accurate and credible results. However, running an algorithm on the same problem multiple times has its computational costs. To address this issue, another trade off is employed in the algorithm.

To this end, every individual that is generated by GE is deemed improvable by default. That is to say, it is assumed that every program is improvable by hybrid optimization. However, as a program is optimized using the meta-heuristic, the

Algorithm 2. Fitness evaluation and hybrid optimization.

```
age=individual.age;

if(age<=params.criticalAge1)
     individual = MSE(individual);
elseif(age<=criticalAge2 && age>criticalAge1)
    r=rand;
    if(r<0.25)
         individual = hybridOptimzation(individual);
    end
else
    individual = linearScaling(individual);
    individual.isEvaluated=1;
end
```

improvement in fitness is noted. If the meta-heuristic further improves the fitness of the program, the program is deemed *improvable* again. If, however, the met-heuristic does not improve the fitness of the program, it is assumed that either the program is not improvable, or it is difficult to improve it. So the program is marked as *not-improvable*. As a result, this program cannot undergo hybrid optimization further.

4.4   Time Bounded Numerical Optimization

To further reduce computational time, another constraint was added to the process of numerical optimization of individuals produced by GE. In this scheme, each of the individuals that undergo numerical optimization is only allowed thirty seconds of exposure to the meta-heuristic for optimization.

4.5   Similarities with Incremental Evolution

The scheme of employing three different fitness functions during the lifetime of an individual is somewhat similar to incremental evolution proposed in [2]. When an individual is a neonate, we evaluate it with simple MSE. As it matures to some extent, by surviving a couple of generations, it is subjected to hybrid optimization, and the fitness is assigned by the meta-heuristic. As it surpasses its eligible age for hybrid optimization it is subjected to MSE_s for further improvement and evaluation.

5   Experimental Setup

5.1   Grammar Used for Experiments

The following grammar was employed in the research. In this, <var> and <const> contain as many variables and constants as the problem requires, respectively.

Table 1. Benchmark Polynomials

Sr. No	No	Polynomial
1	Keijzer1	$f(x) = 0.3x sin(2\pi x)$
2	Keijzer2	$f(x) = 1 + 3x + 3x^2 + x^3$
3	Keijzer3	$f(x, y) = 8/(2 + x^2 + y^2)$
4	Keijzer4	$f(x, y) = x^4 - x^3 + y^2/2 - y$
5	Keijzer5	$f(x, y) = x^3/5 + y^3/2 - y - x$
6	Keijzer6	$f(x_1, \cdots, x_{10}) = 10.59x_1x_2 + 100.5967x_3x_4 - 50.59x_5x_6$ $+ 20x_1x_7x_9 + 5x_3x_6x_{10}$

```
<expr>      ::=       <lexpr><op><expr>
            |    <b−pre−op>(<expr>, <expr>)
                 | <var>
                 | <const>
<lexpr> ::= <expr>
<op>        ::= + | − | .*
<b−pre−op>  ::=  ge_divide
<var>       ::=      X(:,1)
<const> ::=      w(1)
<pop>       ::= ""
```

5.2   Benchmark Problems and Evolutionary Parameters

All experiments had a population size of 500, tournament selection, elitist replacement and adaptive crossover and mutation probabilities. Ramped half-and-half initialisation was used for GP. We used the grammatical counterpart of this scheme for GE that is known as sensible initialization [8]. In order to conduct the GP experiments we used GPLab [10]. The total number of runs in each experiment was equal to 40. Moreover, each run was 50 generations long.

In this research, we have used six different polynomials from the symbolic regression domain. These polynomials have been used in the past in [1] as suitable benchmarks. The polynomials are given in Table 1.

For each of these polynomials, we randomly initialized the independent variables in the interval [−1.5, 1.5]. We generated 100 data points and chose half of them for training and the remaining half for testing on the unseen data.

6   Results

Figures 2 and 3 plot the results of the experiments. The x-axis represents 50 generations throughout. The y-axis represents normalized scores viz a viz MSE. The scores are computed according to Eq. 4.

$$score = \frac{1}{1 + MSE} \qquad (4)$$

Each sampled point in the plots shows an average over 40 independent runs. The 95% confidence limits of the error bars are computed according to equation:

$$\bar{X} \pm 1.96 \frac{\sigma}{\sqrt{n}} \tag{5}$$

where $\bar{X}$ and σ are the mean and standard deviation of the fitness scores of n runs ($n = 40$ in our case). To this end, it implies that one can be 95% confident that the fitness scores from all the runs lie within these error bars. Moreover, a lack of overlap between any two of the modeled schemes means that the corresponding populations are statistically different.

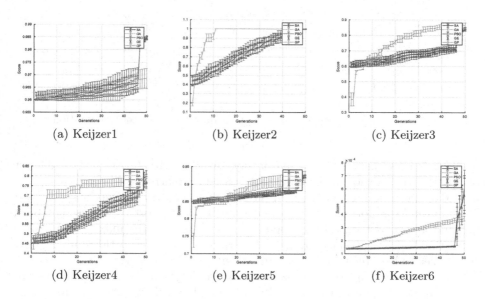

(a) Keijzer1   (b) Keijzer2   (c) Keijzer3

(d) Keijzer4   (e) Keijzer5   (f) Keijzer6

Fig. 2. Mean of best fitness alongwith errorbars for each generation.

A keen look at the test results in Fig. 3 shows that various flavors of GE performed at least as well as GP on 5 out of the 6 chosen benchmark problems. Only on one problem (Fig. 3c) the results of GE were significantly inferior to GP. Moreover, on two problems, Figs. 3a and 3f, all variants of GE performed significantly better than GP. To this end, our results are similar to the ones reported in [1].

7   Conclusions

This paper proposes a novel approach of hybridizing GE with meta-heuristic algorithms. The system uses SA, GA and PSO as different options for hybridizing. Experiments were conducted using several benchmark polynomials for symbolic regression. The results are promising and show improved performance on 5 benchmark problems.

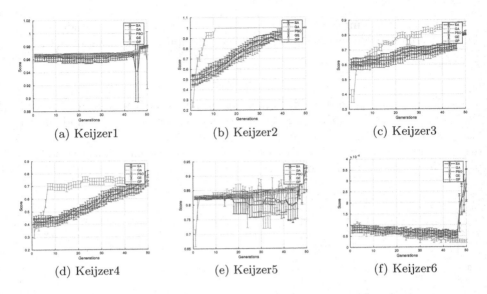

(a) Keijzer1   (b) Keijzer2   (c) Keijzer3

(d) Keijzer4   (e) Keijzer5   (f) Keijzer6

Fig. 3. Mean of test fitness score alongwith errorbars for each generation.

In future, we aspire to test hybrid optimization with GE using a wider set of benchmarks. Although the polynomials used currently have been used to test hybrid optimization techniques using GE [1], the need to employ more specialized benchmarks is important. It can be seen in Fig. 3 for most problems that both GP and all the versions of GE approach a score of "1", the maximum possible score that can be attained by any algorithm. This leaves little room for any specialized hybrid technique to vividly stand out from the rest.

The proposed approach can have widespread applications ranging from symbolic regression to classification and can be employed in evolutionary robotics.

In particular, in [6] an approach was proposed recently to derive controllers for unmanned aerial vehicles using meta-heuristic algorithms. The efficiency of meta-heuristic algorithms, including GP, has been limited so far. By going a step further and hybridizing GP-like systems with other meta-heuristics can lead to better controllers. Hybridizing GE with meta-heuristics is a step ahead in that direction.

Acknowledgements. The authors thank the anonymous referees for their time, comments and helpful suggestions. This work was supported, in part, by Science Foundation Ireland grant 13/RC/2094 and co-funded under the European Regional Development Fund through the Southern & Eastern Regional Operational Programme to Lero - the Irish Software Research Centre (www.lero.ie).

References

1. Azad, R.M.A., Ryan, C.: Comparing methods to creating constants in grammatical evolution. In: Ryan, C., O'Neill, M., Collins, J.J. (eds.) Handbook of Grammatical Evolution, pp. 245–262. Springer, Cham (2018). https://doi.org/10.1007/978-3-319-78717-6_10

2. Barlow, G.J., Oh, C.K., Grant, E.: Incremental evolution of autonomous controllers for unmanned aerial vehicles using multi-objective genetic programming. In: 2004 IEEE Conference on Cybernetics and Intelligent Systems, vol. 2, pp. 689–694. IEEE (2004)

3. Howard, L.M., D'Angelo, D.J.: The GA-P: a genetic algorithm and genetic programming hybrid. IEEE Expert 10(3), 11–15 (1995). https://doi.org/10.1109/64.393137

4. Keijzer, M.: Scaled symbolic regression. Genet. Program Evol. Mach. 5(3), 259–269 (2004). https://doi.org/10.1023/B:GENP.0000030195.77571.f9

5. Mugambi, E.M., Hunter, A., Oatley, G., Kennedy, L.: Polynomial-fuzzy decision tree structures for classifying medical data. Knowl.-Based Syst. 17(2–4), 81–87 (2004). http://www.sciencedirect.com/science/article/B6V0P-4C4VYG9-2/2/8ee7c8541e99bf3c8c22922dad2ebfbf. https://doi.org/10.1016/j.knosys.2004.03.003

6. Raja, M.A., Rahman, S.U.: A tutorial on simulating unmanned aerial vehicles. In: 2017 International Multi-topic Conference (INMIC), pp. 1–6 (2017). https://doi.org/10.1109/INMIC.2017.8289450

7. Raja, M.A., Ryan, C.: GELAB - a matlab toolbox for grammatical evolution. In: Yin, H., Camacho, D., Novais, P., Tallón-Ballesteros, A.J. (eds.) IDEAL 2018. LNCS, vol. 11315, pp. 191–200. Springer, Cham (2018). https://doi.org/10.1007/978-3-030-03496-2_22

8. Ryan, C., Azad, R.M.A.: Sensible initialisation in grammatical evolution. In: Barry, A.M. (ed.) GECCO 2003: Proceedings of the Bird of a Feather Workshops, Genetic and Evolutionary Computation Conference. pp. 142–145. AAAI, Chicago (2003)

9. Ryan, C., Collins, J.J., Neill, M.O.: Grammatical evolution: evolving programs for an arbitrary language. In: Banzhaf, W., Poli, R., Schoenauer, M., Fogarty, T.C. (eds.) EuroGP 1998. LNCS, vol. 1391, pp. 83–96. Springer, Heidelberg (1998). https://doi.org/10.1007/BFb0055930

10. Silva, S., Almeida, J.: Gplab-a genetic programming toolbox for matlab. In: Proceedings of the Nordic MATLAB Conference, pp. 273–278. Citeseer (2003)

11. Topchy, A., Punch, W.F.: Faster genetic programming based on local gradient search of numeric leaf values. In: Spector, L., et al. (eds.) Proceedings of the Genetic and Evolutionary Computation Conference (GECCO-2001). pp. 155–162. Morgan Kaufmann, San Francisco (2001). http://www.cs.bham.ac.uk/~wbl/biblio/gecco2001/d01.pdf

Distribution-Wise Symbolic Aggregate ApproXimation (dwSAX)

Matej Kloska[(✉)] and Viera Rozinajova

Faculty of Informatics and Information Technologies, Slovak University of Technology in Bratislava, Ilkovičova 2, 842 16 Bratislava, Slovak Republic
{matej.kloska,viera.rozinajova}@stuba.sk
https://www.fiit.stuba.sk

Abstract. The Symbolic Aggregate approXimation algorithm (SAX) is one of the most popular symbolic mapping techniques for time series. It is extensively utilized in sequence classification, pattern mining, anomaly detection and many other data mining tasks. SAX as a powerful symbolic mapping technique is widely used due to its data adaptability. However this approach heavily relies on assumption that processed time series have Gaussian distribution. When time series distribution is non-Gaussian or skews over time, this method does not provide sufficient symbolic representation. This paper proposes a new method of symbolic time series representation named distribution-wise SAX (dwSAX) which can deal with Gaussian as well as with non-Gaussian data distribution in contrast with the original SAX, handling only the first case. Our method employs more general approach for symbol breakpoints selection and thus it contributes to more efficient utilization of provided alphabet symbols. The goal is to optimally cover the information space. The method was evaluated on different data mining tasks with promising improvements over SAX.

Keywords: Time series · Kernel density estimator · SAX

1 Introduction

Whenever we work with any kind of event observations taken according to the order of time, we usually talk about time-order sequences also known as time series.

With the raise of big data and streaming technologies we see various fields such as healthcare, finance, security and industry where intelligent analysis and data mining tasks take place. Aforementioned tasks in context of time series are demanding and usually need different approaches due to data instances temporal ordering characteristic. At the same time, time series usually capture feature rich, highly dimensional data which make processing tasks even harder. In our work, the high dimensionality is related to high number of data points in time series. Dimensionality reduction and descriptive forms of time series representation are

© Springer Nature Switzerland AG 2020
C. Analide et al. (Eds.): IDEAL 2020, LNCS 12489, pp. 304–315, 2020.
https://doi.org/10.1007/978-3-030-62362-3_27

recognized as a possible solution for highly performing data mining tasks [6]. A challenging area in the field of effective time series processing is their compact data presentation without sacrificing any significant information [17]. Symbolic representation of time series appears to be the solution to this problem.

The Symbolic Aggregate approXimation algorithm (SAX) [8] is one of the most popular symbolic mapping techniques for time series. SAX as a powerful symbolic mapping technique is widely used due to its data adaptability. It is extensively utilized in sequence classification [13], pattern mining [3], anomaly detection [7] and many other data mining tasks [9,14,15]. However, this approach heavily relies on assumption that processed time series have Gaussian distribution [8]. When time series distribution is non-Gaussian or skews over time, this method does not provide sufficient symbolic representation. This paper proposes a new method named distribution-wise SAX (dwSAX) which can deal also with non-Gaussian data distribution in contrast with the original SAX. Our method employs more general approach for symbol breakpoints selection and thus improves tightness of lower bounding without sacrificing other SAX benefits.

This paper is organized as follows. Section 2 describes original SAX method and possibilities to data distribution estimation. Section 3 introduces dwSAX - our improvement of SAX method. Section 4 contains an experimental evaluation of the proposed method on time series clustering and anomaly detection tasks compared to the original SAX method. Finally, Sect. 5 offers some conclusions and suggestions for future work.

2 Related Work

One of efficient data stream processing problems is their high dimensionality, too high number of data points. Possible solution to this issue is efficient symbolic representation that represents highly dimensional data stream through less dimensional symbolic data stream. In past decades, many different time series representations have been introduced. Lin et al. [8] divided methods into data adaptive (eg. Piecewise Linear Approximation, Singular Value Decomposition, SAX) and non data adaptive (eg. Wavelets, Random Mappings, Discrete Fourier Transformation). Recent research [2,10,11,15,18] shows activities in both method families. In the following sections we discuss fundamentals of original SAX, and techniques for data distribution estimation.

2.1 Symbolic Representation - SAX

SAX is one of the best known algorithms for symbolic time series representation. This method makes it possible to represent any time series of length n using a string of any length w ($w \ll n$) with symbols from predefined alphabet. Looking at the mentioned method, it consists of:

1. *Dimensionality reduction:* applying Piecewise Aggregate Approximation (PAA) [6], it significantly reduces dimensionality and preprocesses time series for further step;

2. *Discretization:* mapping PAA segments into specific symbols from alphabet based on precomputed mapping symbols table.

Concept of this method is illustrated in Fig. 1. Throughout this paper we use common notation used also in original SAX paper [8] which you can find in Table 1.

Table 1. A summarization of common notation used in this paper and the original SAX paper [8].

C	A time series $C = c_1, ..., c_n$ where $c_i \in \mathbb{R}$
$\bar{C}$	A Piecewise Aggregate Approximation of a time series $\bar{C} = \bar{c}_1, ..., \bar{c}_w$
$\hat{C}$	A symbol representation of a time series $\hat{C} = \hat{c}_1, ..., \hat{c}_w$
w	The number of PAA segments representing time series C
a	Alphabet size (e.g., for the alphabet = {a, b, c}, $a = 3$)

Dimensionality Reduction. Intuition based on aforementioned description is to reduce time series from n dimensions into w dimensions. This goal is simply achieved by division of the time series into w equal sized pieces. For each piece, mean value is calculated and this value represents underlying vector of w original values. Total vector of all pieces becomes new reduced representation of the original time series.

More formally, a time series C of length n can be reduced into a w-dimensional time series by a vector $\bar{C} = \bar{c}_1, ..., \bar{c}_w$ where i^{th} element of $\bar{C}$ is calculated as follows [8]:

$$\bar{c}_i = \frac{w}{n} \sum_{j=\frac{n}{w}(i-1)+1}^{\frac{n}{w}i} c_j \tag{1}$$

Discretization. Discretization step replaces PAA segments obtained in the length reduction step by alphabet symbols. Assuming data are normalised before reduction (with zero mean) and have highly Gaussian distribution, the replacement is performed as follows. We first precompute a table of breakpoints. Lin et al. [8] defined breakpoints as sorted list of numbers $B = \beta_1, ..., \beta_{a-1}$ such that the area under a $N(0, 1)$ Gaussian curve from β_i to $\beta_{i+1} = 1/a$ (β_0 and β_a are defined as $-\infty$ and ∞, respectively).

The Gaussian curve enables efficient breakpoints table precomputation, thus the discretization step is trivial in comparison to the single vector lookup operation.

Finally, formal definition of SAX as proposed by Lin et al. [8]: A subsequence C of length n can be represented as a word $\hat{C} = \hat{c}_i, ..., \hat{c}_w$ as follows. Let α_i denote

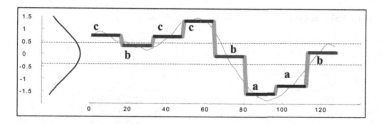

Fig. 1. Concept of Original SAX. Background grey thin line is replaced with bold line segments (PAA). PAA segments are mapped by normal distribution sketched on y axis into symbols, $a = (-\infty; -0.43\rangle, b = (-0.43; 0.43\rangle, c = (0.43, \infty)$ [8].

the ith element of the alphabet, i.e., $\alpha_1 = a$ and $\alpha_2 = b$. Then the mapping from a PAA approximation C to a word $\hat{C}$ is obtained as follows:

$$\hat{c}_i = \alpha_j, \quad iif \quad \beta_{j-1} \leq \bar{c}_i < \beta_j \tag{2}$$

2.2 Techniques for Distribution Estimation

In the previous section we discussed internals of the original SAX method. SAX uses a Gaussian distribution to derive the regional breakpoints resulting in the generation of an equiprobable set of symbols. As we already mentioned, our method wants to make a new SAX method Gaussian distribution requirement free. In this section we want to mention other methods how to estimate data distribution and, based on them, improve SAX by a different way of setting the regional breakpoints. At the beginning, we want to state a common intuition to graphically represent data distribution - histogram plotting from underlying time series data points.

The histogram is former nonparametric density estimator with strong use in explorative data analysis for displaying and summarizing data. Bin width is an important parameter that needs selection prior histogram construction. It is evident that the choice of the bin width has a strong effect on the shape of the resulting histogram. The example for different bin width selection you can find in Fig. 2.

There were proposed many ways to determine optimal bin width $\hat{h}$ with n observed instances such as [16]:

$$\hat{h} = \frac{range\_of\_data}{1 + log_2 n} \tag{3}$$

or alternatively more general formula based on Mean Integrated Squared Error (MISE):

$$\hat{h} = \hat{C} n^{-1/3}, \tag{4}$$

where $\hat{C}$ is any selected statistic. Most known example of above mentioned formula is normal reference rule [12,16]:

$$\hat{h} = 3.49 \hat{\sigma} n^{-1/3}, \tag{5}$$

where $\hat{\sigma}$ is an estimate of the standard deviation.

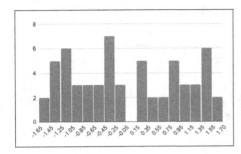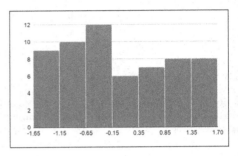

Fig. 2. Comparison of two histograms for the same dataset with bin widths 0.2 and 0.5 respectively. Incorrectly selected bin width causes visually different data distribution.

In past four decades there was a research in the field of distribution estimation using continuous functions - density estimators. Intuition behind kernel estimators is to describe underlying data histogram with smooth continuous line. More formally, given a set of N training data $y_n, n = 1, ..., N$, a kernel density estimator (KDE), with the kernel function K and a bandwidth parameter h, $(h \in R; h > 0)$, gives the estimated density $\hat{f}(y)$ for data y as follows [4]:

$$\hat{f}(y) = \frac{1}{N} \sum_{n=1}^{N} K\left(\frac{y - y_n}{h}\right) \tag{6}$$

Kernel function K should satisfy positivity and integrate-to-one constraints [4]:

$$K(y) \geq 0, \qquad \int_{R^+} K(y)dy = 1 \tag{7}$$

The quality of a kernel estimate depends less on the chosen K than on the bandwidth value h. It is crucial to choose the most suitable bandwidth as a value that is too small or too large will result in not useful estimation. Small values of h lead to undersmoothing estimates while larger h values lead to oversmoothing [5]. Figure 3 illustrates possible cases of incorrectly selected bandwidth parameter h.

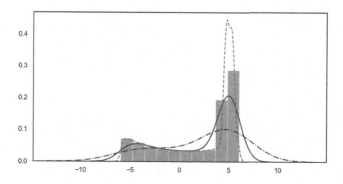

Fig. 3. Implications of different KDE bandwidth parameter selection Gaussian kernel. Optimal (blue line), oversmoothing (green line) and undersmoothing (orange line) bandwidths. (Color figure online)

3 Distribution-Wise SAX (dwSAX)

Formerly proposed SAX method is not well suited for time series with non-Gaussian distribution. If we apply Gaussian distribution lookup vector for breakpoints, we get non-optimal, still feasible, breakpoints. Our modified implementation focuses on elimination of this deficiency. Algorithm 1 illustrates overall main function method flow where input is sequence for symbolization, word length, alphabet size and bandwidth for kernel estimator result is sax representation of input sequence. In the next sections we will discuss specific internals of our method:

1. *Probability density function estimation:* given normalized time series, we need to estimate probability density function to proceed with more precise breakpoints;
2. *Breakpoints vector calculation:* having probability density function, the task is to calculate equiprobable breakpoints covering the domain of pdf.

Data: Sequence, WordLength, AlphaSize, Bandwidth
Result: SAX Repr
PAA Sequence = *ApplyPAA*(Sequence, WordLength);
PDF Estimate = *EstimatePDF*(Sequence);
Breakpoints Vector = *CalculateBreakpoints*(AlphaSize, PDF Estimate);
SAX Repr = *Map*(PAA Sequence, Breakpoints Vector);

Algorithm 1: dwSAX main algorithm

3.1 Probability Density Estimation

With transformed PAA time series we can proceed with probability density estimation. Having precisely estimated probability function will help us in further step breakpoints vector calculation for discretization procedure. Naive solution

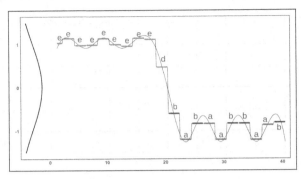

Example of SAX representation.
Resulting symbolic string: *eeeeeeeeedbabaabbaab*

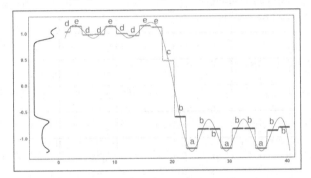

Example of dwSAX representation.
Resulting symbolic string: *deddeddeecbabbabbabb*

Fig. 4. Application of different pdf on breakpoints selection. a) SAX with Gaussian distribution, b) dwSAX with KDE. KDE based breakpoints bring more precise symbolic representation compared with SAX with same alphabet size 5 and word length 2 (PAA = 2).

to this problem seems histograms exploitation as its computational complexity is much lower than the other methods such as KDE. The main drawback of histograms is their discrete representation which is not suitable for breakpoints interpolation. Our method for breakpoints calculation expects continuous probability function suitable for integral calculus with integrate-to-one constraint. KDE appears to be the solution to this problem. This methods needs to specify a kernel function K and a bandwidth parameter h. Selection of appropriate kernel function and bandwidth function depends on data and required precision of overall symbolic representation performance. To our best knowledge, Gaussian kernel gives most relevant results and should be applied as first possible option. The difference between dwSAX and SAX is depicted in Fig. 4.

3.2   Breakpoints Vector Calculation

Having estimated probability function using KDE, we can advance and estimate breakpoints based on probability distribution of time series. The main goal is to efficiently compute those breakpoints as we do not apply only specific Gaussian distribution and its precomputed values table. However, the idea for breakpoints selection is the same - select points from KDE probability density function (pdf) such that they produce equal-sized areas under KDE function curve.

Definition 1. *Let a denote alphabet size, pdf probability density function and β_n, β_{n+1} any two consecutive breakpoints from breakpoints vector B. Then breakpoints vector B is defined as a vector of ordered breakpoints β such that β_n, β_{n+1} follows:*

$$\int_{\beta_n}^{\beta_{n+1}} pdf(y)dy = \frac{1}{a} \tag{8}$$

Discretization process follows the same algorithm as proposed in the original SAX method. For reference see Algorithm 2.

Data: PAA Sequence, Breakpoints Vector B
Result: SAX Representation
foreach *Segment in PAA Sequence* **do**
    for $i \leftarrow 2$ **to** *Length(B)* **do**
        if $B[i-1] \leq Segment$ **and** $Segment < B[i]$ **then**
        | Append(SAX Representation, Alphabet[i])
        end
    end
end

Algorithm 2: dwSAX mapping procedure

---

4   Evaluation

We evaluated our proposed method modification using data mining tasks such as time series clustering and novelty/anomaly detection - we compared it with the former method. As far as we know, there is no similar symbolic representation method that we can confront with, except of the SAX. In the next sections we discuss achieved results with their implications in real life method exploitation.

4.1   Clustering

Clustering is by nature one of the most commonly used data mining tasks. Formally, clustering is the division of data into groups of similar objects [1]. Traditionally, clustering techniques are divided into hierarchical and partitioning method families. For purposes of our evaluation, we decided to employ hierarchical clustering and correctly graphically illustrate results. In this data mining task evaluation we used well known Control Chart dataset[1] with selected Normal, Cyclic, Increasing trend and Decreasing trend charts represetatives.

Hierarchical clustering gives us a brief overview of similarity measures performance. Similarity measure is in case of dwSAX and SAX virtually the same but symbolized sequences of time series are expected to be more closely clustered together within specific subtree. Figure 5 shows resulting dendrograms after applying aglomerative hierarchical clustering with Euclidean distance as similarity measure. Both dendrograms seem to be correct at class level series subtrees (second level subtrees from the bottom of dendrogram). Small differences can be observed at intra-class level subtrees clustering where dwSAX clusters more similar series in a common subtree depicting significantly smaller distance between them.

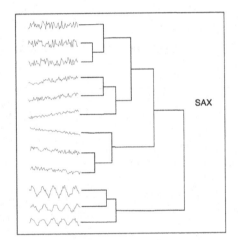 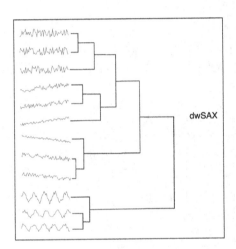

Fig. 5. A comparison of hierarchical clustering of selected Control Chart series. Both methods perform well, dwSAX gives more detailed clustering inside specific series class.

4.2 Novelty/Anomaly Detection

Novelty or anomaly detection is common data mining task, usually applied in data cleaning/preprocessing step or as targeted task. This process consists of learning phase and detection phase itself. In learning phase, we try to infer normal behavior model from observed data and apply it in detection phase. SAX and dwSAX are candidates for improving such kind of detection. Designed detection model consists of symbolization produced by specific SAX method and Markov chain model for encoding normal behavior motifs from produced symbolic representation. During detection phase, we replay window of last N symbols and sign current symbol as anomaly when detection statistic is below specific threshold. We compared performance of mentioned methods on two different time series of

---

[1] https://archive.ics.uci.edu/ml/datasets/Synthetic+Control+Chart+Time+Series.

the same length: a) periodical slightly noisy sine time series, b) periodical time series with strong seasonality. Figure 6 shows the results. In evaluation dataset there were 4 real anomalies and 6 temporal phases with not clear anomalous state. From the detection report we see that both methods are able to help in detection of clear anomalies (4 of 4) in strongly periodical time series. dwSAX slightly outperform SAX in detection of temporal unclear phases (3 of 6).

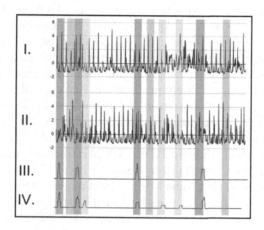

Fig. 6. A comparison of anomaly detection of real life household electricity consumption dataset. I. 1000 points training dataset II. 1000 points test dataset. III. and IV. detection using SAX and dwSAX respectively. Red, orange and grey areas depict commonly detected anomalies, dwSAX only detected anomalies and not detected anomalies respectively. (Color figure online)

5   Conclusion and Future Work

As stated in the Introduction, our main goal was to improve symbolic representation of time series with non-Gaussian data distribution. Lin et al. [8] proposed a superior method for symbolic representation of time series - SAX. Although this approach is interesting, it does not work well for time series with non-Gaussian data distribution. We believe that we have designed an innovative solution for this problem. Our approach extends original SAX by means of dynamically captured data distribution of underlying time series and defining alternative vector of breakpoints for characters mapping. Data distribution estimation at its simplest form could be estimated through well-known and widely applicable histograms. However, this approach suffers from the ease of dynamic computation of breakpoints. An alternative solution, though with high overheads is estimation using continuous, function based, estimator. Density estimators appear to be a solution to this problem. The most common variants of these estimators are kernel density estimators.

This method represents a viable alternative to original SAX method. We compared our method with original SAX in two data mining tasks: clustering of time series and anomaly detection. As stated in the Evaluation, our method was able to improve clustering performance by means of significant lowering objective function over original SAX within the same number of iterations. In anomaly detection, both methods detected major anomalies in provided time series. However, our method detected in addition to major anomalies also the less evident ones. This was achieved by improved breakpoints vector with higher resolution for highly probable values in time series.

The most important limitation lies in unnecessary KDE application in case of highly Gaussain distributed data. Applying both methods in this case will result in very similar symbolic representation. KDE estimates breakpoints similar to precomputed breakpoints from SAX table, but with undoubtly higher computational complexity. On the other had, applying dwSAX without any prior knowledge of data distribution will safely produce efficient symbolic representation. A number of potential shortcomings need to be considered. Firstly, computational complexity of KDE and breakpoints vector recalculations needs to be considered in case of online exploitation. Secondly, the concept drift at its basis is not covered in proposed method, thought KDE with periodical recalculation is able to overcome a skew in data distribution to some extend. The third shotcoming is converned with the fact, that knowledge of breakpoints vector used during discretization is crucial for further operations such as time series indexing. Despite this we believe that our work could be a springboard for research in the field of data distribution-wise symbolic time series representation.

This study has gone some way towards enhancing our understanding of efficient symbolic time series representation. To deepen our research we plan to design online version of our method to tackle computational complexity with hard online processing constraints. Our results are promising and should be validated by a larger sample size time series from real-life environments.

Acknowledgments. This work was partially supported by the project "Knowledge-based Approach to Intelligent Big Data Analysis" - Slovak Research and Development Agency under the contract No. APVV-16-0213 and "International Centre of Excellence for Research of Intelligent and Secure Information-Communication Technologies and Systems - phase II" - ITMS2014+ 313021W404.

References

1. Berkhin, P.: A survey of clustering data mining techniques. In: Kogan, J., Nicholas, C., Teboulle, M. (eds.) Grouping multidimensional data, pp. 25–71. Springer, Heidelberg (2006). https://doi.org/10.1007/3-540-28349-8_2
2. Eghan, R.E., Amoako-Yirenkyi, P., Omari-Sasu, A.Y., Frimpong, N.K.: Time-frequency coherence and forecast analysis of selected stock returns in Ghana using Haar wavelet. J. Adv. Math. Comput. Sci. **30**(5), 1–12 (2019)
3. Fournier-Viger, P., Lin, J.C.W., Kiran, R.U., Koh, Y.S., Thomas, R.: A survey of sequential pattern mining. Data Sci. Pattern Recogn. **1**(1), 54–77 (2017)
4. Hwang, J.N., Lay, S.R., Lippman, A.: Nonparametric multivariate density estimation: a comparative study. IEEE Trans. Signal Process. **42**(10), 2795–2810 (1994)

5. Jones, M.C., Marron, J.S., Sheather, S.J.: A brief survey of bandwidth selection for density estimation. J. Am. Stat. Associ. **91**(433), 401–407 (1996)
6. Keogh, E., Chakrabarti, K., Pazzani, M., Mehrotra, S.: Dimensionality reduction for fast similarity search in large time series databases. Knowl. Inf. Syst. **3**(3), 263–286 (2001)
7. Keogh, E., Lin, J., Fu, A.: HOT SAX: efficiently finding the most unusual time series subsequence. In: Fifth IEEE International Conference on Data Mining (ICDM 2005), pp. 8–pp. IEEE (2005)
8. Lin, J., Keogh, E., Lonardi, S., Chiu, B.: A symbolic representation of time series, with implications for streaming algorithms. In: Proceedings of the 8th ACM SIG-MOD Workshop on Research Issues in Data Mining and Knowledge Discovery, pp. 2–11 (2003)
9. Lin, J., Keogh, E., Wei, L., Lonardi, S.: Experiencing SAX: a novel symbolic representation of time series. Data Min. Knowl. Disc. **15**(2), 107–144 (2007)
10. Mahmoudi, M.R., Heydari, M.H., Roohi, R.: A new method to compare the spectral densities of two independent periodically correlated time series. Math. Comput. Simul. **160**, 103–110 (2019)
11. Sato, T., Takano, Y., Miyashiro, R.: Piecewise-linear approximation for feature subset selection in a sequential logit model. J. Oper. Res. Soc. Japan **60**(1), 1–14 (2017)
12. Scott, D.W.: On optimal and data-based histograms. Biometrika **66**(3), 605–610 (1979)
13. Senin, P., Malinchik, S.: SAX-VSM: interpretable time series classification using sax and vector space model. In: 2013 IEEE 13th International Conference on Data Mining, pp. 1175–1180. IEEE (2013)
14. Shieh, J., Keogh, E.: iSAX: indexing and mining terabyte sized time series. In: Proceedings of the 14th ACM SIGKDD International Conference on Knowledge Discovery and Data Mining, pp. 623–631 (2008)
15. Tamura, K., Ichimura, T.: Clustering of time series using hybrid symbolic aggregate approximation. In: 2017 IEEE Symposium Series on Computational Intelligence (SSCI), pp. 1–8. IEEE (2017)
16. Wand, M.: Data-based choice of histogram bin width. Am. Stat. **51**(1), 59–64 (1997)
17. Wang, X., Mueen, A., Ding, H., Trajcevski, G., Scheuermann, P., Keogh, E.: Experimental comparison of representation methods and distance measures for time series data. Data Min. Knowl. Disc. **26**(2), 275–309 (2013)
18. Yang, S., Liu, J.: Time-series forecasting based on high-order fuzzy cognitive maps and wavelet transform. IEEE Trans. Fuzzy Syst. **26**(6), 3391–3402 (2018)

Intelligent Call Routing
for Telecommunications Call-Centers

Sérgio Jorge[1]([✉])(ID), Carlos Pereira[2](ID), and Paulo Novais[3](ID)

[1] University of Minho, Campus Gualtar, 4710 Braga, Portugal
`A77730@alunos.uminho.pt`
[2] NOS Comunicações, Senhora da Hora, Portugal
`carlos.migpereira@nos.pt`
[3] Algoritmi Center, University of Minho, Campus Gualtar, 4710 Braga, Portugal
`pjon@di.uminho.pt`

Abstract. At telecommunications companies, call-centers have the highest interaction with customers, and the operators' performance is vital because an excellent service satisfies the customer and helps a better operation. Therefore, attempts are made to use customer data, call operator data, and historical service data to improve support. Pairing a customer with an operator who is comfortable with the problem to solve helps companies reducing costs, improves customer service, and increases employee productivity. In this article, we propose an approach based on machine learning and optimization, which predicts the problem for which the customer is calling and routes the call and the customer to the most appropriate call operator. The results show that using large amounts of business data along with innovative algorithms such as LightGBM can improve the customer support performance.

Keywords: Call-center · Customer · Data mining · Machine learning · Optimization · Telecommunications · Workforce

1 Introduction

In a world increasingly connected, companies that can capture data and information from their customers with the highest agility and accuracy stand out. The digital age and the area of Big Data, relative to the enormous amount of data, have helped companies to make better decisions in their businesses.

The telecommunications market has also seen continuous growth since the beginning of the 21st century, as mobile phones and the Internet have become indispensable in our society. A Portuguese communications and entertainment group offers fixed and mobile internet, television, and voice solutions. Therefore, the number of customers (and data) they have is enormous, leading to potentially large profits and challenges in keeping their consumers satisfied. Solving customer problems (doubts, technical or non-technical issues) in a fast and effective way is fundamental and much sought after, not only in this company but in

© Springer Nature Switzerland AG 2020
C. Analide et al. (Eds.): IDEAL 2020, LNCS 12489, pp. 316–328, 2020.
https://doi.org/10.1007/978-3-030-62362-3_28

any organization in the telecom area. In Portugal, according to DECO, a Portuguese consumer protection association, the telecommunications market ranks first in terms of the number of complaints and problems [1]. As one of the largest in the sector in the country, this company seeks to improve the quality and efficiency of its customer service. Thus, advanced methods of prevision and optimization, based on data, are necessary to solve this problem.

Call-centers have evolved considerably in the last years. There are still some scripts and strategies that globally improve support, but the reality is that technology has dramatically changed the way call-centers can work. They can know what transactions the customer made in the last hour, what he asked in the last call to the customer support, information about satisfaction, feeling, propensity for churn, and many other things. Using historical data, it is possible to have a complete view of the customer in the sense that it is possible to see how the customer has already interacted with the company. This scenario makes the customer's process easier since he does not need to explain himself several times. However, it would be even more convenient if the right operator was routed to the right customer. Therefore, the focus on improving call-centers would not only be on customers, but also in operators and service providers. Each operator communicates differently, knows different things, and above all, appeals to a different segment of customers. It is clear that, for example, some operators cannot handle negative calls, while others stand out in such scenarios [2].

The concept of customer relationship management (CRM) is a big thing right now and is defined as a process of creating and maintaining a connection with customers. It is an adverse and complicated process of identifying, attracting, differentiating, and retaining customers. It sometimes integrates the entire supply chain of a company to give more value to the customer at each stage, either through more significant benefits or through cost reduction. The result is higher profit through increased business, with perfect coordination between sales, customer service, marketing, among others [3].

The main goal of this work is to create a predictive modeling and optimization approach that, from customer data provided by a telecom Portuguese company, can allocate customer calls to the most appropriate call-center operator for resolving the issue. Thus, it is first necessary to focus on predicting the customer's call reason through a machine learning approach. After this, an allocation to the most appropriate operator is made, based on optimization techniques. It is intended that the end product results in an improvement in customer service. A high-level representation of the architecture to be developed is shown in Fig. 1. The solution will consist of two modules dependent on each other.

This paper is organized as follows. In the next section, call-centers and some related concepts are explained. The existing literature is also analyzed regarding the problem under analysis. Section 3 is dedicated to the first module of the system, in which the methodology used is explained as well as the learning model to be built and the results obtained from it. Section 4 explains the method and the algorithm used in the second module of the system, where the optimization

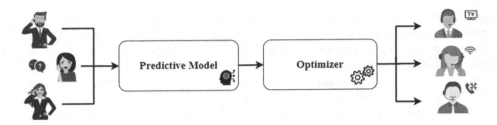

Fig. 1. System Architecture. Customer calls are allocated to the most appropriate call-center operator using predictive modeling and optimization.

is done. Finally, in Sect. 5, we conclude the paper and provide some guidelines for future work.

2   Background and Related Work

2.1   Customer Service

The company, like all telecommunications companies, has a customer support service that is available 24/7, clarifying doubts and providing support for technical difficulties or breakdowns resulting from contracted services. It has several call-centers equipped with qualified operators. When someone calls, they are contacting the operators on the first line who are not necessarily specialized in all areas. When they cannot solve a problem, they transfer the call to another colleague, now on the second line. This transfer not only brings costs to the company because it involves a longer call duration and more than two busy operators, but it also annoys the customer because, in most cases, he has to re-explain the problem to another person. In this way, the company seeks to improve its service by making a prediction, in advance, in search of why the customer is looking for help and support. This enables the company to provide adequate, personalized, fast, and comfortable assistance.

2.2   Data Mining in Telecommunications Industry

Some machine learning solutions have been used as highly useful tools for extracting information hidden in large data repositories. Some well-known examples of this, already explored, are the analysis and detection of breast cancer in the biomedical sector, the credit scores in the financial sector, or even stress assessments, and the detection of emotions [4].

In terms of customer satisfaction, some telecommunication data, such as call details and customer information, can be profitable if used, as it can support the determination of customer behavior and the identification of opportunities to support the objectives of expanding the customer base or reducing churn [5].

Today, for example, there are many approaches to the most significant industry challenge, which is retaining customers. Therefore, customers may switch to other companies for various reasons, such as better services, better prices, better call quality, or fewer billing problems. This problem of customers who shake up and move quickly to other competitors is widespread. Some solutions existing in the market, such as presented by Mishra et al. [6], show the need of a good classifier and the importance of an adequately pre-processed dataset, which removes unnecessary and redundant data. Telecommunications companies collect information about their customers, which is not all relevant.

Useful applications cannot be developed without understanding the various data used in the telecommunications industries. Therefore, the first step in the data mining process is to understand the data. The different types of data used in this industry are mainly grouped into three different types: detailed call data with information such as date, time, and call duration, which can generate data such as average call duration, the average number of calls originated per day, the average number of calls received per day, percentage of calls during the day, or percentage of calls during the week; network data, i.e., telecom networks contain thousands of interconnected components which can generate error or status messages; customer data such as the address, payment history, or contracted service. The information can be used to profile customers and be used for marketing and prevision purposes. The focus of marketing in the telecommunications industry has shifted from identifying new customers to measuring the value of existing customers and making decisions to retain them [7].

2.3   Related Work

In 2011, Abbas Raza Ali stated that it is impossible to train all call operators so that they can handle and respond to all kinds of calls [8]. In that proposed solution, the calls to customer support are related to sales, and it is intended to maximize a score directly related to a sale. If the product is sold at a high price and with reasonable customer satisfaction, then the score is positive. Initially, from historical data and customer and operator information, a dataset is built, which is used to train a set of models in an offline way. Finally, several models are used, and it is verified which model has the best performance to predict the problem. Once the best model is obtained, it is used in order to score online. At each moment, all the customers who are waiting for service and all the available operators are joined. So, if, for instance, 15 customers are waiting and 10 operators are available, a dataset with 150 lines is created, which are subject to the model previously created to predict a score. Finally, for each customer, it checks which operator results in a higher score.

Mehrbod et al. addressed call-center issues with a machine learning approach [9]. The authors made use of three different datasets related to customer data such as age, gender, marital status and occupation, operator information such as age, gender, qualification, recent supervisor evaluation score, and six months of historical data with inbound call information. Besides, they defined the target as the outcome of the call (positive or negative). They applied some data pre-processing, where null data was removed, and then treated the classes' unbalancing since the class referred to positive calls is the majority when compared to negative calls. Using Random Forest, they obtained an accuracy of 0.92 and an AUC value of 0.9.

These works try to solve problems similar to what this paper proposes to solve. In other words, historical customer data and customer information, as well as some operators' statistics, are also used to find the best operator. However, they join the available N operators with the waiting M customers into a single prediction module which returns a positive or negative value, rather than separating into two modules. The problem is the computational load at each moment since a dataset has to be built for each call. Moreover, since the target is positive or negative, they do not ensure the maximum service they can offer, but only one operator that satisfies the customer.

3   Predictive Model

In order to assign the best operator, first it is necessary to predict the reason for each customer's call, as illustrated in Fig. 1. In this section, we present the development of the predictive model to achieve this goal.

3.1   Methodology

In this study, we follow the CRISP-DM method, which describes approaches commonly used by specialists to tackle problems of Data Mining [10]. The biggest advantage of using this method is that it is independent of industry, tools, and data. The methodology phases include Business Understanding, Data Understanding, Data Preparation, Modeling, Evaluation, and Deployment.

3.2   Business Understanding

The objective of the work presented in this paper is to allocate customer calls to the operators better prepared to support them. It requires prior knowledge of the reason for the call and what the customer's problem is. This operation allows the company to provide personalized service (suitable assistance, equipment, or technician), culminating in better customer experience and greater efficiency in the operation of the call-center itself. It can result, for example, in a decrease in service time, or a decrease in the recurrence rate, resulting in an increase in profit for the company.

The data used in this study were retrieved from a telecommunications company in Portugal, which was previously anonymized and filtered in order to comply with privacy regulations. Machine learning aims to develop precise models capable of supporting the decision process, i.e., predicting the customer's problem before the customer says so. To make prediction possible, models take into consideration the customer support records as input values.

Table 1. Target variables and their distribution.

Target class	Before data cleaning	After data cleaning
0 - Churn	1100659 (19%)	160523 (7%)
1 - Equipment	707821 (12%)	60506 (3%)
2 - Billing	290786 (5%)	163695 (7%)
3 - Others	1263669 (21%)	463111 (19%)
4 - Payment	188551 (3%)	144350 (6%)
5 - Internet	583164 (10%)	409821 (17%)
6 - Telephone	115551 (2%)	82726 (3%)
7 - Mobile Phone	45168 (1%)	33157 (1%)
8 - TV	799240 (14%)	630134 (26%)
9 - Services	789823 (13%)	254384 (11%)

3.3   Data Understanding

The original dataset consisted of 5.9M rows, reduced to 2.7M after removing records considered irrelevant or unusable. The primary dataset was the customer support record, which was complemented with other information such as personal data, equipment in use, consumption, contracted services, technical support provided. It resulted in each of the entries being described by a total of 200 variables. In summary, the final set of data is divided into two broad groups, i.e., customer data and historical customer data.

For the target variable, it is essential to consider that the problem is not binary, but multi-class because we want to predict the customer's problem from a set of ten possible problems: television, internet, telephone or mobile phone, payments, and billing, willingness to disconnect or cancel services (churn) and information regarding services and equipment. The remaining problems, as they are many and less predictable, have been aggregated into a class named others. Table 1 lists the target variables.

3.4   Data Preparation

At this phase, it was necessary to prepare the data to be used by the machine learning models. First of all, data integration was done since it was necessary to work with different data sources. This was followed by data cleaning and data transformation, where administrative and derived variables were removed. Also, in the variables, those that were predictably poorly calculated by the system, and those with too many null values were eliminated because they could lead to noise in the models. Initially, it was also decided to remove about 50% of the dataset rows either because they were related to outbound calls or because they were rows related to internal operations, typing errors, or rows created in the company's physical stores. To deal with null values, instead of removing the observations, we decided to make imputations so as not to lose customer information and history.

The most crucial step was the extraction and construction of features. The number of maintenance and installations was first obtained from the dataset of technical interventions. Also, from the contract end date and the customer account creation date, the age of the account and the number of months remaining for the contract end date were determined. We created features indicating how many calls the customer had previously made to the customer support and what the problem was. As mentioned before, observations were deleted for various reasons. However, some of their information was retained in features such as the number of times the customer went to the company's stores, churn requests, number of complaints, among others. The customer's county of residence may, at certain times, suffer from technical problems or peaks in usage, so we built features that indicate the incidence of problems in the county. Finally, some lag features have been built in order to understand the context of the last call. Therefore, we calculated, for example, the percentage of recurrence of the operator, or the average contacts for the last problem.

Some variables are related to information chosen by the call operator in a semi-structured way. Thus, it was necessary to use word embeddings algorithms to extract the context of the call and the reported problem. FastText [11], an extension created by Facebook from the famous Word2Vec [12], was chosen. FastText transforms words and, consecutively, phrases into vectors, which later are the input of a K-Means clustering model. The use of this unsupervised model divided the phrases (vectors) into different clusters.

Finally, the next step was to treat outliers, replacing them with the 95th percentile, and encode the categorical variables into numerical values to be used in the machine learning models. For this, target encoding, a widely used technique, was applied with noise addition in order to avoid overfitting that occurs more often in the standalone technique. This type of encoding was also used, to the detriment of others, to avoid further increase of the dataset size.

3.5   Modeling

This section explores how different ML Models are used after the data was transformed and processed. First of all, the problem was defined as classification. The classification algorithms that we consider are: Logistic Regression (LR), Naive Bayes (NB), Random Forest (RF), Neural Networks (NN), AdaBoost (AB), XGBoost (XGB), and LightGBM (LGBM). Although they are generally considered the best algorithms to apply to similar data, we will describe them briefly to justify the reasons why the algorithms were chosen. LR is a useful and straightforward algorithm that classifies the data by considering outcome variables on extreme ends and tries to make a logarithmic line that distinguishes between them. NB assumes that all dataset variables are naive, i.e., they are not correlated with each other. It is based on the Bayesian theorem, which defines the probability of an event occurring given the probability of another event that has already occurred. Despite its simplicity, it can often perform better than other

Table 2. Combination of alternatives for evaluation.

Scalers	Estimators	Balancing techniques
StandardScaler	ROS (10%) + RUS (30%)	Logistic regression
RobustScaler	ROS (10%) + RUS (50%)	Naive Bayes
	ROS (20%) + RUS (30%)	Random forest
	ROS (20%) + RUS (50%)	Neural networks
	Balanced model class weights	AdaBoost
	None	XGBoost
		LightGBM

algorithms. RF builds multiple decision trees and merges them to get a more accurate and stable prediction. AB, XGB, and LGBM are decision tree-based models with boosting, so several individual models are trained sequentially, and each model learns from the mistakes made by the previous model. Finally, NN is a model inspired at animal's central nervous system.

We decided to evaluate not only of scikit-learn's StandardScaler but also RobustScaler to take extra care with the possible presence of outliers [13]. Due to the class unbalancing verified in Table 1, class balancing was also considered. As we are facing a dataset with high dimensionality, it was impossible to execute over-sampling synthetic algorithms such as SMOTE due to the amount of time it would require. Even so, random over-sampling (ROS) applied to the two major classes (TV problems and Others), and random under-sampling (RUS) applied to the remaining classes were evaluated as well as balanced model class weights.

We divided the dataset into training, validation, and testing sets (70%–15%–15%, respectively). It is essential to note the special care that must be taken with the split so that it does not cause time data leakage. The set of all the various alternatives presented above, which lead to the execution of 46 different models, are summarized in Table 2.

Choosing an adequate evaluation metric for models generated from unbalanced data is a critical task. Accuracy is not a metric suitable for use when working with an unbalanced dataset. Therefore, the choice of a metric that best evaluates performance is quite important in the development of this work. We use the micro-F1 because it is more appropriate for unbalanced classification [14]. We present the Receiver Operating Characteristic (ROC) curve for evaluation purposes. It visually shows the relationship between true positive rate and false positive rate and presents the AUROC, which is a standard metric for unbalanced classification problems as well [15].

3.6 Evaluation

Table 3 presents the results that had the best performances for all the evaluated combinations, without hyperparameter tuning, concerning the validation set, i.e., by combining different scalers with different types of balancing techniques and models (see Table 2) we see the performance in the validation set.

Table 3. Best models obtained.

Estimator	Scaler	Balancing technique	Micro-F1
Logistic regression	StandardScaler	None	0.361
Random forest	StandardScaler	None	0.383
Neural networks	StandardScaler	None	0.385
XGBoost	StandardScaler	None	0.385
LightGBM	**StandardScaler**	None	**0.398**

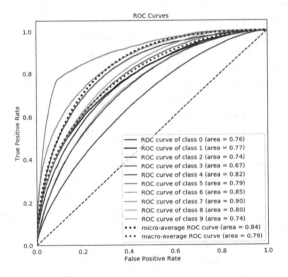

Fig. 2. ROC curves of the best model.

From the results, we observe that the best model was LightGBM. It was therefore selected for hyperparameter tuning to adapt it even more to the data in question. The tuning was achieved through 40 runs of Bayesian Optimization [16], in detriment of other algorithms, due to the enormous amount of data and the long training time. The performance obtained was slightly improved, and the micro-F1 of 0.41 was achieved. To complement this result, the ROC curve for the LightGBM optimized model is shown in Fig. 2, where it is possible to check the performance of the model for predicting each target class, as presented in Table 1, is well above the diagonal.

The top 10 features most important were mainly customer-related data such as time since the beginning of the contract, customer age, customer location, and time since the last contact. These features can serve as guidelines for other companies to implement such a system as well.

The LightGBM optimized model was finally tested with the test set, and the same result was obtained, i.e., micro-F1 of 0.41. It is important to achieve the best performance possible for the predictive model. It will greatly influence the system's performance since calls will be routed to the most appropriate call operator according to the prediction.

4 Operator Allocation

Once the reason for the customer's call is known, an optimization to allocate the call to the best operator to resolve the issue can be made. Some simulation techniques are used to recreate a past scenario. Next, we describe our optimization module, as illustrated in Fig. 1.

4.1 Method

A traditional company's call-center can be represented by a queue system, proper of the queue theory, in which no intelligent call distribution is available, and any operator can be confronted with all problems. Thus, assuming unlimited capacity, infinite population, FIFO attendance discipline, we can represent the system by M/M/S queue system, in which S operators can attend at every moment. Customers are impatient because they can get tired of waiting, and each operator can only serve one person at a time.

In order to simulate this system, we need to set the operators' working periods and recreate the customers' calls. To make it reliable, customers should give up when they have been waiting for a long time, and operators should take short breaks between different calls. To proceed with the optimization, we used the SimPy library [17], which allows the simulation of events simply and intuitively.

Initially, it was necessary to estimate the number of operators available at each moment. This estimate was extracted from the data, i.e., from what really happened. For example, if an operator answered a call at 4:00 p.m., that means he was working at that time. Thus, the timetable of each operator was calculated. Then, the service time of each operator was calculated as well, based on the calls answered in the past for each of the problems. Thus, for example, on a given day, John's service time in television-related problems is calculated and estimated by the median duration of the calls related to television he answered until that day.

After these steps, the simulation is executed, and customers are received in the system, now knowing why they are calling. Among the available operators, the fastest to resolve similar issues in the past is the one which the call is allocated. At the end of the call, the operator becomes available again to answer a new call. We compare the simulated service time of our optimized approach against the current system.

Table 4. Comparison between the two solutions.

Model	Mean service time	Utilization rate
Actual solution	7.9 min	15%
Proposed solution	4.7 min	9%

4.2 Results

The solution to this problem increases customer satisfaction. In practice, it happens because a customer wants to be served as quickly as possible (time is money), and an efficient service shows the customer the professionalism of the company. Also, from the company's side, it allows for the reduction of costs because it is possible to answer more calls in the same time or even by fewer operators. Operators would also be more satisfied with carrying out their functions because they feel more useful and fulfilled when answering calls in which they are prepared to help in the best way.

Table 4 shows the comparison between the solution proposed in this paper with the current solution, in which a customer is assigned to a random operator. We can observe that an improvement can be achieved. These results, however, depend considerably on the effectiveness of the predictive model. The simulation shows that the proposed solution allows reducing the service time and the workload of the operators by approximately 40%, allowing not only better management of resources, but also higher customer satisfaction.

5   Conclusion and Future Work

The telecommunications market has experienced enormous growth in recent years, attracting several customers. For this reason, competitiveness is tremendous, and companies try to invest in their infrastructure to improve their relationship with their customers. Most of the efforts in recent years have been focused on churn prediction, which is a late stage of customer dissatisfaction. In this article, an attempt is made to predict the reason for a customer's potential dissatisfaction when calling the customer support to improve internal and external operations. As AI applications are growing in the industry and the many data they have is available for use at anytime and anywhere, it is hoped that the application of such technological advances will lead to a better relationship with customers and their satisfaction.

In this article, we devised a predictive model that anticipates the problem for which the customer is calling and then, resorting to optimization procedures, routes the call and the customer to the most appropriate call operator. The results of our predictive model were generally satisfactory, with micro-F1 at 0.41 and AUROC at 0.84. They give some confidence in using these types of models and data in the support process. The optimization in a simulated scenario indicated possible improvement in the order of 40%, making the service more efficient and faster.

Despite everything, some improvements can be achieved by collecting more data and further optimizing models. Some work on data integration and variable extraction is necessary and computationally expensive to obtain the results shown here. Therefore, the task of deploying this solution into production, the last phase of CRISP-DM, is still the next step to check whether the improvement is proven to be obtainable.

Acknowledgements. This work has been supported by FCT – Fundação para a Ciência e Tecnologia within the R&D Units Project Scope: UIDB/00319/2020.

References

1. DECO: Telecomunicações sempre no topo das reclamações (2019). https://www. deco.proteste.pt/institucionalemedia/imprensa/comunicados/2019/telecons-1705. Accessed 10 May 2020
2. Fly, A.: 5 Best Practices for AI- and Data-Driven Call Centers. https:// towardsdatascience.com/5-best-practices-for-ai-and-data-driven-call-centers-647406b4234b. Accessed 7 June 2020
3. Hassan, R.S., Nawaz, A., Lashari, M.N., Zafar, F.: Effect of customer relationship management on customer satisfaction. Procedia Econ. Financ. **23**(October 2014), 563–567 (2015)
4. Carneiro, D., Novais, P., Pêgo, J.M., Sousa, N., Neves, J.: Using mouse dynamics to assess stress during online exams. In: Onieva, E., Santos, I., Osaba, E., Quintián, H., Corchado, E. (eds.) HAIS 2015. LNCS (LNAI), vol. 9121, pp. 345–356. Springer, Cham (2015). https://doi.org/10.1007/978-3-319-19644-2_29
5. Tundjungsari, V.: Business intelligence with social media and data mining to support customer satisfaction in telecommunication industry. Int. J. Comput. Sci. Electron. Eng. **1**(1), 1–4 (2013)
6. Mishra, K., Rani, R.: Churn prediction in telecommunication using machine learning. In: 2017 International Conference on Energy, Communication, Data Analytics and Soft Computing, ICECDS 2017 (2012), pp. 2252–2257 (2018)
7. Marwaha, R.: Data mining techniques and applications in telecommunication industry. Artif. Intell. Med. **16**(1), 1–2 (2014)
8. Ali, A.R.: Intelligent call routing: optimizing contact center throughput. In: Proceedings of the 11th International Workshop on Multimedia Data Mining, MDMKDD 2011 in Conjunction with SIGKDD 2011, pp. 34–43 (2011)
9. Mehrbod, N., Grilo, A., Zutshi, A.: Caller-agent pairing in call centers using machine learning techniques with imbalanced data. In: Proceedings of the 2018 IEEE International Conference on Engineering, Technology and Innovation, ICE/ITMC 2018 (2018)
10. Pete, C., et al.: CRISP-DM 1.0. CRISP-DM Consortium, p. 76 (2000)
11. Bojanowski, P., Grave, E., Joulin, A., Mikolov, T.: Enriching word vectors with subword information (2016)
12. Mikolov, T., Corrado, G., Chen, K., Dean, J.: Efficient estimation of word representations in vector space, pp. 1–12, January 2013
13. Buitinck, L., et al.: API design for machine learning software: experiences from the scikit-learn project. In: ECML PKDD Workshop: Languages for Data Mining and Machine Learning, pp. 108–122 (2013)

14. Scikit-learn: scikit-learn: f1_score. https://scikit-learn.org/stable/modules/generated/sklearn.metrics.f1_score.html. Accessed 19 Aug 2020

15. He, H., Garcia, E.A.: Learning from imbalanced data. IEEE Trans. Knowl. Data Eng. **21**(9), 1263–1284 (2009)

16. Nogueira, F.: Bayesian optimization: open source constrained global optimization tool for Python (2014). https://github.com/fmfn/BayesianOptimization. Accessed 9 June 2020

17. Team, S.: SimPy: Discrete-Event Simulation for Python. https://simpy.readthedocs.io/en/latest/. Accessed 9 June 2020

Entropy Based Grey Wolf Optimizer

Daniel Duarte[1] ⓘ, P. B. de Moura Oliveira[1,2](✉) ⓘ, and E. J. Solteiro Pires[1,2] ⓘ

[1] Department of Engineering, University of Trás-os-Montes and Alto Douro (UTAD),
Vila Real, Portugal
`danielduarte777@hotmail.com, {oliveira,epires}@utad.pt`
[2] INESC-TEC Technology and Science, Campus da FEUP, Porto, Portugal

Abstract. Recently Shannon's Entropy has been incorporated in nature inspired metaheuristics with good results. Depending on the problem, the Grey Wolf Optimization (GWO) algorithm may suffer from premature convergence. Here, an Entropy Grey Wolf Optimization (E-GWO) technique is proposed with the overall aim to improve the original GWO performance. The entropy is used to track the GWO swarm diversity, comparing the distance values between the Alpha in relation to the Beta and Delta wolves. The aim of the E-GWO variant is to improve convergence and prevent stagnation in local optima, since ideally restarting the swarm agents will prevent this from happening. Simulation results are presented showing that E-GWO restarting mechanism can achieve better results than the original GWO algorithm for some benchmark functions.

Keywords: Grey Wolf Optimization · Shannon entropy · Nature inspired algorithms

1 Introduction

Grey Wolf Optimizer (GWO) originally proposed by [1] is one of many algorithms that are inspired in mother nature. This inspiration comes from many sources, such as physics, chemistry, evolution, and the behavior of a large range of living beings [2]. Swarm algorithms [3] are inspired on the highly social and intelligent behavior of animals living and working together as groups, such as a colony of ants or a swarm of bees. GWO is one of those swarm algorithms, as it takes inspiration in the way grey wolves live, communicate, and interact together in a pack. Since the GWO algorithm proposal [1] several variants were proposed, such as the following examples: Mean Grey Wolf Optimizer (MGWO), by [4], in which some algorithm equations were changed; Modified Grey Wolf Optimizer (mGWO) developed by [5], with the purpose of balancing exploitation and exploration to increase precision and Minkowski Based Grey Wolf Optimizer (MGWO) developed by [6]. There are variants of other algorithms that incorporate entropy, such as the variant proposed by [7] where entropy is used to evaluate the self-organizing properties of swarm algorithms and is applied to the SPARROW-SNN adaptive flocking algorithm, and the variant proposed by [8] where entropy is used on Particle Swarm Optimization (PSO) to measure and improve the algorithm's convergence, by analyzing the entropy's signal as a criterion for reinitializing the swarm.

© Springer Nature Switzerland AG 2020
C. Analide et al. (Eds.): IDEAL 2020, LNCS 12489, pp. 329–337, 2020.
https://doi.org/10.1007/978-3-030-62362-3_29

Entropy in information theory, while not unrelated, is different from entropy in thermodynamics. Information entropy measures the uncertainty or surprise that is naturally a part of random variables. One of GWO's problems is to maintain the swarm diversity. In this paper a new GWO variant incorporating entropy is proposed. The overall aim of the proposed GWO variation is to introduce multiple restarts based on an entropy criterion. It will be shown that the proposed algorithm obtains better results due to a wider exploration and better exploitation, by preventing the algorithm becoming "trapped" in local optima. Balancing exploration and exploitation in an algorithm is important and will lead to better results, as described by [9].

Since the population will change throughout the running of the algorithm, the population will have a higher diversity. This diversity is important, because an algorithm with a diversity that is too low may not have optimal results. There are many ways to analyze the diversity of a population, as analyzed by [10] in their work. While entropy itself does not necessarily mean diversity, it can be used to measure diversity in a population [11].

The remaining of this paper is organized as follows: in Sect. 2 basic entropy concepts are presented, Sect. 3 reviews the GWO basic natural mechanisms and inspiration, Sect. 4 explains the proposed Entropy Based GWO variant and in Sect. 5 the results of the tests are displayed and analyzed. Finally, Sect. 6 concludes the paper and outline further work.

2 Entropy

Entropy is a concept that can now be applied to many different fields [12] but was initially used by Rudolf Clausius when attempting to find a mathematical expression to describe the transformations of a body through heat exchange [13]. It can be interpreted as a measure of a system irreversibility. Entropy can be expressed by many concepts, such as disorder and chaos [8]. In information theory, entropy (also called information entropy) measures the average amount of uncertainty or information associated to random variables, due to the innate possibilities of their outcomes. It was first described by Claude Shannon [14] in 1948. Shannon used entropy as a part of his communication theory, by describing a data communication system as having a channel, a receiver, and data source, and addressing the main problem of any communication system: the identification of data sent through the channel. Entropy may also be perceived as the expected value of information obtained through a stochastic data source. This means that if a value with low probability to occur is produced by a data source, this event has more information contained in it than when an event with high probability happens [15]. The entropy H(X) of a random variable X can be described by:

$$H(X) = -\sum_i P_X(x_i) \log(P_X(x_i)) \tag{1}$$

where x_i represents the possible values of X and $P_X(x_i)$ the probability of them happening.

3   Grey Wolf Optimizer (GWO) Overview

3.1   GWO Natural Inspiration

GWO is an algorithm inspired on the individual and collective behavior of grey wolves, developed by [1] in 2014. More specifically, it is inspired on the grey wolves' social structure and hunting mechanisms [16]. They are predators at the top of the food chain, and they live in packs of 5 to 12 wolves on average with a very strong dominance hierarchy that is divided into four groups (see Fig. 1) [17]:

- Alpha (α), which consists of a male and a female, are the leaders of the pack. They have important responsibilities such as making decisions related to hunting, the packs sleeping place and when to wake up. These are the dominant wolves because the rest of the pack must respect their decisions. The Alpha wolves are not necessarily the physically strongest wolves, but those who can better lead their pack.
- Beta (β), composed of wolves that help the Alpha wolves in their decision making and other activities. They must respect the Alpha wolves but are able to give orders to wolves that are lower in the hierarchy. In the case an Alpha wolf death, the most likely candidate to replace him would be a Beta wolf.
- Delta (δ), consisting of wolves inferior to the Alpha and Beta wolves, but superior to the Omega. These wolves main tasks are taking care of weak or sick wolves, protect the pack and warn about dangers.
- Omega (ω), the lower group in the hierarchy. These wolves are the scapegoat of the pack and are the last to eat. The other wolves unleash their rage and frustrations on the Omega, which makes them useful because this pleases the pack and allows the dominance hierarchy to be maintained.

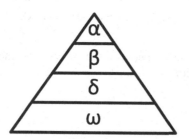

Fig. 1. Dominance hierarchy of a pack of wolves [1].

Grey Wolves also show smart and social behaviors when hunting. The hunting process can be divided into three parts [18]:
- Find, follow, and approach the prey;
- Chase, circle and tire the prey until it stops moving;
- Attack the prey.

3.2 GWO Algorithm Description

In the GWO Algorithm (Fig. 2), the first step is the population initialization (a pack of grey wolves), where each wolf has a random position, a possible solution to the problem, and "hunts" a "prey" (attempts to find a better solution). The step of chasing the prey can be represented by [1]:

$$D = |C \cdot X_P(t) - X(t)|$$ (2)

$$X(t + 1) = X_p(t) - A \cdot D$$ (3)

```
Initialize the grey wolf population P;
Initialize parameters;
Calculate each wolf's fitness;
Store the 3 best wolves;
While exit conditions aren't met:
       For each wolf
              Update position;
       End
       Update parameters;
       Calculate each wolf's fitness;
       Update the 3 best wolves;
End
Return best wolf;
```

Fig. 2. A pseudocode for GWO.

where D represents the distance between the wolves and their prey, $X(t)$ is the wolf position, $X_p(t)$ a prey position and A and C are obtained from the following expressions:

$$A = 2a \cdot r_1 - a$$ (4)

$$C = 2 \cdot r_2$$ (5)

where r_1 and r_2 are randomly generated numbers in between 0 and 1, and a is a parameter that gradually decreases from 2 to 0. The purpose of these parameters is to balance between search exploitation and exploration and to avoid the early convergence of the algorithm. Similarly, to the social grey wolves hierarchy, the three best solutions are X_α, X_β e X_δ. These solutions are stored and then the wolves update their positions, according to the average position of the best wolves in each group. These update processes can be described by the following equations:

$$X_1 = X_\alpha - A_1 \cdot (D_\alpha)$$ (6)

$$X_2 = X_\beta - A_1 \cdot (D_\beta) \tag{7}$$

$$X_3 = X_\delta - A_1 \cdot (D_\delta) \tag{8}$$

$$X(t + 1) = \frac{X_1 + X_2 + X_3}{3} \tag{9}$$

4 Entropy Based Grey Wolf Optimizer

The proposed variant for the Grey Wolf Optimizer (E-GWO) algorithm functions similarly to the original GWO. It evaluated entropy in every iteration based on the distances of the Beta and Delta wolves compared to the Alpha wolves. The distance between Beta and Delta wolves is not so relevant, since the best results are always assigned to the Alpha wolves positions. Therefore, this distance is not incorporated into this GWO variant. If the entropy value "stagnates" for a determined number of iterations, then the search agents are restarted. To prevent an excessive number of algorithm restarts, since that would hinder both exploration and exploitation, after a restart occurs there is a pre-established number of iterations in which the search agents are not allowed to be restarted. Restarting allows the algorithm not to get "trapped" in local optima and allows for a wider search and helps in convergence. E-GWO is depicted in Fig. 3.

```
Initialize the grey wolf population P;
Initialize parameters;
Calculate each wolf's fitness;
Store the 3 best wolves;
While exit conditions are not met:
        For each wolf
                Update position;
        End
        Update parameters;
        Calculate each wolf's fitness;
        Update the 3 best wolves;
        Calculate and store entropy;
        If iteration > pause
            Compare entropy to past entropy values;
            If number of equal values> threshold
                    Restart positions;
            End
        End
End
Return best wolf;
```

Fig. 3. Entropy GWO (E-GWO) pseudocode.

5 Simulation Results

The proposed E-GWO algorithm was tested on sub-set of the functions (see Table 1) used in the original GWO paper [1]. The criterion adopted for restarting the GWO search agents is a pre-defined number of iterations being reached without any changes in the entropy value. By changing the restarting iterations threshold, the number of restarts will increase or decrease and depending on the function to be optimized, better or worse results can be obtained. Therefore, it may be necessary to perform a proper adjustment of this restarting iterations threshold according to the objective function to achieve the best possible results.

Table 1. Function set used to test the E-GWO

Function	Dim	Range	F_{min}
$F_6(x) = \sum\limits_{i=1}^{n} ([x_I + 0.5])^2$	30	$[-100, 100]$	0
$F_7(x) = \sum\limits_{i=1}^{n} i x_i^4 + random(0, 1)$	30	$[-1.28, 1.28]$	0
$F_8(x) = \sum\limits_{i=1}^{n} -x_i \sin\left(\sqrt{|x_i|}\right)$	30	$[-500, 500]$	-417.9729×5
$F_9(x) = \sum\limits_{i=1}^{n} [x_i^2 - 10|\cos(2\pi x_i + 10)|$	30	$[-32, 32]$	0
$F_{10}(x) = -20\exp(-0.2\sqrt{\frac{1}{n}\sum\limits_{i=1}^{n} x_i^2} - \exp(\frac{1}{n}\sum\limits_{i=1}^{n}\cos(2\pi x_i)) + 20 + e$	30	$[-32, 32]$	0
$F_{11}(x) = \frac{1}{4000}\sum\limits_{i=1}^{n} x_i^2 - \prod\limits_{i=1}^{n}\cos\left(\frac{x_i}{\sqrt{i}}\right) + 1$	30	$[-600, 600]$	0
$F_{12}(x) = \frac{\pi}{n}\{10\sin(\pi y_1) + \sum\limits_{i=1}^{n-1}(y_i - 1)^2[1 + 10\sin^2(\pi y_{i+1})] + (y_n - 1)^2\}$ $+ \sum\limits_{i=1}^{n} u(x_i, 10, 100, 4)$ $y_i = 1 + \frac{x_i+1}{4}$ $u(x_i, a, k, m) = \begin{cases} k(x_i - a)^m x_i > a \\ 0 - a < x_i < a \\ k(-x_i - a)^m x_i < -a \end{cases}$	30	$[-50, 50]$	0
$F_{13}(x) = 0.1\left\{\sin^2(3\pi x_1) + \sum\limits_{i=1}^{n}(x_i - 1)^2[1 + \sin^2(3\pi x_i + 1)] + (x_n - 1)^2\left[1 + \sin^2(2\pi x_n)\right]\right\}$ $+ \sum\limits_{i=1}^{n} u(x_i, 5, 100, 4)$	30	$[-50, 50]$	0

Some E-GWO results are presented in Figs. 4, 5, 6, 7 and Tables 2, 3 obtained from 10 different GWO and E-GWO runs and showing average evolution for the best value and average values. Each run was executed through 20000 iterations in both algorithms. These functions were selected as they enable to show E-GWO performance improvements compared to the original GWO.

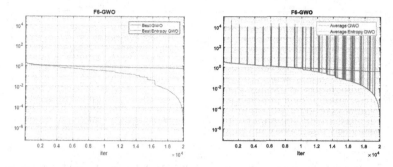

Fig. 4. Average of best value and average value in GWO vs. E-GWO for function F_6

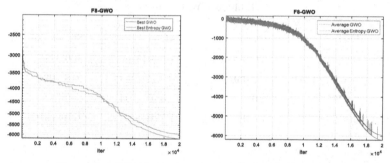

Fig. 5. Average of best value and average value in GWO vs. entropy E-GWO for function F_8.

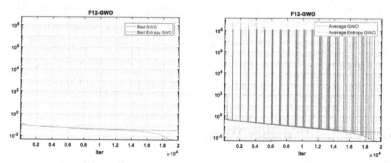

Fig. 6. Average of best value and average value in GWO vs. entropy E-GWO for function F_{12}.

From the E-GWO evolution curves presented in Figs. 4, 5, 6 and 7, it is possible to see spikes in the average graph. These correspond to a E-GWO restart with the corresponding increase of the average values. This also causes an increase in the mean and standard deviation in some functions, but overall, for these functions, the GWO entropy variant converges more quickly and achieves better results than the original GWO algorithm.

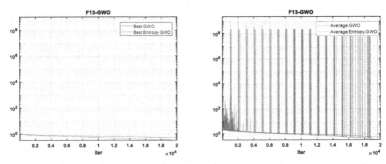

Fig. 7. Average of best value and average value in GWO vs. E-GWO for function F_{13}.

Table 2. GWO results

Function	Best	Worst	Mean	Standard deviation
F6	0.4697	6.73E + 04	18.0845	800.3127
F8	−6.25E + 03	−2.05E + 03	−4.72E + 07	918.508
F12	0.0256	6.02E + 08	9.03E + 04	5.75E + 06
F13	0.4353	1.14E + 09	1.19E + 07	1.19E + 07

Table 3. Entropy based GWO results

Function	Best	Worst	Mean	Standard deviation
F6	0.0226	6.93E + 04	17.3457	793.9201
F8	−6.42E + 03	−2.32E + 03	−4.89E + 03	936.2025
F12	0.0179	5.05E + 08	8.35E + 04	5.20E + 06
F13	0.2276	1.23E + 09	1.86E + 05	1.18E + 07

6   Conclusion and Further Work

Eight functions were tested with the proposed Entropy Based Grey Wolf Optimizer algorithm. Although restarting the search agents greatly worsens the average for a few iterations, the best value keeps improving, resulting in better results and faster convergence when compared to the original GWO algorithm. Some functions responded well to the variant, providing a significative improvement, while others provided only small improvements, and some of them did not improve at all. These improvements are also greatly dependent on the parameters used for restarting the search agents, and the number of "pause" iterations where no restarts can happen, as these control the number of restarts that happen. This is crucial because for example, if there are too many restarts, the algorithm may not have time to converge properly. If all these parameters are properly adjusted this variant achieves better results than the original GWO algorithm.

More research is necessary to try to devise an adaptive mechanism to automatically determine the E-GWO restarting threshold.

References

1. Mirjalili, S., Mirjalili, S.M., Lewis, A.: Grey wolf optimizer. Adv. Eng. Softw. **69**, 46–61 (2014)
2. Fister Jr., I., Yang, X.-S., Fister, I., Brest, J., Fister, D.: A brief review of nature-inspired algorithms for optimization. Elektrotehniski Vestnik/Electrotechnical Review. **80**(3), 116–122 (2013)
3. Chu, S.-C., Huang, H.-C., Roddick, J.F., Pan, J.-S.: Computational Collective Intelligence. Technologies and Applications, vol. 6922 (2011)
4. Singh, N., Singh, S.B.: A modified mean grey wolf optimization approach for benchmark and biomedical problems. Evol. Bioinform. **13**(1), 1–28 (2017)
5. Mittal, N., Singh, U., Sohi, B.S.: Modified grey wolf optimizer for global engineering optimization. Appl. Comput. Int. Soft Comput. Article ID 7950348, 16 (2016)
6. Khanum, R., Jan, M. Aldegheishem, A., Mehmood, A., Alrajeh, N., Khanan, A.: Two new improved variants of grey wolf optimizer for unconstrained optimization digital object identifier https://doi.org/10.1109/access.2019.2958288
7. Folino, G., Forestiero, A.: Using entropy for evaluating swarm intelligence algorithms. In: González, J.R., Pelta, D.A., Cruz, C., Terrazas, G., Krasnogor, N. (eds.) Nature Inspired Cooperative Strategies for Optimization (NICSO 2010) Studies in Computational Intelligence, vol. 284, pp. 331–343. Springer, Heidelberg (2010). https://doi.org/10.1007/978-3-642-12538-6_28
8. Pires, E.J.S., Machado, J.A., Oliveira, P.B.M.: PSO evolution based on a entropy metric. In: 18th International Conference on Hybrid Intelligent Systems (HIS 2018), Porto, Portugal, 13–15 December 2018
9. Črepinšek, M., Liu, S.-H., Mernik, M.: Exploration and exploitation in evolutionary algorithms: a survey. ACM Comput. Surv. **45**(3), Article 35, 33 (2013)
10. Tang, E.K., Suganthan, P.N., Yao, X.: An analysis of diversity measures. Mach. Learn. **65**, 247–271 (2006)
11. Jost, L.: Entropy and diversity. Oikos **113**, 2 (2006)
12. Solteiro Pires, E.J., Tenreiro Machado, J.A., de Moura Oliveira, P.B.: PSO evolution based on a entropy metric. In: Madureira, A., Abraham, A., Gandhi, N., Varela, M. (eds.) HIS 2018. AISC, vol. 923, pp. 238–248. Springer, Cham (2020). https://doi.org/10.1007/978-3-030-14347-3_23
13. Camacho, F., Lugo, N., Martinez, H.: The concept of entropy, from its origins to teachers. Revista Mexicana de Física E **61**(2015), 69–80 (2015)
14. Shannon, C.E.: A mathematical theory of communication. Bell Syst. Tech. J. **27**, 379–423 (1948)
15. https://machinelearningmastery.com/what-is-information-entropy/. Accessed 1 June 2020
16. Heidari, A.A., Pahlavani, P.: An efficient modified grey wolf optimizer with Lévy flight for optimization tasks. Appl. Soft Comput. **60**, 115–134 (2017)
17. Teng, Z.-J., Lv, J.-I., Guo, L.-W.: An improved hybrid grey wolf optimization algorithm. Soft. Comput. **23**, 6617–6631 (2019)
18. Luo, K.: Enhanced grey wolf optimizer with a model for dynamically estimating the location of the prey. Appl. Soft Comput. J. **77**, 225–235 (2019)

Mapping a Clinical Case Description to an Argumentation Framework: A Preliminary Assessment

Ana Silva$^{(\boxtimes)}$ ⓘ, António Silva ⓘ, Tiago Oliveira ⓘ, and Paulo Novais ⓘ

Algoritmi Centre/Department of Informatics, University of Minho, Braga, Portugal
{ana.paula.silva,asilva}@algoritmi.uminho.pt,
tiago.jose.martins.oliveira@gmail.com,
pjon@di.uminho.pt

Abstract. Medical reasoning in the context of multiple co-existing diseases poses challenges to healthcare professionals by demanding a careful consideration of possible harmful interactions. Computational argumentation, with its conflict resolution capabilities, may assist medical decisions by sorting out these interactions. Unfortunately, most of the argumentation work developed for medical reasoning has not been widely applied to real clinical sources. In this work, we select ASPIC+G and formalise a real clinical case according to the definitions of this argumentation framework. We found limitations in the representation of a patient's evolution and the formalisation of clinical rules which can be inferred from the context of the clinical case.

Keywords: Conflict resolution · Computational argumentation · Medical reasoning

1 Introduction

In medical reasoning, a healthcare professional establishes a connection between observable phenomena and medical concepts that explain such phenomena [1,2]. This process involves the integration of clinical information, medical knowledge and contextual factors. The flow of reasoning is guided by medical knowledge which consists of the set of heuristics that use evidence and observations as antecedents and conclude diagnoses and/or next steps. A particularly challenging context for medical reasoning is that of multimorbidity [3], characterised by the co-existence of multiple health conditions in a patient. The difficulties posed by multimorbidity are mainly related to drug-drug and drug-disease interactions. The fist occurs when different drugs prescribed to address different health conditions interact and cause harm to the patient while the latter occurs when a drug prescribed of a health condition causes the aggravation of another existing condition. Hence, the health care professionals must consider not only the observations and clinical evidence in order to recommend treatments to a patient,

© Springer Nature Switzerland AG 2020
C. Analide et al. (Eds.): IDEAL 2020, LNCS 12489, pp. 338–349, 2020.
https://doi.org/10.1007/978-3-030-62362-3_30

but also the interactions such recommendations may produce. Another dimension of this process that must be taken into consideration is the preferences of the health care professional and the patient about the recommendations themselves and about the overall goal of the treatment.

Applications of argumentation theory and argumentation frameworks in medical reasoning are not new. We can look as back as early as when Fox and Sutton [4] proposed the PROforma model to find an initial work conveying the usefulness of computational argumentation in a medical setting. However, since then, there have not been works in which the proposed argumentation frameworks are tested and assessed in clinical cases that originate from a different source and with a different structure from the ones used to develop said argumentation frameworks. As such, in this work, we qualitatively assess an argumentation framework, called ASPIC+G [5], proposed for medical reasoning in a context of multimorbidity by applying it to a clinical case extracted from MIMIC III [6], a freely accessible database developed by the MIT Lab for Computational Physiology, comprising the identified health data associated with intensive care unit admissions. Case descriptions include demographics, vital signs, laboratory tests, medications, and medical notes. We focus on the medical notes to identify the clinical information necessary to instantiate the ASPIC+G argumentation framework. The contributions of this work are (i) an analysis of a goal-driven argumentation framework, ASPIC+G, to reason with conflicting medical actions; (ii) an analysis of how information elements in a real clinical case are conveyed and represented in the framework; (iii) the identification of limitations ASPIC+G and an outline of possible to mitigate them. This work does fill in the gap in the literature mentioned above but is intended as a first step in closing that gap.

The paper is organised as follows. In Sect. 2, we provide an overview of argumentation works applied to medical reasoning and highlight the differences between these works and ASPIC+G. Section 3 contains a brief description of the most important components of ASPIC+G, including the representation of clinical information elements. In Sect. 4, we describe the selected clinical case to assess the framework and in Sect. 5 we formalise it in order to build an argumentation theory for medical reasoning. Section 6 conveys the limitations found during representation. Finally, Sect. 7 presents conclusions and future work directions.

2   Related Work

The PROforma modelling language [4] is an executable model language aimed at executing clinical guideline recommendations as tasks. The argumentation component in this work resides in the representation of the core components of decision tasks as arguments for or against a candidate solution. This approach focuses on selecting a task within a single clinical guideline, which, in principle, consists of a set of consistent tasks and does not feature conflicts amongst recommendations. No preferences are considered in argument aggregation in this work.

Another example of reasoning within a single clinical guideline is the approach in [7]. The proposed framework was implemented in the COGENT modelling system and encompasses situations of diagnostic reasoning and patient management. Similarly to PROforma, there are actions that have associated beliefs concerning their effects on the patient state.

A success case in the use of argumentation for medical reasoning is the work in [8]. Therein clinical trials are summarised using argumentation. The framework produces and evaluates arguments that establish the superiority of a treatment over another. These arguments provide conclusions pertaining to a set of outcome indicators and it is possible to establish preferences over these outcomes.

In [9], the authors use argumentation schemes to solve conflicts in recommendations for patient self-management. The argument scheme used is an adaptation of the sufficient condition scheme for practical reasoning which produces an argument in support for each possible treatment. This type of argument leads to the goal to be realised.

Existing approaches generally focus mainly on reasoning within a single set of recommendations [4,7] or do not consider a multimorbidity setting with conflicting recommendations and goals with different priorities [8,9]. Alternatively, the ASPIC+G approach [5] aims to capture these dimensions of clinical reasoning, hence the interest in observing how ASPIC+G handles a case that is different from the one disclosed in that work.

3   ASPIC+G

In this section, we provide an overview of the ASPIC+G argumentation framework as defined in [5] and describe the steps taken to evaluate the framework in light of a real clinical case, extracted from MIMIC III [6].

3.1   Framework Definition

ASPIC+G is an argumentation framework developed to formalise conflict resolution in a medical setting of multimorbidity and compute aggregated consistent sets of clinical recommendations. An argumentation theory in ASPIC+G is a tuple $\langle \mathcal{L}, \mathcal{R}, n, \leqslant \mathcal{R}_d, \mathcal{G}, \leqslant_{\mathcal{G}} \rangle$, where:

- $\mathcal{L}$ is a logical language closed under negation ($\neg$).
- $\mathcal{R} = \mathcal{R}_s \cup \mathcal{R}_d$ is a set of strict ($\mathcal{R}_s$) and defeasible ($\mathcal{R}_d$) rules of the form $\phi_1, \ldots, \phi_n \rightarrow \phi$ and $\phi_1, \ldots, \phi_n \Rightarrow \phi$ respectively, where $n \geq 0$ and $\phi_i, \phi \in \mathcal{L}$;
- n is a partial function s.t.[1] $n : \mathcal{R} \rightarrow \mathcal{L}$;
- $\leqslant_{\mathcal{R}_d}$ is a partial pre-order over defeasible rules $\mathcal{R}_d$, denoting a preference relation, with a strict counterpart $<_{\mathcal{R}_d}$ given by $X <_{\mathcal{R}_d} Y$ iff $X \leqslant_{\mathcal{R}_d} Y$ and $Y \not\leqslant_{\mathcal{R}_d} X$;
- $\mathcal{G} \subseteq \mathcal{L}$ is a set of goals that the arguments will try to fulfil s.t. $\forall\, \theta \in \mathcal{G}$, there exists a rule $\phi_1, \ldots, \phi_n \rightarrow \phi$ in $\mathcal{R}_s$ or $\phi_1, \ldots, \phi_n \Rightarrow \phi$ in $\mathcal{R}_d$ s.t. $\phi = \theta$;

---

[1] s.t.: such that.

- $\leqslant_{\mathcal{G}}$ is a total pre-order on $\mathcal{G}$, denoting *preferences* over goals, with $<_{\mathcal{G}}$ given by $\phi <_{\mathcal{G}} \psi$ iff $\phi \leqslant_{\mathcal{G}} \psi$ and $\psi \nleqslant_{\mathcal{G}} \phi$, and $\simeq_{\mathcal{G}}$ given by $\phi \simeq_{\mathcal{G}} \psi$ iff $\phi \leqslant_{\mathcal{G}} \psi$ and $\psi \leqslant_{\mathcal{G}} \phi$.

Argument construction and argument relations, such as attack and defeat, follow the well-established definitions set by ASPIC+ [10]. One feature provided by ASPIC+G on top of the reasoning mechanisms of ASPIC+ is goal-driven reasoning applied on preferred extensions in order to select a top preferred extension. Let $F = (\mathcal{A}, \mathcal{D}, \mathcal{G}, \leqslant_{\mathcal{G}}, \mathcal{F})$ – where $\mathcal{A}$ is a set of arguments, $\mathcal{D} \subseteq \mathcal{A} \times \mathcal{A}$ is a binary relation of defeat, $\mathcal{G}$ is a set of goals, $\leqslant_{\mathcal{G}}$ is a preference order over goals, and $\mathcal{F}$ is a binary relation of fulfilment s.t. $\mathcal{F} \subseteq \mathcal{A} \times \mathcal{G}$ – be an ASPIC+G argumentation framework and S, a finite set of goals, a preferred extension of F. S is a *top preferred extension of F* iff for every preferred extension S' of F, $\mathrm{Goal}(S') \trianglelefteq_{\mathcal{G}} \mathrm{Goal}(S)$, where S' is another finite set of goals, and defining the goal set ordering, denoted by the operator $\trianglelefteq_{\mathcal{G}}$, as: $S' \trianglelefteq_{\mathcal{G}} S$ iff $S' = \emptyset$ or $\exists g \in (S \setminus S')$ such that $\forall g' \in (S' \setminus S)$, $g' \leqslant_{\mathcal{G}} g$. In the context of a clinical decision, this top preferred extension would be the set of treatments selected to be applied to a patient.

3.2 Clinical Information Elements

In the original mapping of clinical information elements to the ASPIC+G three main types of clinical information elements are considered:

- $\mathbb{A}$: the set of all clinical actions which are up for recommendation;
- $\mathbb{E}$: the set containing contraries for all possible effects;
- $\mathbb{S}$: the patient state containing conditions manifested by a patient.

An action $\mathsf{A_x} \in \mathbb{A}^2$ is represented as tuple $\langle \mathsf{t_{x,a}}, \mathsf{O_{x,a}}, \mathsf{P_{x,a}} \rangle$ in which

- $\mathsf{t_x}$ is the treatment conveyed by the action;
- $\mathsf{O_x} = \{(e_1, \mathsf{C_1}, \lambda_1), \ldots, (e_n, \mathsf{C_n}, \lambda_n) : n > 0\}$ is a set of outcomes in which each outcome is a tuple $(e_i, \mathsf{C_i}, \lambda_i), i \in \{1, \ldots, n\}$ brought about by treatment $\mathsf{t_x}$, where: e_i is a description of an effect; $\mathsf{C_i} = \{c_1, \ldots, c_m : m \geq 0\}$ is a set with patient-specific conditions unifiable with the patient state $c_j, j \in \{1, \ldots, m\}$ that enable the occurrence of effect e_i over treatment t_x; λ_i is the impact of an effect e_i, if e_i is a positive effect, then $\lambda_i = \oplus$, otherwise, if it is a negative effect, $\lambda_i = \ominus$.
- $\mathsf{P_x} = \{p_1, \ldots, p_n : n \geq 0\}$ denotes pre-conditions and contains constraints for the application of a treatment $\mathsf{t_x}$.

As an example, let us consider an action that recommends the administration of metformin (*met*) with the intended effect of decreasing glucose levels (*gd*). However, metformin has an undesired side effect which is the acceleration of chronic kidney disease (*ackd*) in patients who have this health condition (*ckd*). Additionally, let us now consider an alternative action which

---

² Here we omit the second index.

recommends the administration of sulfunylurea ($sulf$) to also decrease glucose levels (gd). If a patient takes $sulf$ he should not take met as these drugs have the same effect and their combination could potentially cayuse harm to the patient. The first action would then be represented as $A_1 = \langle met, \{(gd, \emptyset, \oplus), (ackd, \{ckd\}, \ominus)\}, \{\neg sulf, \neg met\}\rangle$. Similarly, the second action would be represented as $A_2 = \langle sulf, \{(gd, \emptyset, \oplus), \emptyset\}, \{\neg sulf, \neg met\}\rangle$. We see the pre-conditions used in these actions prevent the simultaneous application of A_1 and A_2.

The next component of multimorbidity management is a set containing the contraries of effects $\mathbb{E} = \{C_1, \ldots, C_n : n \geq 0\}$ where each $C_i, i \in \{1, \ldots, n\}$, is a tuple (e_j, e_k) s.t. $\exists\ A_x = \langle t_x, O_x, P_x\rangle, A_y = \langle t_y, O_y, P_y\rangle \in \mathbb{A}$, s.t. $(e_j, C_j, \lambda_j) \in O_x$ and $(e_k, C_k, \lambda_k) \in O_y$. Expanding the earlier example, let us consider that it would be possible to delay chronic kidney disease by taking another form of medication in addition to one of the previous actions. Considering this new addition to the example effect contraries would take the form $\mathbb{E} = \{(dckd, ackd)\}$.

As for the patient state $\mathbb{S}$, it is defined as a set $\mathbb{S} = \{s_1, \ldots, s_n : n \geq 0\}$ where each $s_i \in S$ is a condition observed or diagnosed in the patient. In the running example, this set would consist of $\mathbb{S} = \{ckd\}$ since there is only one condition the patient is known to have.

3.3   Instantiating the Argumentation Framework

To construct an argumentation theory based on clinical information elements $\mathbb{A}$, $\mathbb{E}$, and $\mathbb{S}$, it is necessary to construct a set of rules $\mathcal{R}$ which will be the backbone of the argumentation theory. The construction of these rules obeys the following specifications:

- $\mathcal{R} = \mathcal{R}_d \cup \mathcal{R}_s$ are respectively defeasible and strict rules in which:
  - $\mathcal{R}_d = \mathcal{R}_1 \cup \mathcal{R}_2$ where $\mathcal{R}_1 = \{\Rightarrow t_{x,a} \mid \exists A_{x,a} = \langle t_{x,a}, O_{x,a}, P_{x,a}\rangle \in \mathbb{A}\}$ and $\mathcal{R}_2 = \{t_{x,a}, c_1, \ldots, c_n \Rightarrow e_z \mid \exists A_{x,a} = \langle t_{x,a}, O_{x,a}, P_{x,a}\rangle \in \mathbb{A}, (e_z, \{c_1, \ldots, c_n\}, \oplus) \in O_{x,a}, n \geq 0\}$;
  - $\mathcal{R}_s = \mathcal{R}_3 \cup \mathcal{R}_4 \cup \mathcal{R}_5 \cup \mathcal{R}_6$ where $\mathcal{R}_3 = \{t_{x,a}, c_1, \ldots, c_n \to e_z \mid \exists\ A_{x,a} = \langle t_{x,a}, O_{x,a}, P_{x,a}\rangle \in \mathbb{A}, (e_z, \{c_1, \ldots, c_n\}, \ominus) \in O_{x,a}, n \geq 0\}$, $\mathcal{R}_4 = \{t_{x,a} \to \neg t_{y,b} \mid \exists\ A_{x,a} = \langle t_{x,a}, O_{x,a}, P_{x,a}\rangle, A_{y,b} = \langle t_{y,b}, O_{y,b}, P_{y,b}\rangle \in \mathbb{A}, \neg t_{y,b} \in P_{x,a}\}$, $\mathcal{R}_5 = \{e_j \to \neg e_k \mid (e_j, e_k) \in \mathbb{E}\ \text{or}\ (e_k, e_j) \in \mathbb{E}\}$, and $\mathcal{R}_6 = \{\to s \mid s \in \mathbb{S}\}$.

The clinical information elements of our running example would produce the following rules:

- $\mathcal{R}_d = \{\Rightarrow sulf,\ \Rightarrow met\} \cup \{sulf \Rightarrow gd,\ met \Rightarrow gd\}$;
- $\mathcal{R}_s = \{met, ckd\} \cup \{sulf \to \neg met\} \cup \{ackd \to \neg dckd\} \cup \{\to ckd\}$;
- $\mathcal{R} = \mathcal{R}_d \cup \mathcal{R}_s$.

There are some important notes about this mapping provided in the original ASPIC+G paper [5]. Disputable facts such as treatments up for selection are represented as defeasible rules with empty antecedents. Additionally, there are two different representations of treatment/effect relationships. A treatment/effect relationship which features a positive effect is represented as a defeasible rule whereas a relationship featuring a negative effect is represented as a strict rule.

It is possible to place preferences on defeasible rules that convey treatments, which allows the users to express preferences for one treatment over another. Adding a scenario in which a patient would manifest a preference for $sulf$ over met would yield the following partial pre-order $\leqslant_{\mathcal{R}_d}$: $(\Rightarrow met) <_{\mathcal{R}_d} (\Rightarrow sulf)$.

In ASPIC+G, it is also possible to specify goals and a total order over these goals which are used in reasoning to select the top preferred extension containing the consistent set of treatments. Goals are set amongst the positive effects of treatments. In the running example, the set of goals would be $\mathcal{G} = \{gd, dckd\}$. Let us add the information that the patient would prioritise $dckd$ over gd. To convey this preference, the total pre-order over treatment goals is used as follows $\leqslant_{\mathcal{G}}$: $gd <_{\mathcal{G}} dckd$.

4   Case Example

The clinical case selected to map into an ASPIC+G argumentation theory was retrieved from the MIMIC-III database, a publicly-available critical care database [6]. The case was selected from discharge summary reports because they depict complex multimorbidity clinical cases that allow us to test several aspects of ASPIC+G. The case provides a detailed description of the clinical process, which make it possible to easily follow up the inherent medical reasoning.

The clinical case description concerns an 81-year-old female who was admitted to the Medicine service, complaining about a gastrointestinal (GI) haemorrhage (bright red blood per rectum, noticed by blood in her stool and red blood on the paper towel). The patient had an allergy to Penicillins, Vicodin, Cipro and Polysporin. She presented a medical history of atrial fibrillation (A-Fib), diverticulosis and myasthenia gravis. Ex-smoker (quit 25–30 years ago), denied alcohol and drugs. Her father had congestive heart failure, her mother died of myocardial infarction and her siblings had pulmonary fibrosis.

The patient arrived at the hospital after calling her Primary Care Provider, who checked an International Normalized Ratio (INR) which was elevated to 8.0. She went to the hospital where hematocrit (HCT) was 29.0 and INR was 6.1.

On admission, the patient presented sclera anicteric and tachycardia.

During the hospital stay, pericardial effusion was detected and the patient had hypoxia. Lastly, after 5 days of hospitalisation, the patient and her family decided to take comfort measures only (CMO). She was given Morphine and she passed away with her family at her side on the morning of the day after. Below are the medical notes recorded by healthcare professionals.

- **Pericardial effusion:** After a chest x-ray (CXR) it was *noted a left pleural effusion*. Subsequently, in a chest computed tomography (CT) showed *a small* *to moderate left effusion with a significant pericardial effusion and ascending thoracic aorta aneurysm. (...) A formal echo was performed on the dilated aortic root with probable small sinus of Valsalva aneurysm of the right coronary cusp and severe aortic regurgitation, a moderate circumferential pericardial effusion without frank tamponade. The cardiology recommended CT with contrast due to concern for dissection. The CT showed dissection of the thoracic ascending aorta with possible rupture into the pericardium. A CT surgery was consulted who recommended surgery but the patient declined. She was medically managed with heart rate (HR) and blood pressure control but continued to decline over the next several days with increasing oxygen requirements eventually requiring 100% on a non-rebreather.*

- **Atrial fibrillation with Rapid Ventricular Response (RVR):** *Patient has a history of atrial fibrillation on Coumadin diagnosed a year ago. Coumadin was held in the setting of GI bleed. She was given 5 mg IV Metoprolol (x2) and rates slowed to the 120 s. She then had a likely vagal episode and became acutely bradycardic with a 6-s pause. The episode resolved spontaneously and the patient reverted back to atrial fibrillation. The patient was transitioned to 25 mg Metoprolol three times a day (TID) but remained in A-Fib with RVR. She was placed intermittently on Diltizem drip and converted to normal sinus rhythm (NSR) with rates in the 60 s. When the decision was made to make her CMO, these were discontinued.*

- **GI haemorrhage (bright red blood per rectum):** *The patient presents with bright red blood per rectum in the setting of an elevated INR of 8.0 and HCT of 29.0. Her hematocrit dropped to 24.2. She received 2 units of packed red blood cells (PRBC) and had an appropriate rise in hematocrit. The INR was corrected with 2 units fresh frozen plasma (FFP) and 5 mg vitamin K x 2. Gastroenterology was consulted who felt this was likely a lower GI bleed, most likely diverticular or arteriovenous malformation (AVM), exacerbated in the setting of elevated INR. Also on the differential diagnosis are haemorrhoids and malignancy. She did have a colonoscopy 2 year ago that did not show any polyps, making malignancy unlikely. The patient's hematocrit stabled and she had no further episodes of bleeding.*

- **Hypoxia:** *Patient with intermittent desaturation. Etiology likely multifactorial secondary to pericardial effusion and poor reserve with underlying myasthenia. The CXR did not show any evidence of acute infection. She was placed on the nasal cannula to maintain oxygen saturation >92%. Please see above, but the patient had increasing oxygen requirements eventually requiring 100% on a non-rebreather. At that time the patient decided to made CMO.*

- **Myasthenia gravis:** *The patient was recently diagnosed with myasthenia after 3 years of progressive symptoms. She had a positive anti-acetylcholine receptor antibody and was started on Pyridostigmine with little improvement. She was recently started on Mestinon. Neurology was consulted regarding diagnosis and treatment and concern for underlying malignancy with paraneoplastic syndrome. They recommended monitoring vital capacity and negative inspiratory force. Neurology weighed in regarding possible surgery and advised that the patient may have a slower recovery coming off of the vent.*

From this clinical case, the information was acquired and treated. Succinctly, the clinical case can be described as:

- GI haemorrhage (*gihem*): to normalize INR (*dinr*) values (which were high), she was given FFP (*ffp*) and 5 mg vitamin K (*vitk*); to increase HCT (*ihct*) was given PRBC (*prbc*).
- Atrial fibrillation (*afib*) with Rapid Ventricular Response (RVR): to prevent a stroke (*ps*), Coumadin (*cou*) was administered; to control heart rate (*chr*), she was given Metoprolol (*met*), but remained in A-Fib with RVR. She was placed intermitantly on Diltizem (*dil*) drip and converted to NSR. These were discontinued when the decision was made to make her CMO (*cmo*).
- Hypoxia (*hyp*): to increase the oxygen intake (*ioi*), patient was ventilated (*vent*).
- Myasthenia gravis (*mg*): She had a positive anti-acetylcholine receptor antibody (*aara*). To relieve symptoms (*rsmg*), the patient took Pyridostigmine (*pyr*) and Mestinon (*mes*). Neurology weighed a possible surgery (*mgsurg*), but the patient may have a slower recovery by leaving ventilation (*vent*).
- Pericardial effusion (*pe*): in a CXR (*cxr*) was noted a left pleural effusion (*lpe*). In a chest CT (*cct*), was remarkable for a small to moderate left effusion with a significant pericardial effusion (*spe*) and ascending thoracic aorta aneurysm (*ataa*). An echo (*echo*) was performed on dilated aortic root with probable small sinus of Valsalva aneurysm (*ssva*) of the right coronary cusp and severe aortic regurgitation (*sar*), a moderate circumferential pericardial effusion (*mcpe*) without frank tamponade. CT with contrast (*ctc*) due to concern for dissection and the CT showed dissection of the thoracic ascending aorta (*dtaa*) with possible rupture into pericardium (*rip*). To treat pericardial effusion (*tpe*), a surgery (*pesurg*) was proposed, but patient refused. The patient and her family made the decision to be made CMO (*cmo*). She was given Morphine (*mor*) and she passed away with her family at her side.

5   Case Mapping

Considering the multimorbidity clinical case mentioned in Sect. 4, in this section, we perform mapping of its components to ASPIC+G framework. From this case example we have the following actions in $\mathbb{A}$:

$A_1 \langle ffp, \{(dinr, \emptyset, \oplus)\}, \emptyset \rangle;$
$A_2 \langle vitk, \{(dinr, \emptyset, \oplus)\}, \emptyset \rangle;$
$A_3 \langle prbc, \{(ihct, \emptyset, \oplus)\}, \emptyset \rangle;$
$A_4 \langle cou, \{(ps, \emptyset, \oplus)\}, \emptyset \rangle;$
$A_5 \langle met, \{(chr, \emptyset, \oplus)\}, \emptyset \rangle;$
$A_6 \langle dil, \{(chr, \emptyset, \oplus)\}, \emptyset \rangle;$
$A_7 \langle cxr, \emptyset, \emptyset \rangle;$
$A_8 \langle cct, \emptyset, \emptyset \rangle;$

$A_9\langle echo, \emptyset, \emptyset\rangle$;
$A_{10}\langle ctc, \emptyset, \emptyset\rangle$;
$A_{11}\langle pesurg, \{(tpe, \emptyset, \oplus)\}, \emptyset\rangle$;
$A_{12}\langle mor, \{(cmo, \emptyset, \oplus)\}, \{\neg dil, \neg pesurg, \neg mgsurg\}\rangle$;
$A_{13}\langle vent, \{(ioi, \emptyset, \oplus)\}, \emptyset\rangle$;
$A_{14}\langle aara, \emptyset, \emptyset\rangle$;
$A_{15}\langle pyr, \{(rsmg, \emptyset, \oplus)\}, \emptyset\rangle$;
$A_{16}\langle mes, \{(rsmg, \emptyset, \oplus)\}, \emptyset\rangle$;
$A_{17}\langle mgsurg, \{(rsmg, \emptyset, \oplus)\}, \{\neg vent\}\rangle$.

Next step, comprising defining a set of contrary effects $\mathbb{E}$; however, the clinical case example doesn't provide any contrary effect. Thus, the effect contraries are: $\mathbb{E} = \emptyset$.

After, we define the state of the patient as a set of conditions manifested by the patient. We consider $\mathbb{S} = \{lpe, spe, ataa, ssva, sar, paara, mcpe, rip, dtaa\}$.

The set of rules $\mathcal{R}$ that we produce from the clinical case, are as follows:

- $\mathcal{R}_d = \{\Rightarrow ffp, \Rightarrow vitk, \Rightarrow prbc, \Rightarrow cou, \Rightarrow met, \Rightarrow dil, \Rightarrow cxr, \Rightarrow cct, \Rightarrow echo, \Rightarrow ctc, \Rightarrow pesurg, \Rightarrow mor, \Rightarrow vent, \Rightarrow aara, \Rightarrow pyr, \Rightarrow mes, \Rightarrow mgsurg\} \cup \{ffp \Rightarrow dinr, vitk \Rightarrow dinr, prbc \Rightarrow ihct, cou \Rightarrow ps, met \Rightarrow chr, dil \Rightarrow chr, pesurg \Rightarrow tpe, mor \Rightarrow cmo, vent \Rightarrow ioi, pyr \Rightarrow rsmg, mes \Rightarrow rsmg, mgsurg \Rightarrow rsmg\}$;
- $\mathcal{R}_s = \{cmo \rightarrow \neg dil, mgsurg \rightarrow \neg vent, \rightarrow cmo, cmo \rightarrow \neg pesurg\} = \{cmo \rightarrow \neg dil, mgsurg \rightarrow \neg vent, cmo \rightarrow \neg mgsurg, cmo \rightarrow \neg pesurg\}$;
- $\mathcal{R} = \mathcal{R}_d \cup \mathcal{R}_s$;
- $\leqslant \mathcal{R}_d$: $(\Rightarrow pesurg) <_{\mathcal{R}_d} (\Rightarrow mor), (\Rightarrow dil) <_{\mathcal{R}_d} (\Rightarrow mor)$;
- $\mathcal{G} = \{dinr, ihct, ps, chr, tpe, cmo, ioi, rsmg\}$;
- $\leqslant_{\mathcal{G}}$: $dinr \simeq_{\mathcal{G}} ihct \simeq_{\mathcal{G}} ps \simeq_{\mathcal{G}} chr \simeq_{\mathcal{G}} tpe \simeq_{\mathcal{G}} rsmg <_{\mathcal{G}} ioi <_{\mathcal{G}} cmo$.

Based on the argument construction rules of ASPIC+G [5] and the goal set described in Sect. 3.3, the arguments $\mathcal{A}$ and goals $\mathcal{G}$ are as follows:

- $\mathcal{A} = \{A_1 :\Rightarrow ffp, A_2 : A_1 \Rightarrow dinr, B_1 :\Rightarrow vitk, B_2 : B_1 \Rightarrow dinr, C_1 :\Rightarrow prbc, C_2 : C_1 \Rightarrow ihct, D_1 :\Rightarrow cou, D_2 : D_1 \Rightarrow ps, E_1 :\Rightarrow met, E_2 : E_1 \Rightarrow chr, F_1 :\Rightarrow dil, F_2 : F_1 \Rightarrow chr, G_1 :\Rightarrow cxr, H_1 :\Rightarrow cct, I_1 :\Rightarrow echo, J_1 :\Rightarrow ctc, L_1 :\Rightarrow pesurg, L_2 : L_1 \Rightarrow tpe, M_1 :\Rightarrow mor, M_2 : M_1 \Rightarrow cmo, M_3 : M_2 \rightarrow \neg dil, M_4 : M_2 \rightarrow \neg pesurg, M_5 : M_2 \rightarrow \neg mgsurg, N_1 :\Rightarrow vent, N_2 : N_1 \Rightarrow ioi, O_1 :\Rightarrow aara, P_1 :\Rightarrow pyr, P_2 : P_1 \Rightarrow rsmg, Q_1 :\Rightarrow mes, Q_2 : Q_1 \Rightarrow rsmg, R_1 :\Rightarrow mgsurg, R_2 : R_1 \Rightarrow rsmg, R'_2 : R_2 \rightarrow \neg vent\}$;
- $\mathcal{G} = \{G_1 : dinr, G_2 : ihct, G_3 : ps, G_4 : chr, G_5 : tpe, G_6 : cmo, G_7 : ioi, G_8 : rsmg\}$

The preferred extensions and their respective goals are as follows:

- $\mathcal{S}_1 = \{$ A$_1$, A$_2$, B$_1$, B$_2$, C$_1$, C$_2$, D$_1$, D$_2$, E$_1$, E$_2$, F$_1$, F$_2$, G$_1$, H$_1$, I$_1$, J$_1$, L$_1$, L$_2$, N$_1$, N$_2$, O$_1$, P$_1$, P$_2$, Q$_1$, Q$_2$, R$_1$, R$_2$, R$'_2\}$, Goal($\mathcal{S}_1$) = {G$_1$, G$_2$, G$_3$, G$_4$, G$_5$, G$_7$, G$_8\}$;

– $S_2 = \{$ A_1, A_2, B_1, B_2, C_1, C_2, D_1, D_2, E_1, E_2, G_1, H_1, I_1, J_1, M_1, M_2, M_3, M_4, M_5, N_1, N_2, O_1, P_1, P_2, Q_1, $Q_2\}$, Goal(S_2) = $\{G_1$, G_2, G_3, G_4, G_6, G_7, $G_8\}$;

The argumentation theory has two preferred extensions: S_1 - S_2. Admitting the goal ordering established earlier, we calculate the goal set order. There is only one extension S_2 that achieves goal G_6 (the most preferred goal over all the set of goals). Thus, S_2 is the top preferred extension. This means that the decision was made to apply CMO by:

– Administer Morphine;
– Administer FFP and vitamin K to decrease INR;
– Have the patient receive PRBC to increase HCT;
– Administer Coumadin to prevent a stroke;
– Administer Metoprolol to control heart rate;
– Increase the oxygen intake by venting the patient;
– Administer Pyridostigmine and Mestinon to relief symptoms of myasthenia gravis.

6 Discussion

During the mapping of the clinical case, some limitations of the framework were found, particularly concerning temporal events. For example, mapping symptoms over time or even monitoring the evolution of a disease. Moreover, during the construction of rules to ASPIC+G, indirect rules can be derived from a particular clinical case. For instance, it's clearly stated in the mentioned clinical case of Sect. 4 that when the decision was made to CMO, this leads to skipping pericardial effusion surgery. However, there is another rule that we can indirectly set, which is CMO leads to skipping myasthenia gravis surgery. ASPIC+G has a limitation in deriving these indirect rules since it does not provide a formal process of deriving them.

Other information, which is usually relevant in the construction of clinical cases, is the social and family history, which in this framework is also not possible to map. These topics are important to assist health professionals in the diagnosis process and to identify risk factors.

Complementary medicine was mapped as a treatment of an action, with no outcomes and no pre-conditions. This information can be important to validate or discard diseases, helping medical reasoning. It would be important to upgrade ASPIC+G in order to be able to get more information: the reason that led to complementary medicine and the conclusions (if the disease was validated or not, for instance).

To address some of the limitations mentioned above we propose a workflow for extracting the components of the arguments and/or the rules, as depicted in Fig. 1. Each clinical case is processed using machine learning techniques (specifically natural language processing techniques) for automatic extraction of treatment, outcomes/effects and the patient status. The main goal will be to use this information to feed an argumentation framework. At the same time, this information can be used in an ontology and its relationships can feed the same framework.

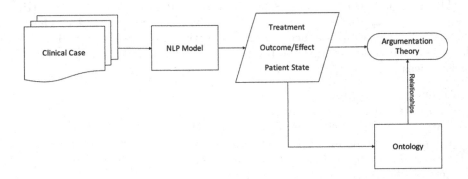

Fig. 1. Workflow to extract argumentation components.

7   Conclusions and Future Work

ASPIC+G has the potential to positively affect clinical behaviour since it provides a framework that properly assesses different conflicting solutions that can arise when treating multimorbidity patients. It provides a formal process to computationally represent clinical actions and their respective components (treatment/effect relations, contrary effects, patient state, and so forth), which provides the necessary information and expressiveness to reasoning in complex scenarios such as the multimorbidity clinical cases. Moreover, its argumentation system offers an alternative to Multiple Criteria Decision Making (MCDM), which allows merging clinical recommendations and use patient-specific goal preferences over treatments to produce the best solution. Furthermore, it has the additional advantage of being more explanatory. However, some points of improvement have been identified that can complete the framework, namely, the inclusion of social and family history, adding information about complementary medicine and map temporal events.

As future work, it is intended to provide an automated mechanism for the extraction of argumentation components (clinical actions, rules and arguments) to include in an argumentation tool. To this end, we will use deep/machine learning techniques [11] such as long short-term memory (LSTM), for natural language processing.

Acknowledgments. This work has been supported by FCT - Fundação para a Ciência e a Tecnologia within the R&D Units project scope UIDB/00319/2020. The work of Ana Silva is also supported by a Portuguese doctoral grant, SFRH/BD/143512/2019, issued by FCT in Portugal.

References

1. Costa, R., Neves, J., Novais, P., Machado, J., Lima, L., Alberto, C.: Intelligent mixed reality for the creation of ambient assisted living. In: Neves, J., Santos, M.F., Machado, J.M. (eds.) EPIA 2007. LNCS (LNAI), vol. 4874, pp. 323–331. Springer, Heidelberg (2007). https://doi.org/10.1007/978-3-540-77002-2_27
2. Cook, D.A., Sherbino, J., Durning, S.J.: Management reasoning: beyond the diagnosis. Jama **319**(22), 2267–2268 (2018)
3. Fraccaro, P., Casteleiro, M.A., Ainsworth, J., Buchan, I.: Adoption of clinical decision support in multimorbidity: a systematic review. JMIR Med. Inform. **3**(1), e4 (2015)
4. Sutton, D.R., Fox, J.: The syntax and semantics of the pro forma guideline modeling language. J. Am. Med. Inform. Assoc. **10**(5), 433–443 (2003)
5. Oliveira, T., Dauphin, J., Satoh, K., Tsumoto, S., Novais, P.: Goal-driven structured argumentation for patient management in a multimorbidity setting. In: Dastani, M., Dong, H., van der Torre, L. (eds.) CLAR 2020. LNCS (LNAI), vol. 12061, pp. 166–183. Springer, Cham (2020). https://doi.org/10.1007/978-3-030-44638-3_11
6. Johnson, A.E.W., et al.: Mimic-iii, a freely accessible critical care database. Sci. Data **3**(1), 160035 (2016). https://doi.org/10.1038/sdata.2016.35
7. Grando, M.A., Glasspool, D., Boxwala, A.: Argumentation logic for the flexible enactment of goal-based medical guidelines. J. Biomed. Inform. **45**(5), 938–949 (2012)
8. Hunter, A., Williams, M.: Aggregating evidence about the positive and negative effects of treatments. Artif. Intell. Med. **56**(3), 173–190 (2012)
9. Kokciyan, N., et al.: Towards an argumentation system for supporting patients in self-managing their chronic conditions. In: Workshops at the Thirty-Second AAAI Conference on Artificial Intelligence (2018)
10. Modgil, S., Prakken, H.: The ASPIC+ framework for structured argumentation: a tutorial. Argument Comput. **5**, 31–62 (2014)
11. Carneiro, D., Novais, P., Pêgo, J.M., Sousa, N., Neves, J.: Using mouse dynamics to assess stress during online exams. In: Onieva, E., Santos, I., Osaba, E., Quintián, H., Corchado, E. (eds.) HAIS 2015. LNCS (LNAI), vol. 9121, pp. 345–356. Springer, Cham (2015). https://doi.org/10.1007/978-3-319-19644-2_29

Learning User Comfort and Well-Being Through Smart Devices

David Sousa[1]([⊠]), Fábio Silva[1,2]([⊠]), and Cesar Analide[1]([⊠])

[1] ALGORITMI Center, Department of Informatics, University of Minho, Braga, Portugal
a78938@alunos.uminho.pt, fas@estg.ipp.pt, analide@di.uminho.pt
[2] CIICESI, ESTG, Politécnico do Porto, Felgueiras, Portugal

Abstract. This article aims to provide a large-scale study, without geographical restrictions, on how people's habits can influence their comfort and well-being. In this sense, sensing techniques are used, through smart devices, such as the smartphone, with the main objective of collecting information about the user and the environment that surrounds him. The collected data are subsequently processed, and several models of deep learning are built that aim to predict the well-being and comfort in the different environments in which a person is inserted. However, due to the pandemic, the main focus has been changed and the main objective is to understand if it is possible to predict comfort and well-being in the quarantine.

Keywords: Comfort · Deep learning · Supervised learning · Well-being

1 Introduction

According to the Oxford dictionary habit is defined as "a thing that you do often and almost without thinking, especially something that is hard to stop doing". Now by the quick interpretation of this concept, we can infer that there are patterns that we are running throughout our lives. Recognizing these standards and acting accordingly can give people a better quality of life.

In this sense, concepts such as smart devices, intelligent environments, are crucial in building such systems. These bring systems that are very useful as they are able to feel the environment around them.

Well-being and comfort have had a major impact on society. In the case of well-being, higher levels mean less risk of disease and more longevity [2]. Consequently, comfort is also a vogue concern. Universities, hospitals, institutions spend most of their time promoting increased comfort and consequently reducing discomfort. Several studies address that employees with higher comfort have a better performance on work [9]. However, it is quite difficult to understand what these terms are and how they may be affected. In fact, one of the main questions that come up when we talk about monitoring comfort and well-being is: What to measure? The environment condition alone is correlated to well-being and comfort status [17]. The heterogeneity of the environment and of the population makes difficult to perceive what may be affecting us in a given place.

C. Analide et al. (Eds.): IDEAL 2020, LNCS 12489, pp. 350–361, 2020.
https://doi.org/10.1007/978-3-030-62362-3_31

However, these days it is increasingly easy to explore these terms. Effectively more and more people have access to smart devices. These devices make the use of sensors easy and inexpensive since, for example, when we buy a smartphone it comes with a huge variety of sensors. This will allow some globalization, making possible to gather more heterogeneous information and give the possibility to answer this and other questions. In short, this article aims to demonstrate the main results obtained in the search for a solution that allows foreseeing comfort and well-being.

2 State of Art

As one of the focus of this project is comfort and well-being it is important to define these concepts in order to clarify their meaning. In this sense, on the next sub-chapters, is presented a little about these and other concepts crucial to the understanding of the present work.

2.1 Well-Being and Comfort

On the one hand, the term welfare is quite difficult to define, and its refinement has been the subject of many studies over the years. It is a term that is often used by the community, but not always with a correct connotation. In fact, because of the associated confusion, we do not always know the meaning of being well. Several authors seek to define this concept. Elena Alatarsteva and Galina Barysheva [1] subdivide this term into two strands: one objective and one subjective. In the case of the objective strand, well-being could be characterizing by wage levels, residence conditions, educational opportunities, social quality, the environment, security and civil rights. In the case of the subjective strand, well-being is conceptualized only as an internal state of an individual. However, this is not the only vision about the definition of this concept. Through the consulting of various articles published by authors from different branches it is possible to arrive at the definition of this concept more specifically, that is, by categorizing it into different classes. Some of this classes may be Community Well-being [14], Economic Well-being, Emotional Well-being [4], Physical Well-being [14], Social Well-being [3] and Work Well-being [15]. There are several ways to categorize well-being, however, in general, they all end up touching the same points.

In the case of comfort, looking for and maintaining it is common to all people. Many people quickly identify loss of comfort and work diligently to restore it.

In our days, the term comfort is often seen related to the marketing of products like chairs, cars, clothing, hand tools, and a lot of others [20]. Evidence from the literature suggests that comfort, as a concept, has historical and more recent relevance for nursing [19]. It is, therefore, a concept that needs further exploration to increase knowledge in this area because it is difficult to reach a consensus on its definition. In order to establish the definition of comfort, this article will use the meaning present in the Oxford dictionary. According to this, comfort is a broader concept that can be defined from a physical and a psychological perspective. In the first case, it is seen as a "state of physical ease and freedom from pain or constraint". In the second it is defined as: "The easing or alleviation of a person's feelings of grief or distress". However, despite all these definitions, they all note the existence of factors that influence comfort. Some papers show that different activities during measurements can influence comfort, concluding that characteristics of the environment and the context, can change how people feel [20]. Although comfort is often associated with a synonym for well-being, a clear difference can be demonstrated between both concepts. By the previous definitions, it is noticeable that comfort is most used to classify the atmosphere that surrounds the human being. This implies that comfort is associated with momentary aspects. However, well-being arises characterized by an exhaustive variety of groups, which makes it associated with a long-term context. By way of example, a person may find himself comfortable but unhappy (and vice versa). In this sense, well-being is referred to in many studies as being progressively improved, while comfort is something that can be improved on time. As we can see, the mental health organization in the UK said that "it is important to realize that well-being is a much broader concept than moment-to-moment happiness" [5]. Although comfort and well-being are distinct terms, this does not imply that they exist separately. Indeed, both terms arise intrinsically linked. Comfort at a certain level contributes to well-being.

2.2 Learning Techniques

One of the areas of artificial intelligence is machine learning which is known for the capacity to automatically learn and improve from experience without the need of being explicitly programmed [18]. With the definition of machine learning comes another, deep learning. Deep learning is a subset of machine learning. It is considered the next evolution of machine learning algorithms. The designation deep comes because of the many layers that usually this type of technique has. The main difference between these two areas is that deep learning can automatically discover the features to be used while in machine learning the features have to be provided manually [7].

Another term that arises with deep learning is the term of artificial neural network. These types of algorithms were drawn to simulate the way the human brain analyzes and processes information [6]. There are many types of neural networks that can be used depending on what is necessary. In this work, these mechanisms can be useful to make predictions. When we speak of the term predictions, we speak of the outputs that are achieved from the training that a given algorithm had on a given dataset. In many cases, the term does not refer to future issues. It can be used, for example, to predict if there has been fraud in a given transaction that has already happened. Predictions are very important as they allow highly accurate guessing, thus helping in various aspects. However, for good guessing happen an algorithm has to learn. In fact, machine and deep learning are bounded by the concept of learning. There are different types of learning: supervised learning, unsupervised learning and reinforcement learning. Each of these types has several machines and deep learning algorithms that, depending on problem, can be applied. Although there many types of learning, this article will focus on supervised learning. In this type of learning is given to the machine what each data block means. This is, the model is trained on a labeled dataset [11]. A label dataset is one that has both parameters: input and output parameters. In this case, learning is based on regression and classification techniques. If the output variable is a category, such as "yes" or "no", it is a classification problem. If the output variable is a real value, such as "weight" it is a regression problem. In this article, will be useful, a category technique because one of the main purposes is to identify well-being and comfort levels which can be represented by classes.

There are many algorithms that can be applied in this work. One of these algorithms that can be really useful explore the concept Recurrent Neural Networks (RNN). This is because RNN "can model sequence of data so that each sample can be assumed to be dependent on previous ones" [12]. One of the most popular versions of RNN is Long Short-Term Memory (LSTM) Network which is used to process and predict time series given time lags of unknown duration [12]. In fact, comfort and well-being can be dependent on what happened in a previous time.

2.3 Related Work

Generally speaking, there are several solutions we use every day that are related to the concept of data collection and intelligent suggestions. Take for example the case of virtual assistants such as Siri, Cortana, and others. This type of system collects as much information as possible about the user and the environment around them, and from there suggests and performs actions that simplify people's lives. The use of smart devices for data collection is not new, and there are a lot of applications in this field. One of the main applications is, for example, context recognition [13]. In the case of comfort and well-being, there are advances with the use of smart devices sensors. One example of an application that has this purpose is *BeWell* [10]. It is an application designed to make people care about certain concepts of their well-being, concepts that are often overlooked. This is the case of, for example, lack of sleep, hygiene or high social isolation. The authors believe that this situation is caused by an absence of adequate tools for effective self-management of overall well-being and health. The main idea of this application is to be capable of monitoring multiple dimensions of human behavior, encompassing physical, mental and social dimensions of well-being. Comparing to what is presented in this article, this application will be useful as a support for this work since the main objective is similar to what is presented. However, this application only makes use of some device sensors whereby only a few pre-established well-being metrics are evaluated.

In another way, in the field of comfort, there are equally some studies and consequently some applications. One example of that is the *ThermOOstact* [16]. This is an application that has with main purpose study the thermal comfort inside a University Campus. Nevertheless, it is really specific since only is used concepts associated with temperature and only works on the University of California occupants. The purpose of this article is more comprehensive because a lot of different data will be collected and according to that data will be possible to explore a wide range of concepts. This brings a great advantage because it allows the creation of a broader application. Furthermore, the proposal application will allow to explore comfort and well-being at the same time.

3 Architecture

According to the system architecture, it is possible, in addition to fundamental components such as the database, to divide the application into two strands. A strand oriented only to data collection so that the data can be stored and processed later and another strand more related to the building of predictive models. In Fig. 1 could be seen this process.

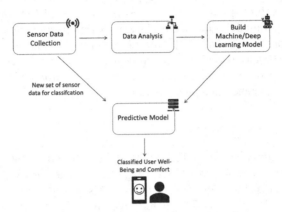

Fig. 1. Application workflow

3.1   Data Collector Application Strand

Since for built machine learning algorithms data is needed, an application that collects data has to be built. With an application developed for smart devices, it is possible to use inherent sensors of this type of system. Some information that could be extracted is the number of steps, ambient temperature, light level, magnetometer values, gyroscope values, accelerometer values, humidity, and location. However other information can be collected, like the environment type (indoor or outdoor), Wi-Fi information (SSID, RSSID, and Mac address of the provider), battery percentage, call number and duration, during a day, the number of events for the day and activity performed by the user by the time that the data is collected. Some of this information is sensible and for more security, it has to be saved through hash key, furthermore, no information identifying the user could be saved. In addition, it is intended to use APIs so that the information gathered was even wider. This includes, for example, the OpenWeather API that allows collecting a vast amount of data associated with weather conditions (such as weather state, wind speed, ultraviolet radiation index, pressure, and temperature) and the Foursquare API that allows capturing location-related data (such as local category). In addition to these, an API to capture data about air quality (such as AQI and the dominant polluter) can also be used.

One decision to make is how users will be asked about their comfort and well-being. In this sense, to avoid being too intrusive, two situations were taken into account. On the one hand, the questions about comfort could be made by context analysis. That is, when it is detected that the user has moved, questions will be asked more frequently. Another situation is that these questions can appear at specific times. In this case, it could be given to the user the possibility to choose how often the questions occur. Three options could be taken into account: low, medium, or high frequency. In the case of the well-being question, it is intended to appear only less frequently. This is due to what was defined in the previous section. Indeed, it is assumed that well-being throughout the day will not have as many variations as comfort. In these cases, the notification could show up every two hours. On the other hand, the user could also be allowed to answer a question about comfort or well-being voluntarily without waiting for a notification. To make both processes faster a response scheme through a smiley face could be used (Fig. 2). This idea came about over an adaptation of the Likert scale, commonly used in opinion polls. As the following figure demonstrate, the questionnaire for well-being is more complex to be possible to associate the different type of data collected to different types of well-beings (defined in state of art). These questionnaires were designed with the aim of being short answer so that users can answer quickly and without effort.

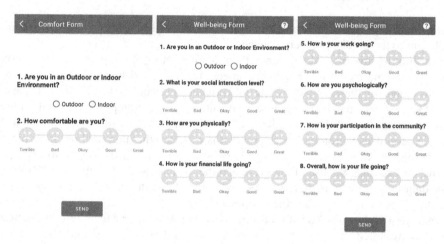

Fig. 2. Application questions

3.2 Learning Workflow Strand

After an initial analysis of the data, the basic procedures of machine learning and, more particularly deep learning, can be used to build predictive models. Such procedures result from the application of frameworks that arise from different studies. One example of an approach was the one described by Yufeng Guo [8]. This involves, after the data collection, a pre-processing phase. This is a complex phase, as it involves a set of steps (missing values treatment, data-type conversions, data normalization, among others) where various strategies could be explored. Then the objective is to choose a model and train it, trying to improve it in relation to a base value. Finally, it is necessary to evaluate the model by making optimizations regarding its hyper-parameters. The studies applied below could be applied to comfort and well-being.

4   Case Study

The case study that will be presented next involves an application created for Android, with the assumptions stated above, with the participation of ten users who used the application during the period from 15 March 2020 to 20 April 2020. Users are located in Portugal and due to this factor, the data collection period coincided with a mandatory quarantine period due to the global pandemic. In this sense, there was a change in the final objective of the work so that it fit the conditions that were felt. In short, the collected dataset contains 1954 lines and a total of 43 features. In the following steps, we summarize some of the steps taken. This work of building a model was done in the *Python* programming language, using some libraries such as *TensorFlow* and *sklearn*. Some of the data processing done was:

- **Elimination of some irrelevant variables:** Variables like the user's key, which is not relevant to the prediction, were deleted;
- **Handling of missing values:** Since there were a lot of missing values, we started by using some methods for their treatment. What seemed more adequate was the replacement of these values by the immediately previous value. This was due to the fact that once the data is captured sequentially, this technique could result in less introduction of variability in the dataset. In addition, in the case of some features, such as the case of meteorology, for example, it is expected that they will not have major variations over the course of a day. However, other techniques were also explored here, for example, the use of the mean and the median. Since the number of lines in the dataset is not very high, it was decided not to eliminate lines with missing values in order not to further reduce the sample number;
- **Treatment of the encoding of non-numeric variables:** Several features of the dataset were represented by strings. Since deep learning models only accept numbers, it was necessary to proceed with some type of pre-processing. For this purpose, the One Hot Encoder method was used;
- **Data Normalization:** In order to reduce the discrepancy, some techniques for framing the values were also applied, among them the Standardization and Normalization methods were tested;
- **Target Encoding:** Since the target presents values in a certain way sorted from 1 to 5, a label encoding technique was used to normalize these values (thus transforming these values into classes 0 to 4).

We want to study if the previous data are relevant to the prediction of physical well-being. For that case were used RNN, more in particular LSTM. As this is a neural network where the order of the data is quite relevant, no shuffle has been done. Although, other special precautions regarding the way the data had to be processed were taken. In addition to pre-processing, the data were transformed according to a sliding window principle. Thus, the previous data are organized so that they serve as input to predict the target of the next line. Since we want to classify the well-being some precautions have to be taken. As this is a multiclass classification problem, the loss function used was therefore *categorical_crossentropy*. Furthermore, in the final layer, a *softmax* activation function was used. Here we have to take into account the type of this activation function and the loss function, as incorrect use of these can lead to false results. Some studies were carried out in order to understand what is the method that could give best results. Essentially two approaches were built:

1. **LSTM using data from a single user**
   a. Only one user was used for training the rest were used for testing.
2. **LSTM without separating by user**
   b. There was no separation of each user (only two users are saved for the test);
   c. The transformation into a sliding window occurred in all data, so the data was sorted by user key and date-time in order to reduce inconsistencies.

The following are some of the results obtained in both approaches. It is although important to refer that the following experiments were made on physical well-being. However, the study could be extended to other well-being types.

First, a study was done on what should be the best treatment for missing values. The results could be seen summarized on Table 1.

Table 1. Nan treatment experiments

Nan treatment	Approach type	Train accuracy	Train loss	Test accuracy	Test loss
Using only forward propagation	1	78.96%	0.5802	64.42%	1.2970
	2	77.04%	0.7422	28.70%	1.8702
Using mean values	1	78.96%	0.6429	64.42%	1.2372
	2	41.99%	1.0705	30.26%	2.0822
Using median values	1	78.96%	0.5680	64.42%	1.3350
	2	41.56%	1.0673	30.26%	2.0683

Comparing the two approaches is possible to see that the first one, gives in a first instance better result than the second one. This could be for the fact that the first one, is simpler and the data is less. However, test accuracy does not have big changes in the two approaches. In these experiments seem like using the mean values to solve the Nan problem is the most appropriate in the first approach (less loss) and the median is better in the second approach.

With the use of values normalization techniques, it was possible to obtain a big improvement in the accuracy of the second approach, as shown in Table 2. However, in the first one, no improvement has been achieved. Despite all, generally, a value normalization is better. The *MinMaxScaler* technique was the chosen one.

Since the number of data is low, the next step was to test resampling mechanisms. The method varies depending on the approach. In the first one, the data was resampling in an interval of 15 min because it was the minimum time that notification is thrown.

Table 2. Value normalization experiments

Value normalization	Approach type	Train accuracy	Train loss	Test accuracy	Test loss
Standardization (*StandardScaler*)	1	100%	9.6612e − 04	23.93%	3.3128
	2	99.57%	0.0147	65.91%	2.1979
Normalization (*MinMaxScaler*)	1	83.23%	0.4657	64.42%	1.2368
	2	98.97%	0.0265	88.52%	0.6357

In the case of the second approach, the resampling was done taking into account some considerations. First, the data were grouped by user, after that the resampling was applied for each user individually. This is, for every user was generated samples that correspond to the interval of 15 min. However, in this case, it was not possible to improve the accuracy. The results could be consulted in Table 3.

Table 3. Resample experiments

Resample type	Approach type	Train accuracy	Train loss	Test accuracy	Test loss
Without resample	1	83.23%	0.4657	64.42%	1.2368
	2	98.97%	0.0265	88.52%	0.6357
With resample	1	99.70%	0.0140	82.16%	0.9723
	2	99.61%	0.0133	62.91%	0.8457

The final step was to validate and tuning the model. In these approaches the objective was to experiment some combinations in order to find a good fit. The number of layers, the number of neurons, the windows size, epochs, batch size, among other, were tested together. It is, therefore, important to refer that a good fit is not that one that has high accuracy. In fact, after the application of some of these techniques, it was possible to improve the results and consequently combat problems such as underfitting and overfitting which can lead to a leak of performance on the final model despite the accuracy value.

In the first approach, the model seems to converge after a few Epochs. The learning curves of the best model could be seen in Fig. 3. The windows size used was 8 and the batch size was 6.

Despite these results, this approach has some problems. The main one is in choosing a user who is the most possible representative of the problem. For example, if user A is used for training and user B is used for testing the model will fail if the classes in B are not the same as in A (user A has chosen 1 and 2 for physical well-being, and user B has chosen 2 and 3, as the model was not trained in class 3 this could be a problem). Furthermore, each user individually has few rows of data. This makes the model be more fallible in a real scenario since it will depend on that one user used for the training. The

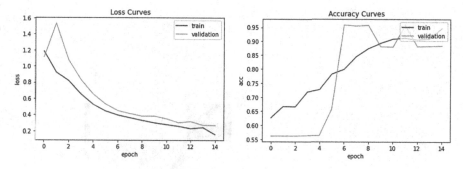

Fig. 3. Approach 1: learning curves

next image (Fig. 4) represents two confusion matrixes from two different users. The one on the right (Fig. 4b) has a higher accuracy than the one on the left (Fig. 4a). This could be explained for what was said before.

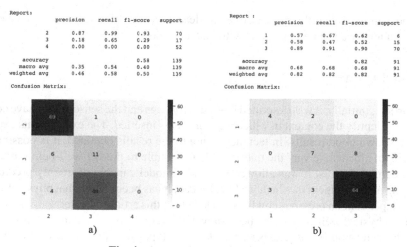

Fig. 4. Approach 1: confusion matrixes

The validation of the model and consequently the final accuracy was obtained by using all the remaining users from the dataset. The mean accuracy obtained was 84.22%.

In a search for finding a good model the same steps following before were applied in the second approach. The final results could be seen on Fig. 5.

The windows size used was 12. The best batch size found was 14. Although this approach would have the same problem as the first one, since the number of data is bigger the probability of that problem happen is lower.

Despite all this, this approach has also some other problems. First, by joining the data of all users in the same file, we may be creating some inconsistencies mainly about the creation of sliding windows. Second, the classes are unbalanced, this combined with

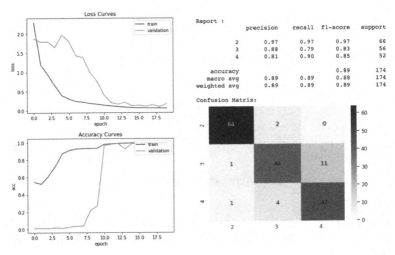

Fig. 5. Approach 2: final model

the impossibility to shuffle the dataset could lead to a leak of performance. However, the second approach seems to be one with better results.

5 Conclusion

In this article, initially, it was described how smartphones and the sensors they have could be used to identify the context in which a given user is inserted. The experiments carried out have shown good results. In fact, according to the results obtained, it is possible to answer in an affirmative way the question "Is it possible to predict well-being in times of a pandemic?". Thus, it was allowed to build a model capable of mainly predicting well-being at the moment in which we live. Indeed, it has been shown that physical well-being is dependent on previous conditions. Despite in this article only some of the results regarding physical well-being have been shown, the process is more comprehensive and allows the extension of these experiences to other types of well-being. In the case of comfort, it is intended to apply the same studies.

In the future, and after having overcome the contingencies of the times in which we live, we hope to proceed to a new collection of data and thus not only be able to predict well-being and comfort during quarantine but also for different environments and different spaces. Furthermore, it is intended to create a more comprehensible study and in that way fight against the problems mentioned. Such a study proves to be quite useful both from a personal point of view, as well as from a more professional point of view. From a personal point of view, in a future work, notifications could be launched that provide advice to users so that they can consciously act on their well-being and comfort. In the professional case, on the other hand, the development of an application that can cope with changes in the conditions of comfort and well-being in a given space can be useful. So, the next steps are to improve this study and apply it to help the needs of different populations.

Acknowledgments. This work has been supported by FCT - Fundacao para a Ciencia e Tecnologia within the R&D Units Project Scope: UIDB/00319/2020.

References

1. Alatartseva, E., Barysheva, G.: Well-being: subjective and objective aspects. Proc. – Soc. Behav. Sci. **166**, 36–42 (2015)
2. CDC: Well-being concepts (2018). https://www.cdc.gov/hrqol/wellbeing.htm. Accessed: 15 Oct 2019
3. Davis, T.: What is well-being? definition, types, and well-being skills (2019). https://www. psychologytoday.com/us/blog/click-here-happiness/201901/what-is-well-being-definition-types-and-well-being-skills. Accessed 10 Oct 2019
4. Eid, M.: The Science of Subjective Well-Being. Guilford Press, New York (2010)
5. Foundation, M.H.: What is wellbeing, how can we measure it and how can we support people to improve it? (2015). https://www.mentalhealth.org.uk/blog/what-wellbeing-how-can-we-measure-it-and-how-can-we-support-people-improve-it. Accessed 11 Nov 2019
6. Frankenfield, J.: Artificial neural networks (ANN) defined (2018). https://www.investopedia. com/terms/a/artificial-neural-networks-ann.asp. Accessed 20 Nov 2019
7. Garbade, M.: Clearing the confusion: Ai vs machine learning vs deep learning differences (2018). https://towardsdatascience.com/clearing-the-confusion-ai-vs-machine-learning-vs-deep-learning-differences-fce69b21d5eb. Accessed: 26 Jan 2020
8. Guo, Y.: The 7 steps of machine learning (2017). https://towardsdatascience.com/the-7-steps-of-machine-learning-2877d7e5548e. Accessed 30 Apr 2020
9. Haynes, B.P.: The impact of office comfort on productivity (2008). http://shura.shu.ac.uk/4593/1/Haynes__Impact_Office_Comfort_2008.pdf. Accessed 15 Oct 2019
10. Lane, N., et al.: Bewell: a smartphone application to monitor, model and promote well-being. In: Proceedings of the 5th International ICST Conference on Pervasive Computing Technologies for Healthcare, January 2011. https://doi.org/10.4108/icst.pervasivehealth.2011.246161
11. Marsland, S.: Machine Learning an Algorithmic Perspective. CRC Press, Boca Raton (2015)
12. Mittal, A.: Understanding RNN and LSTM (2019). https://towardsdatascience.com/understanding-rnn-and-lstm-f7cdf6dfc14e. Accessed 14 Jan 2020
13. Otebolaku, A.M., Andrade, M.T.: User context recognition using smartphone sensors and classification models. J. Netw. Comput. Appl. **66**, 33–51 (2016). http://www.sciencedirect.com/science/article/pii/S1084804516300261
14. Rath, T., Harter, J.: The Economics of Wellbeing. Gallup press, Washington DC (2010)
15. Rothmann, S.: Job satisfaction, occupational stress, burnout and work engagement as components of work-related wellbeing. SA J. Ind. Psychol. **34**, 11–16 (2008). http://www.scielo. org.za/scielo.php?script=sci_arttext&pid=S2071-07632008000300002&nrm=iso
16. Sanguinetti, A., Pritoni, M., Salmon, K., Morejohn, J.: Thermoostat: occupant feedback to improve comfort and efficiency on a university campus, August 2016
17. Silva, F., Analide, C.: Tracking context-aware well-being through intelligent environments. ADCAIJ: Adv. Distrib. Comput. Artif. Intell. J. **4**(2) (2015). http://campus.usal.es/~revistas_trabajo/index.php/2255-2863/article/view/ADCAIJ2015426172
18. Team, E.S.: What Is machine learning? a definition (2017). https://expertsystem.com/machine-learning-definition/. Accessed 20 Nov 2019
19. Tutton, E., Seers, K.: An exploration of the concept of comfort. J. Clin. Nursing **12**(5), 689–696 (2003)
20. Vink, P., Hallbeck, S.: Editorial: comfort and discomfort studies demonstrate the need for a new model. App. Ergon. **43**, 271–276 (2011). https://doi.org/10.1016/j.apergo.2011.06.001

A Deep Learning Approach to Forecast the Influent Flow in Wastewater Treatment Plants

Pedro Oliveira[1](✉) [iD], Bruno Fernandes[1] [iD], Francisco Aguiar[3],
Maria Alcina Pereira[2] [iD], Cesar Analide[1] [iD], and Paulo Novais[1] [iD]

[1] ALGORITMI Centre, University of Minho, Braga, Portugal
poliveira199208@gmail.com,bruno.fmf.8@gmail.com,
{analide,pjon}@di.uminho.pt
[2] Centre of Biological Engineering, University of Minho, Braga, Portugal
alcina@deb.uminho.pt
[3] Águas do Norte, Guimarães, Portugal
f.aguiar@adp.pt

Abstract. For the management and operation of a Wastewater Treatment Plant (WWTP), the influent flow is one of the most important variables. Hence, this paper presents an evaluation of multiple Deep Learning models to forecast the influent flow in WWTPs for the next three days, taking into account previous influent observations as well as historical climatological data. Long Short-Term Memory networks (LSTMs) and one-dimensional Convolutional Neural Networks (CNNs), following a channels' last approach, were conceived to tackle this time series problem. The best candidate LSTM model was able to forecast the influent flow with an approximate overall error of $200\,m^3$ for the three forecast days. On the other hand, the best candidate CNN model presented a slightly higher error, being outperformed by LSTM-based models. Nonetheless, CNNs, which are typically applied in the computer vision domain, also showed interesting performance for time series forecasting.

Keywords: Deep Learning · Wastewater Treatment Plants · Influent flow · Long Short-Term Memory Networks · Convolutional Neural Networks

1   Introduction

Population growth worldwide has lead, directly and indirectly, to an increase in the pollution levels on our planet [1]. In fact, liquid and solid wastes are produced in industrial activities, which, when discharged into rivers, can pollute many of the world's water sources. Access to clean water and the disposal of effluents represents a fundamental challenge for all authorities connected to the water supply. Currently, in developed countries, wastewater is treated using Wastewater Treatment Plants (WWTPs) to ensure that it is returned to its natural habitat in the best conditions [2].

© Springer Nature Switzerland AG 2020
C. Analide et al. (Eds.): IDEAL 2020, LNCS 12489, pp. 362–373, 2020.
https://doi.org/10.1007/978-3-030-62362-3_32

To guarantee a high-level of water quality, it is necessary to monitor several WWTPs' parameters. Such monitoring leads to the detection of failures in the WWTPs and, consequently, to an improvement in terms of quality and in the reduction of maintenance risks. The influent flow in a WWTP has a key impact on its operation and management. One of the cases in which the forecast of the influent flow has a great impact is on the use of electricity [3]. By forecasting this flow, it is possible to optimize the selection and programming of the pumps to be used in the wastewater treatment, thus being able to reduce the electricity used. Some factors need to be taken into account in such forecasts, including climatological conditions and the characteristics of the WWTP itself.

This manuscript aims to evaluate multiple Deep Learning models to forecast the influent flow in WWTPs for the next three days, taking into account previous influent observations as well as historical climatological data. Long Short-Term Memory networks (LSTMs) and one-dimensional Convolutional Neural Networks (CNNs) are to be conceived, tuned and evaluated. Hence, the remainder of this paper is structured as follows: the next section summarizes the state of the art regarding influent flow forecasting in WWTPs. The third section aims to present the materials and methods, where the used datasets are addressed, explored and pre-processed. The fourth and fifth sections present a description of the conducted experiments and a discussion of the obtained results, respectively. The sixth and final section examines the main conclusions of this study and presents future perspectives.

2   State of the Art

Many studies have been carried out taking into account the forecast of influent flow in WWTPs [4–8]. Zhang et al. (2018) carried out a study with the objective of reducing the overflow in a WWTP, making a real-time forecast of the influent flow in the WWTP [4]. The authors divided their study into two parts: the first in the development of a hydraulic system to identify free space in the WWTP inlet tubes, and finally, the development of Recurrent Neural Network (RNN) models to predict the overflow of these tubes in rainy situations. The three models used were Elman [5], Nonlinear autoregressive exogenous model (NARX) [6] and LSTMs. The data was collected from the Regnbyge platform. As metrics for evaluating the models, the authors used R-Squared (R^2) and Nash-Sutcliffe Efficiency (NSE). The data were normalized between 0 and 1. In the case of the Elman and NARX models, the data was partitioned into training, test and validation data. On the other hand, in the LSTM models, the data were partitioned in training and testing. In all models, there was a implemented a tuning process. The candidate models, concerning the used metrics, presented a very similar performance. Despite this, the authors concluded that the LSTM model is more effective in long-term dependencies, dealing better with dynamic flow changes.

Another study by Zhou et al. (2019) was based on the proposal of a model using Random Forest for the daily forecast of influent flow in two WWTPs [7].

The data used by the authors were collected from Hydromantis Environmental Software Solution and the data related to climate through Weather Canada. Regarding the used features, the date and time were one-hot encoded to reflect patterns of time sequences. A tuning process was carried out concerning the number of trees and the number of resources tested in each division for the model developed. To compare the performance of the developed model, the authors compared it with an AutoRegression Integrated Moving Average (ARIMA) and Multilayer Perceptron (MLP) models, using the same architecture. The metrics used to evaluate the performance were R^2, NSE and Mean Absolute Percentage Error (MAPE). The data were divided into training and testing. The authors concluded that, in general, the Random Forest model can predict wastewater inputs competently. Regarding the comparison with the ARIMA model, although in one of the stations the results were not satisfactory, the MAPE value is low. Regarding the MLP model, the Random Forest model cannot capture extreme values as well, but the results were generally satisfactory. One of the points highlighted by the authors is that the range of forecast results in the Random Forest model is determined by the range of training data, thus not being able to make a forecast that exceeds that range.

Szelag et al. (2017) developed a study whose objective was to compare the application of different non-linear methods to predict the flow of sewage in a WWTP for the next day [8]. The authors compare four models: Random Forest, K-nearest neighbour (KNN), Support Vector Machines (SVM) and Kernel Regression. As input, the models received precipitation values, the water levels of the Wislok river and WWTP sewage inlets, between 2005 to 2008. The input variables were normalized by the min-max transformation and selected using a matrix relevant correlation. Regarding the evaluation metrics of both models developed, Mean Absolute Error (MAE) and MAPE were used. The models were tested in 12 experiments with different inputs. The authors concluded that in about 75% of the experimented cases the SVM method was more effective than the others. The Kernel Regression never managed to be the best model, in any of the experiments. The authors conclude that experiments with the largest number of input variables presented better results both for MAE and MAPE.

3   Materials and Methods

The materials and methods used in the conducted experiments are detailed in the following lines.

3.1   Data Collection

The influent flow dataset used in this study is based on real-world data and was made available by a Portuguese water company. Climatological data include temperature, precipitation and humidity values and were collected through APIs calls to the Open Weather Maps platform. The present study analyzes observations belonging to the period between January, 2016, to May, 2020.

3.2   Data Exploration

The data used in the conducted experiments come from two distinct datasets. The first corresponds to daily historical data regarding influent flow from one Portuguese WWTP. This dataset consists of two features as described in Table 1, and contains a total of 1610 timesteps. The same table also presents the most important features present in the climatological dataset, which consist of a total of 25 features, with a grand total of 76891 hourly timesteps.

Table 1. Available features in the used datasets.

#	Features	Description	Unit
	Influent Flow Dataset		
1	*date*	Timestamp	date & time
2	*flow*	Accumulated influent flow value	m^3
	Weather Dataset		
1	*dt_iso*	Timestamp	date & time
2	*temp*	Temperature	°C
3	*feels_like*	Human perception of climate	°C
4	*temp_min*	Minimum temperature	°C
5	*temp_max*	Maximum temperature	°C
6	*pressure*	Atmospheric pressure	hPa
7	*humidity*	Humidity percentage	%
8	*wind_speed*	Wind speed value	m/s
9	*wind_deg*	Wind direction	Degrees
10	*rain*	Rain volume	mm
11	*clouds_all*	Cloudiness percentage	%

No missing values were present in the dataset. However, being this a time series problem, we were also required to pay attention to missing timesteps. After an analysis of both datasets, it was found that no timestep was missing and the time series was complete. To identify the existence of outliers, the z-score method was used. It was possible to understand that in some timesteps there was an insertion error in the influent flow. The identified values were corrected.

To understand the variation of the influent flow over the different years, the average monthly variation of the flow was analyzed. Figure 1 illustrates the mentioned variation. It is possible to verify that the biggest peaks of influent flow are verified in 2019. In that same year, the values have higher monthly values than in other years. In addition, it can be seen that the influent flow tends to decrease every year after October, except in 2017. In that year, the flow continues to grow until December.

Then, it was found that all features in both datasets assumed a non-Gaussian distribution, taking into account the Kolmogorov-Smirnov test with $p < 0.05$.

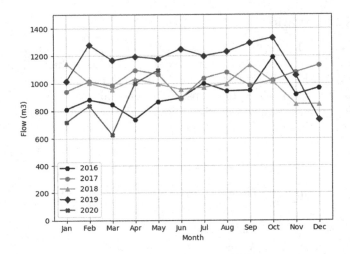

Fig. 1. Monthly variation of influent flow over the years.

The non-parametric Spearman's rank correlation coefficient was used to analyze the correlation between several pairs of features. No interesting correlations were found.

3.3 Data Preparation

The first step performed to prepare the data was to apply a feature engineering process [16–18], in both datasets, to create 3 more features from the timestamps (*year*, *month* and *day*). The climatological data show hourly average records for all features, except the temperature, which also presented minimum and maximum values. Since we aim to work with a daily dataset, climatological observations were aggregated per *day*, *month* and *year*, being grouped to depict the minimum, maximum and average values for all features.

The influent *flow* values were presented as accumulated values. To obtain daily values for this feature, it was necessary to perform a mathematical calculation on each timestep. For this, the influent *flow* value of the previous day is subtracted from the value of the current timestep. Since the first timestep of the dataset does not have a previous known value, it was removed.

We now have two daily datasets, the climatological and the influent flow ones. They were joined in a single dataset using, as key, the *year*, *month* and *day*. The target feature is the *flow* one (in m^3). Hence, to select the input features for the candidate models, a F-test analysis was performed with an *alpha* = 0.05. This test showed that the features with the highest correlation with the influent flow were the *year*, the *average_humidity* and the *maximum_rain*. Hence, all other features were discarded from the dataset.

The data used in this study were normalized to fit the interval $[0, 1]$, for CNN-based models, and $[-1, 1]$, for LSTM ones. After all the performed treatment, the final dataset (Table 2) contains 1609 daily timesteps with a shape of $(1609, 4)$.

Table 2. Features present in the final dataset.

#	Features	Observation Example
0	*timestep index*	2019-04-23
1	*flow*	527
2	*year*	2019
3	*average_humidity*	77.917
4	*maximum_rain*	3.81

To create a supervised problem, a sliding window method was conceived to create pairs of input/label, i.e., the X and the y, depending on the number of timesteps that make a sequence. For instance, if the number of timesteps that make a sequence is set to 7 days, than the final shape of the models' input is $(1601, 7, 4)$ while the labels' shape is $(1601, 1)$. As explained in subsequent sections, the number of timesteps that make a sequence was tuned in order to experiment input sequences of 7, 14 and 21 days.

3.4   Evaluation Metrics

To assess how effective the candidate models were, two error metrics were used. The first corresponds to the Root Mean Squared Error (RMSE), a metric that translates how well the model is executed, realizing the difference between the predicted and the real values. This metric also highlights outliers present in the data. The formula for this metric is as follows:

$$\text{RMSE} = \sqrt{\frac{\sum_{i=1}^{n}(y_i - \hat{y}_i)^2}{n}} \qquad (1)$$

Regarding the second metric, the MAE was the one used. This aims to measure the average magnitude of errors in a set of forecasts, regardless of their direction. In other words, the MAE is the average over the verification sample of the absolute values of a set of differences between the forecast and the observation.

$$\text{MAE} = \frac{1}{n}\sum_{i=1}^{n}|y_i - \hat{y}_i| \qquad (2)$$

3.5   LSTMs

LSTMs are a type of architecture belonging to RNNs. In a RNN, the evolution of the state depends on the current entry, as well as the entry of previous timesteps. LSTMs are used to learn from past experiences, solving problems that emerge when using typical RNNs [9,10].

LSTMs take a similar approach to data in the computer's memory. Information can be stored, written or read from a cell of this type of network. In addition, this type of architecture can prevent the propagation of the error over time and in different layers. This type of technique allows the network to continue its learning process through several "steps" in time [11]. In this study, the LSTM architectures used by the candidate models were structured in blocks of an LSTM plus a Dropout layer to control overfitting.

3.6   CNNs

CNNs are a type of network specially developed and tuned for visual recognition as the extraction of features is done by the network itself [12]. This type of deep neural network is usually divided into two parts, the feature extractor, which can be composed of convolution and reduction layers, and the classifier, composed of fully connected layers, as in a typical Artificial Neural Network. CNNs consist of a set of layers that extract features from the input images through successive convolutions and poolings.

Lately, it is possible to unscramble the use of CNNs for time series problems [13–15]. In this study, a one dimensional CNN is used to forecast the influent flow of a particular WWTP. A channels' last approach was followed, both in the Conv1D and Pooling layers. This type of approach aims to reduce the number of timesteps throughout the network, keeping the number of filters intact. The shape of a Conv1D layer follows the format $(timesteps, filters)$. In CNNs for time series forecasting, the kernel size is used to define the length of the window of timesteps that will be affected by each filter. The output shape of each Conv1D layer is calculated using the following equation:

$$(Timesteps - KernelSize) + 1 \tag{3}$$

Through the analysis of the aforementioned calculation, it is verified that to obtain a more complex model (a deeper one) in terms of Conv1D layers, it is necessary to have a lower kernel size value or use an higher amount of timesteps to build a sequence.

Regarding the pooling layer, AveragePooling1D is used in this study. In this layer, a channels' last approach was also followed. Again, this type of approach aims to reduce the number of timesteps, keeping the number of filters intact. Figure 2 describes an example of the channels' last approach with a Conv1D layer and an AveragePooling1D layer, with an example input of 7 timesteps and 4 features.

4   Experiments

To achieve the goal of forecasting the influent flow at WWTPs, it was necessary to develop several candidate LSTM and CNN models. In these experiments, no statistical models were used, since in some studies the best performance of LSTMs related to this method has already been shown [9]. Concerning CNNs,

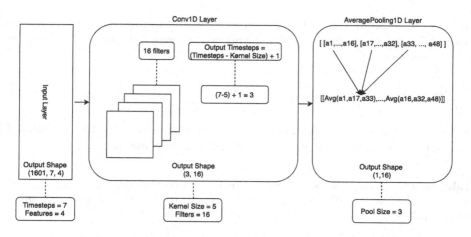

Fig. 2. Example of a channels' last approach in a CNN model.

one dimensional layers were used, i.e., Conv1D and AvergePooling1D. Several experiments were carried out for each of the candidate models to find the best combination of hyperparameters for each model type. The candidate models aim to forecast the influent flow for the next three days, using a multi-variate multi-step recursive approach.

Table 3. LSTM vs CNN hyperparameters' searching space.

Parameter	CNN	LSTM	Rationale
Epochs	[1, 20]	[300, 500]	-
Timesteps	[7, 14, 21]	[7, 14, 21]	Input days
Batch size	[16, 32]	[16, 23]	Batch of 2 to 3 weeks
LSTM layers	-	[3, 4]	Number of LSTM layers
CNN layer	[1,2,3,4,5,6,7,8,9]	-	Number of CNN layers
Activation	[ReLU, linear, sigmoid]	[ReLU, tanh]	Activation function
Pool Size	[2,3]	-	Iin the Pooling Layers
Neurons	[32, 64, 128]	[32, 64, 128]	For each layer
Filters	[8,16,32,64]	-	Number of filters
Dropout	[0.0, 0.5]	[0.0, 0.5]	For Dense Layers
Learning rate	callback	callback	Keras callback
Multisteps	3	3	3 days forecasts
CV Splits	3	3	Time series cross-validator

The performance of the candidate models was evaluated based on error metrics. The search for the best combination of hyperparameters is described in Table 3, which also depicts the searched values for each hyperparameter.

In each experiment, a cross-validator suitable for a time series problem was used, in particular, the TimeSeriesSplit. For each split, the RMSE and MAE metrics are calculated, both in terms of training and testing. The mean value of both error metrics is then calculated and used to evaluate the performance of the candidate models.

5   Results and Discussion

The obtained results are presented in Table 4, which depicts the top-3 results of the best LSTMs and CNNs candidate models. Error values are normalized.

Table 4. LSTM vs CNN top-3 models. Legend: a. timesteps; b. batch size; c. n° of layers; d. n° of neurons/filters; e. pool size; f. kernel size; g. dropout; h. activation; i. RMSE; j. MAE; k. time (in seconds).

#	a.	b.	c.	d.	e.	f.	g.	h.	i.	j.	k.
				LSTM candidate models							
50	7	23	3	32	-	-	0.5	ReLU	**0.168**	**0.132**	**188**
14	7	16	4	32	-	-	0.5	ReLU	0.169	0.134	562
110	7	16	4	32	-	-	0.5	ReLU	0.17	0.133	824
				CNN candidate models							
315	7	16	3	32	2	2	0.5	linear	**0.078**	**0.062**	**11.5**
369	7	16	4	32	3	2	0.5	linear	0.079	0.063	15
356	7	16	4	16	3	2	0	linear	0.08	0.063	14.5

Regarding the best candidate LSTM model, it obtained a normalized RMSE of 0.168 and a MAE of 0.132. On the other hand, the best CNN candidate model presented a normalized RMSE of 0.078 and a MAE of 0.062. It should be noted that LSTMs input data was normalized to fit the interval $[-1, 1]$, while CNNs were normalized into the interval $[0, 1]$. Hence, the comparison of both values is only possible using non-normalized data. In fact, non-normalized errors depict the better performance of LSTMs as an error of 0.132 corresponds to $200\,m^3$. On the other hand, an error of 0.062 for the best CNN model corresponds to $232.7\,m^3$.

LSTMs showed better performance to forecast three future days when using data about the last seven days (the last week). In fact, all the best candidate models, regardless of being LSTMs or CNNs, used information about the last week to predict the next three days. Moreover, regarding the best LSTM models, it is worth noting that the values of the hyperparameters are quite homogeneous in the three best candidates. On the other hand, CNNs present more heterogeneous hyperparameter values. The best overall model, an LSTM one, uses a batch size of three weeks, 3 hidden layers of 32 neurons each, a dropout rate of 50% and ReLU as activation function. It is also interesting to note that CNNs take significantly less time to train when compared to LSTMs. Figure 3 presents two random forecasts of three days for the best LSTM and CNN candidate models.

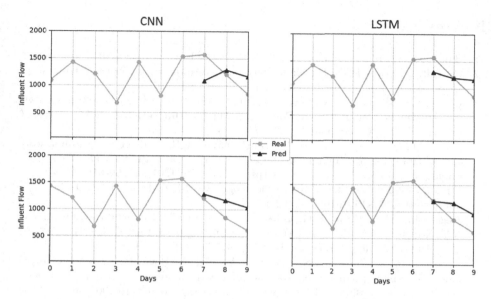

Fig. 3. Two random forecasts of three days performed by the best LSTM and CNN candidate models.

6 Conclusions

The influent flow in a WWTP is a crucial variable in its operation. Hence, this study focused on the use of Deep Learning models, in this case LSTMs and CNNs, to forecast the influent flow for the next three days. Multiple experiments were carried out on both models, through the combination of several hyperparameters. The best model for influent flow prediction was based on LSTMs, presenting a RMSE of 0.168 and a MAE of 0.132. These results demonstrate that it is possible to predict, with satisfactory results, the influent flow in a WWTP by taking into account historical data about the influent flow as well as climatological data. It was interesting to note that CNN-based models can also be applied to time series problems, presenting promising performance. It should also be noted that the number of epochs required, before reaching overfitting, is much lower in CNNs, which is reflected in the time these models take to train when compared to LSTM ones.

Concerning future work, we aim to investigate if a greater number of samples improves the results obtained by CNN-based models. We also aim to investigate how does the performance of these models change if a uni-variate approach is followed, i.e., how do the models behave if they only receive the influent flow as input feature.

Acknowledgments. This work is financed by National Funds through the Portuguese funding agency, FCT - Foundation for Science and Technology within project DSAIPA/AI/0099/2019. The work of Bruno Fernandes is also supported by a Portuguese doctoral grant, SFRH/BD/130125/2017, issued by FCT.

References

1. Baus, D.: Overpopulation and the Impact on the Environment (2017)
2. Metcalf, L., Eddy, H.P., Tchobanoglous, G.: Wastewater Engineering: Treatment, Disposal, and Reuse, vol. 4. McGraw-Hill, New York (1979)
3. Di Fraia, S., Massarotti, N., Vanoli, L.: A novel energy assessment of urban wastewater treatment plants. Energy Conversion Manag. **163**, 304–313 (2018). https://doi.org/10.1016/j.enconman.2018.02.058
4. Zhang, D., Martinez, N., Lindholm, G., Ratnaweera, H.: Manage sewer in-line storage control using hydraulic model and recurrent neural network. Water Resources Manag. **32**(6), 2079–2098 (2018). https://doi.org/10.1007/s11269-018-1919-3
5. Elman, J.L.: Finding structure in time. Cogn. Sci. **14**(2), 179–211 (1990). https://doi.org/10.1207/s15516709cog1402_1
6. Siegelmann, H.T., Horne, B.G., Giles, C.L.: Computational capabilities of recurrent NARX neural networks. IEEE Trans. Syst. Man Cybern. Part B (Cybern.) **27**(2), 208–215 (1997). https://doi.org/10.1109/3477.558801
7. Zhou, P., Li, Z., Snowling, S., Baetz, B.W., Na, D., Boyd, G.: A random forest model for inflow prediction at wastewater treatment plants. Stochastic Environ. Res. Risk Assess. **33**(10), 1781–1792 (2019). https://doi.org/10.1007/s00477-019-01732-9
8. Szelag, B., Bartkiewicz, L., Studziński, J., Barbusiński, K.: Evaluation of the impact of explanatory variables on the accuracy of prediction of daily inflow to the sewage treatment plant by selected models nonlinear. Arch. Environ. Protect. **43**(3), 74–81 (2017). https://doi.org/10.1515/aep-2017-0030
9. Fernandes, B., Silva, F., Alaiz-Moretón, H., Novais, P., Analide, C., Neves, J.: Traffic flow forecasting on data-scarce environments using ARIMA and LSTM networks. In: Rocha, Á., Adeli, H., Reis, L.P., Costanzo, S. (eds.) WorldCIST'19 2019. AISC, vol. 930, pp. 273–282. Springer, Cham (2019). https://doi.org/10.1007/978-3-030-16181-1_26
10. Sagheer, A., Kotb, M.: Time series forecasting of petroleum production using deep LSTM recurrent networks. Neurocomputing **323**, 203–213 (2019). https://doi.org/10.1016/j.neucom.2018.09.082
11. DiPietro, R., Hager, G.D.: Deep learning: RNNs and LSTM. In: Handbook of Medical Image Computing and Computer Assisted Intervention, pp. 503–519. Academic Press (2020). https://doi.org/10.1016/B978-0-12-816176-0.00026-0
12. LeCun, Y., et al.: Handwritten digit recognition with a back-propagation network. In: Advances in Neural Information Processing Systems, pp. 396–404 (1990)
13. Zhao, B., Lu, H., Chen, S., Liu, J., Wu, D.: Convolutional neural networks for time series classification. J. Syst. Eng. Electron. **28**(1), 162–169 (2017). https://doi.org/10.21629/JSEE.2017.01.18
14. Borovykh, A., Bohte, S., Oosterlee, C.W.: Dilated convolutional neural networks for time series forecasting. J. Comput. Finance (2018, Forthcoming)
15. Koprinska, I., Wu, D., Wang, Z.: Convolutional neural networks for energy time series forecasting. In: 2018 International Joint Conference on Neural Networks (IJCNN), pp. 1–8. IEEE (2018). https://doi.org/10.1109/IJCNN.2018.8489399
16. Carneiro, D., Novais, P., Pêgo, J.M., Sousa, N., Neves, J.: Using mouse dynamics to assess stress during online exams. In: Onieva, E., Santos, I., Osaba, E., Quintián, H., Corchado, E. (eds.) HAIS 2015. LNCS (LNAI), vol. 9121, pp. 345–356. Springer, Cham (2015). https://doi.org/10.1007/978-3-319-19644-2_29

17. Costa, A., Rincon, J.A., Carrascosa, C., Julian, V., Novais, P.: Emotions detection on an ambient intelligent system using wearable devices. Future Gen. Comput. Syst. **92**, 479–489 (2019)
18. Lima, L., Novais, P., Costa, R., Cruz, J.B., Neves, J.: Group decision making and Quality-of-Information in e-Health systems. Logic J. IGPL **19**(2), 315–332 (2011)

Automatic Multispectral Image Classification of Plant Virus from Leaf Samples

Halil Mertkan Sahin$^{(\boxtimes)}$ (ID), Bruce Grieve (ID), and Hujun Yin (ID)

Department of Electrical and Electronic Engineering, The University
of Manchester, Manchester M13 9PL, UK
halil.sahin@postgrad.machester.ac.uk,
{bruce.grieve,hujun.yin}@manchester.ac.uk

Abstract. Multispectral imaging has become a useful means of sensing in increasing number of fields, including precision agriculture due to its many advantages such as being non-invasive and extracting additional information than the visible band. Combined with machine learning, such systems can further extract complex relationships among extracted spectral information to facilitate intelligent agriculture. This paper focuses on making the process of multispectral data analysis automatic for classifying virus infection in plants, specially cassava plants infected with brown streak virus and tomato plants infected with mottle virus. Multispectral images are sampled from cassava and tomato plants with a designed portable device and then passed on to the proposed automatic analysis process. Conventional methods often depend on time-consuming manual processes of cropping valid patches from sampled leaf images and then averaging the patches to obtain the spectral reflectance signatures of the leaves, prior to the classification stage. The developed automatic process can automatically extract valid leaf pixels from scanned leaf images and subsequent spectral information and integrate with the classification method for optimal detection of infected plants. It has not only reduced processing time and errors significantly but also make the entire process more optimal. Extensive experiments on cassava brown streak virus and tomato mottle virus have been conducted and results demonstrated the intended advantages of the developed process. Support vector machines with RBF kernel have been shown to perform well for the classification of uninfected, and infected classes. Experiments show that the application can offer less time-consuming automatic analysis of the captured data.

Keywords: Multispectral data analysis · Classification · Precision agriculture · Virus detection

1 Introduction

Hyperspectral imaging (HSI) and multispectral imaging (MSI) are becoming increasingly popular because of their advantages over visual imaging along recent developments in the technology. Unlike colour or greyscale images, HSI/MSI images extend the visible regions (400–700 nm wavelengths) of the electromagnetic spectrum (EM) and comprise

© Springer Nature Switzerland AG 2020
C. Analide et al. (Eds.): IDEAL 2020, LNCS 12489, pp. 374–384, 2020.
https://doi.org/10.1007/978-3-030-62362-3_33

of additional sub-bands of wavelength outside the visible bands, which may bring out finer details with respect to the subject under investigation. For instance, the infrared area of the spectrum may contain chemical element or structure information about the target [1], by virtue of the harmonics from the mid-infrared molecular vibrational spectroscopy. Similarly, such data in the near-infrared (NIR) area of the EM spectrum can also reveal information about biological and organic samples [2].

The main differences between HSI and MSI systems are the number of spectral bands and their bandwidths. Whilst HSI systems have consecutive narrow spectral bands across the spectrum of interest, MSI systems have only a few or tens spectral bands in certain wavelengths. MSI systems have gained a wide range of applications due to their efficiency. One major application field is agriculture and there are many different purposes in agriculture and food industry. For example, HSI/MSI systems are used for plant health monitoring [3], quality evolution of plant products [4], mapping crop yield for precision agriculture [5], and quality evaluation of beef and pork [6]. Medical imaging [7], compositional information of mineral mixtures [8], security, surveillance and target acquisition [9] are also among a variety of examples of applications of HSI/MSI systems in different areas.

HSI/MSI systems are used to obtain valuable spectral information about the target objects in finer wavelengths than available conventional visual inspection. As a consequence interpreting the resulting data is not intuitive. Therefore, a classification algorithm is needed for quantitatively distinguishing data samples from different classes of stressed crops such as those infected by a pathogen versus uninfected or control samples. Often the support vector machine (SVM) is the preferred classifier for MSI image analysis due to its efficiency [10]. As an example of MSI systems in agriculture; ripeness of fruits is of a principal indicator for the optimal harvest time and heavily relies on human experiences and judgment. However, automatic determination of ripeness is nontrivial. In [11], 180 healthy strawberries, with three different classes (60 ripe, 60 mid-ripe and 60 unripe) were harvested by locally experienced growers. These samples were then captured by a laboratory based HSI system, by two different camera systems with 380–1030 nm and 874–1734 nm spectral ranges. A conveyor belt was used for line scan configuration. Besides the spectral information, spatial properties were extracted using grey level co-occurrence matrix (GLCM). Then SVM with radial basis function kernel (RBF) was employed to build classification models, achieving 85% classification accuracy.

The main reason behind developing the current work is to reduce the time spent producing the datasets for classification purpose. Spectral reflectance of each wavelength is often obtained by extracting patches from an MSI image of a leaf or using the image of the whole leaf. However, working with a whole leaf image will require leaf segmentation and removal of noise due to veins, defects and stomata and is also computational more demanding. On the other hand, cropping out patches is a manual task and highly time consuming and requiring experienced user assistance. It may also miss useful information as patches may not cover the entire leaf. The developed application can also automatically capture patches from a leaf by simple user interactions such as defining the number of patches wanted.

The rest of the paper is organised as follows. The background section (Sect. 2) provides a short summary of MSI systems and support vector machines (SVMs). Also, essential equipment of MSI systems are briefly presented in this section. Information about the used MSI device, the dataset, and the preparing process of the dataset are respectively revealed in materials and methods section (Sect. 3). Experiment and results are given in Sect. 4, followed by conclusion in the final section.

2 Background

2.1 Multispectral Imaging Systems

Digital image analysis with different imaging modalities started around in the 60 s. Greyscale values were used in early studies of image analysis. Greyscale images are single band images produced using the reflection of the light intensity over the electromagnetic spectrum. After 1980 s, colour images gained importance for image analysis. Unlike greyscale images, colour images are produced using red, green, and blue (RGB) bands of the EM spectrum of intensities. While greyscale or colour images provide spatial information, spectroscopy gives spectral information about the image. Further, MSI systems provide both spatial and spectral information. These combinations can be represented as a three-dimensional (3D) data cube. The first two elements of the data cube gives spatial information in the targeted image, while the third dimension of the data cube is the obtained spectral information [1, 12]. Figure 1 illustrates an example of a multispectral data cube or hypercube. A hypercube is made by voxels, with each voxel containing pixel intensity values [13]. Scanning configuration is an important feature to understand how the hypercube is produced. There are four methods: point scan, line scan, area scan, and single shot method. The first two are spatial based scan configurations, while area scan method is spectral based scan configuration. Unlike first three approaches, the single shot method provides spectral and spatial based approach. It is

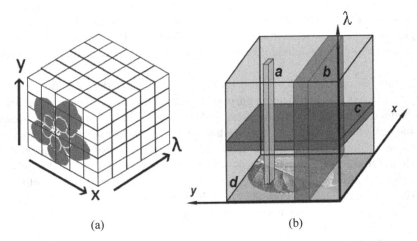

(a)   (b)

Fig. 1. A demonstration of hypercube (a) and scanning configuration (b). (Color figure online)

important to note that spatial based scanning may have registration problems because of the movements of the targets [12–15]. These configurations can be seen in Fig. 1.

In addition to different scanning configurations, there are three other crucial components: illumination light, a device for wavelength selection, and a detector [13, 16].

The light source of any MSI system is crucial as it affects what purpose MSI systems will be used for. The spectral composition and optical power of the illumination used will dictate the wavelength range and sensitivity of the overall system. There are many different light sources used for illumination of target objects. Halogen lamps are one of the most common light sources for HSI/MSI systems. Visible (VIS) and near-infrared (NIR) spectral regions can be covered using this kind of lamps, which is also low-cost. However, halogen lamps generate high heat that may cause a problem (burn or alter) to the target objects. Multiple light emitting diodes (LEDs) can also cover the same spectral region as halogen lamps. LEDs present many advantages with respect to halogen lamps. Consumption of low energy, producing low heat, long lifespan, their small size, and fast response time are some of the advantages of LEDs. However, the materials used for LEDs may affect the bandwidth of the light and LEDs have narrow wavebands. LEDs and halogen lamps are used for reflectance of light and transmittance of light in imaging modalities. Beside LED and halogen lamps, there are many more light sources used for HSI/MSI systems, such as, lasers, tunable sources [14, 16–18]. Also, additional illumination sources can improve quality of capturing targets with decreasing signal noise level. In [19], the effect of an additional illumination source was studied on sugar beet leaves. The additional 400–500 nm (blue light) LED array caused an improvement in the sensitivity of hyperspectral plant images. These wavelengths are important to catch earlier symptoms of plant diseases.

The wavelength dispersion devices are one of the essential (and the most important) equipment of any HSI/MSI systems. These devices divide the broadband spectrum into small wavelengths portions. There are many different types of optical and electro–optical wavelength dispersion devices. For instance, bandpass filter wheels are the most basic type of devices. These filter devices are practical but also it has got some drawbacks, such as mechanical vibration, wrong registration, and limited spectral range. Imaging spectrographs and tunable filters are a variety of examples in terms of the wavelength dispersion devices [14, 17, 18].

The last core part of any HSI/MSI systems is a detector (camera), which reads the valuable information carried by the light. The charge-coupled device (CCD) and complementary metal-oxide-semiconductor (CMOS) are two main types of detectors. Silicon (Si) or indium gallium arsenide (InGaAs) is a material used for producing CCD image sensors. InGaAs CCD image sensors are relatively expensive, but these sensors cover longer wavelengths of EM, e.g. between 900 and 1700 nm for a standard InGaAs image sensor. On the other hand, CCD image sensors are used for VIS and short-wavelength NIR regions (400–1000 nm) of EM. CMOS image sensors have some advantages in contrast with CCD image sensors. For instance, CMOS sensors are suitable for high speed required applications, also they are cheap, small sensor size, and low power consumption. However, CMOS image sensors provide lower sensitivity and lower dynamic range than CCD types of image sensors [14, 17, 18].

The advantage of MSI is that it can obtain a large amount of valuable information about its target. This huge amount of information can be obtained quickly, although it depends on the configuration of the MSI system. The target object needs minimal time to be ready to be captured. MSI systems offer a wide range of applications due to its non-invasive, non-destructive and chemical-free structure. On the other hand, HSI/MSI systems are expensive, a real disadvantage for customers. The captured images need enormous space to store and the process of these images may be computationally intensive. HSI/MSI systems also need to be calibrated every time in order to get accurate data from the target; therefore, trained specialists are often required [14, 15, 20].

2.2 Support Vector Machine (SVM)

SVM is one of the most popular supervised classifiers in machine learning. SVMs provide highly successful performances in a range of applications with few training samples. Classification process generally starts with dividing the dataset into training and testing sets. A small portion of the training set is usually used as the validation set for fine-tuning parameters of the classifier. The aim of the SVM is to find separating hyperplanes with large margins, from the training set. The produced model, as a result of SVM, is used to predict the target value with the test set [22, 23]. The basic kernels used in SVMs are shown below.

$$\text{Linear: } K\left(x_i, x_j\right) = x_i \cdot x_j \tag{1}$$

$$\text{Polynomial: } K\left(x_i, x_j\right) = (\gamma x_i \cdot x_j + r)^d, \gamma > 0 \tag{2}$$

$$\text{Radial basis function: (RBF) : } K\left(x_i, x_j\right) = exp\left(-\gamma \left\|x_i - x_j\right\|^2\right) \tag{3}$$

$$\text{Sigmoid: } K\left(x_i, x_j\right) = tanh(\gamma x_i \cdot x_j + r) \tag{4}$$

In these equations; (x_i, y_i) indicates features and classes, $i = 1, \ldots, l$ where $x_i \in R^n$ and $y \in \{1, -1\}^l$. Unlike linear kernel, other three kernels have kernel parameters such as weight, coefficient and degree of the polynomial kernel, represented by γ, r, and d respectively [22].

In addition to these parameters, C parameter is the cost parameter that is related with soft margins. RBF kernel is one of the most preferred kernels since it provides non-linear hyperplane with C and γ parameters. These two parameters are highly important for the result of non-linear SVM. Therefore, Grid Search method can be used to estimate C and γ parameters with cross-validation [24].

3 Materials and Methods

In this study, an in-house designed multispectral imaging system has been used to capture target leaves. This imaging device utilises 15 different wavebands, ranging from 395–880 nm, covering the visible region and the near-infrared region. The acquisition process starts with calibration to ensure uniform illumination across the different wavebands and the calibration files are saved for future use. Then the device is ready to capture multispectral images of leaves and store the captured image files.

Cassava and tomato plants have been used in the experiments and generated datasets contain samples from Infected, Mock, and Uninfected/Control classes from a number of plants. Infected is defined as controlled inoculation of plant samples with viral titre bound upon gold nanoparticles [25], Mock as per infected but with no viral titre, and Control as an untreated healthy plant specimen. Cassava plants have been inoculated with the Ugandan cassava brown streak virus (UCBSV) (Source: NC-State University, NC, USA, L. Hanley Bowdoin *et al.*), while tomato plants with the mottle virus (ToMoV). The device has captured leaf images on different days after the inoculation. The main aim behind capturing images in different days is to detect early symptoms of viruses or responses from the plants.

There are four steps to prepare a dataset for the machine learning algorithm. 1) Captured multispectral image files are extracted and read to the application. 2) Calibration is checked and further calibration is performed in required, based on the calibration image file. 3) Image processing (leaf segmentation and vein removal) is performed to extract the average reflectance values of corresponding wavelengths. Processed images can be saved as image files. Patch-based extraction can be performed. For this type of extraction, the user has to determine the number and size of patches. 4) SVM-classifier is trained, and results displayed on infected vs. uninfected, mock vs infected and mock vs uninfected. The flow chart shows the steps of the designed application in Fig. 2.

The developed application interface is shown in Fig. 3. The red lamp indicates the status of the step and it turns into green when the process is done.

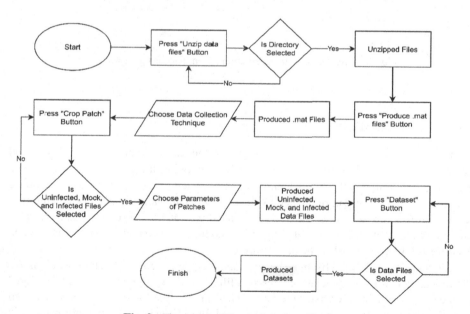

Fig. 2. Flowchart of the designed application.

Software environment: Processor: Intel(R) Core(TM) i5-8365U CPU @ 1.60 GHz, 1901 MHz, 4 Core(s), 8 Logical Processor(s), Installed Physical Memory (RAM): 8 GB, 240 GB SSD hard disk and Microsoft Windows10 system. Experiment platform: Matlab.

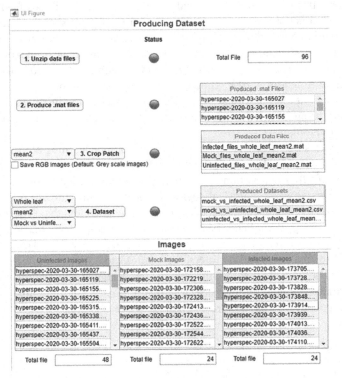

Fig. 3. Screenshot of designed application. (Color figure online)

3.1 Image Processing

First, calibration aims to ensure the images across the various bands are comparable and have uniform isotropic illumination across the area being imaged. It can be achieved manually or automatically. In the hand-held device used in this study (source: in-house manufactured, currently unpublished by the authors), calibration is performed automatically, before scans are taken, and as a result, a calibration file is produced and saved for any future use. Both multispectral image files and their related calibration file have been used for producing a mask in this application.

Then leaf is segmented from the background with veins removed as they would make the reflectance noisy. Laplacian of Gaussian filter has been applied, followed by global image threshold using the Otsu's method [21]. Then, morphological operations are used to polish the result of the segmentation process. Table 1 shows the segmentation and vein removal algorithm. Segmented images have been used for extracting spectral reflectance

Table 1. Leaf segmentation and vein removal algorithm

Segmentation and Vein Removal Algorithm
Requirements: IMG_hyp: Multispectral files, 15 channels in total **Start:** IMG_filtered: Laplacian of Gaussian filter (IMG_hyp) IMG_threshold: Global image threshold using Otsu's method (IMG_filtered) IMG_removed_vein: Morphological operations (IMG_threshold) **end**

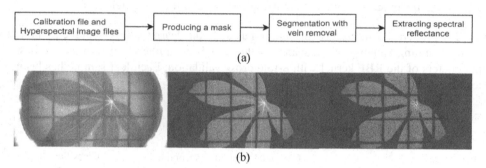

(a)

(b)

Fig. 4. Leaf image segmentation and vein removal. (a) Steps of image processing procedure. (b) Results: from left to right, input image, segmented mask image using calibration file, and vein removed image.

which can be mean value of the entire image. Image processing steps can be seen in Fig. 4(a) and typical results are illustrated in Fig. 4(b).

4 Experiments and Results

Classification performance and computational cost have been evaluated and compared on two different data processing techniques on two different plant datasets. Table 2 shows the datasets used in the experiments. The datasets have been produced using an equally selected number of uninfected and infected leaves. 6 infected and 6 uninfected samples for tomato leaves and 24 infected and 24 uninfected samples for cassava leaves have been used. However, the optical power associated with Band 6 (700 nm) was insufficient to achieve low-noise imaging in the Cassava plants' scans; therefore, data of Band 6 was excluded in the classification.

Patch-Based Analysis: Cassava leaf images were sampled by cropping patches. Tomato leaves are significantly smaller than cassava leaves; therefore, they could not be selected for cropping purpose. The cropping patches have increased the total computational cost of the application around 190 s more because the algorithm has cropped 6 patches (30 by 30 pixels) from each leaf sample. Each 6 patches have been used to calculate the mean value of them. Randomly selected patches and their classification results are highly related with their location. The classification accuracies of five independent runs between uninfected vs infected were 77.08%, 89.58%, 62.5%, 83.33%, and 83%, respectively. As can be seen, there are large variabilities among these runs.

Segmented Leaf-Based Analysis: With the developed automatic segmentation of the leaf, classification can be done based on the whole (segmented) leaves. Results are shown in Table 2. As can be seen from the results, using the spectral reflectance of the whole leaf provides better accuracies than using the patches, though good segmentation and noise removal are key. More importantly, the results seem more stable. The training times are also given as computational cost for each step of the entire process. Inference time for a new leaf with the trained model is 0.66 s, which indicates the possibility of integrating the processing with the imaging device for a real-time field test.

The produced datasets have been tested using Scikit-learn toolbox with the Python-3 programming language. Grid search method has been applied to determine the best parameters of the RBF kernel with 8-fold cross-validation. Each leaf sample has been tested against the rest of the samples.

Table 2. Classification results of uninfected vs infected with entire segmented leaves.

Parameters		Computational cost (seconds)					Classification accuracy
Data collection techniques	Plant						
		Z	M	S	D	Total	
Mean2	Cassava (96 files)	77	199	187	3	466	81.3%
Sum/NNZ	Cassava (96 files)	77	199	190	4	470	87.5%
Mean2	Tomato (72 files)	52	109	128	4	293	95.8%
Sum/NNZ	Tomato (72 files)	52	109	127	3	291	95.8%

Computational cost involves four different steps which are extracting zip files (Z), producing.mat files (M), obtaining spectral reflectance information (S), and producing datasets (D).

5   Conclusions

This paper presents an automatic application for multispectral image classification of plant virus from leaf samples. An efficient image processing procedure has been derived to segment leaf areas and remove veins and SVM classifier has been used to classify infected and control samples. The experimental results on two different plants have shown promising results, indicating that the prototype multispectral machine may provide high classification accuracy for early stage (pre-visual symptomatic) detection of viral titre response within plant foliage. This automatic process can be completed in a few minutes, compared to tens of minutes by conventional methods. Further trials will be required for reliability evaluation. Also, in these experiments, only spectral information has been used. Future study may also explore extracting spatial information from leaf samples as well as integrating the analysis to the imaging device.

Acknowledgments. Authors would like to thank Prof. Linda Hanley-Bowdoin and her team at the North Carolina State University for the provision of plant samples and related biological analysis. Halil Mertkan Sahin would also like to acknowledge the Scholarship provided by the Ministry of National Education of the Republic of Turkey.

References

1. Prats-Montalbán, J.M., de Juan, A., Ferrer, A.: Multivariate image analysis: a review with applications. Chemom. Intell. Lab. Syst. **107**(1), 1–23 (2011). https://doi.org/10.1016/j.chemolab.2011.03.002
2. Geladi, P., Burger, J., Lestander, T.: Hyperspectral imaging: calibration problems and solutions. Chemom. Intell. Lab. Syst. **72**(2), 209–217 (2004). https://doi.org/10.1016/j.chemolab.2004.01.023
3. Lee, W.S.: Plant health detection and monitoring. In: Park, B., Lu, R. (eds.) Hyperspectral Imaging Technology in Food and Agriculture. FES, pp. 275–288. Springer, New York (2015). https://doi.org/10.1007/978-1-4939-2836-1_11
4. Tallada, J.G., Bato, P.M., Shrestha, B.P., Kobayashi, T., Nagata, M.: Quality evaluation of plant products. In: Park, B., Lu, R. (eds.) Hyperspectral Imaging Technology in Food and Agriculture. FES, pp. 227–249. Springer, New York (2015). https://doi.org/10.1007/978-1-4939-2836-1_9
5. Yang, C.: Hyperspectral imagery for mapping crop yield for precision agriculture. In: Park, B., Lu, R. (eds.) Hyperspectral Imaging Technology in Food and Agriculture. FES, pp. 289–304. Springer, New York (2015). https://doi.org/10.1007/978-1-4939-2836-1_12
6. Konda Naganathan, G., Cluff, K., Samal, A., Calkins, C., Subbiah, J.: Quality evaluation of beef and pork. In: Park, B., Lu, R. (eds.) Hyperspectral Imaging Technology in Food and Agriculture. FES, pp. 251–273. Springer, New York (2015). https://doi.org/10.1007/978-1-4939-2836-1_10
7. Lu, G., Fei, B.: Medical hyperspectral imaging: a review. J. Biomed. Opt. **19**(1), 010901 (2014). https://doi.org/10.1117/1.jbo.19.1.010901
8. Zaini, N., van der Meer, F., van der Werff, H.: Determination of carbonate rock chemistry using laboratory-based hyperspectral imagery. Remote Sens. **6**(5), 4149–4172 (2014). https://doi.org/10.3390/rs6054149
9. Yuen, P.W.T., Richardson, M.: An introduction to hyperspectral imaging and its application for security, surveillance and target acquisition. Imaging Sci. J. **58**(5), 241–253 (2010). https://doi.org/10.1179/174313110X12771950995716
10. Burger, J.E., Gowen, A.A.: Classification and prediction methods. In: Park, B., Lu, R. (eds.) Hyperspectral Imaging Technology in Food and Agriculture. FES, pp. 103–124. Springer, New York (2015). https://doi.org/10.1007/978-1-4939-2836-1_5

11. Zhang, C., Guo, C., Liu, F., Kong, W., He, Y., Lou, B.: Hyperspectral imaging analysis for ripeness evaluation of strawberry with support vector machine. J. Food Eng. **179**, 11–18 (2016). https://doi.org/10.1016/j.jfoodeng.2016.01.002
12. ElMasry, G., Sun, D.W.: Principles of hyperspectral imaging technology. In: Hyperspectral Imaging for Food Quality Analysis and Control, pp. 3–43 (2010)
13. Geladi, P.L.M., Grahn, H.F., Burger, J.E.: Hyperspectral imaging: background and equipment. In: Techniques and Applications of Hyperspectral Image Analysis, pp. 1–15 (2007)
14. Wu, D., Sun, D.W.: Advanced applications of hyperspectral imaging technology for food quality and safety analysis and assessment: a review - part I: fundamentals. Innov. Food Sci. Emerg. Technol. **19**, 1–14 (2013). https://doi.org/10.1016/j.ifset.2013.04.014
15. Dale, L.M., et al.: Hyperspectral imaging applications in agriculture and agro-food product quality and safety control: a review. Appl. Spectrosc. Rev. **48**(2), 142–159 (2013). https://doi.org/10.1080/05704928.2012.705800
16. Gowen, A.A., O'Donnell, C.P., Cullen, P.J., Downey, G., Frias, J.M.: Hyperspectral imaging - an emerging process analytical tool for food quality and safety control. Trends Food Sci. Technol. **18**(12), 590–598 (2007). https://doi.org/10.1016/j.tifs.2007.06.001
17. Qin, J.: Hyperspectral imaging instruments. In: Hyperspectral Imaging for Food Quality Analysis and Control, 1st edn, pp. 129–172. Elsevier Inc. (2010)
18. Gowen, A.A., Feng, Y., Gaston, E., Valdramidis, V.: Recent applications of hyperspectral imaging in microbiology. Talanta **137**, 43–54 (2015). https://doi.org/10.1016/j.talanta.2015.01.012
19. Mahlein, A.K., Hammersley, S., Oerke, E.C., Dehne, H.W., Goldbach, H., Grieve, B.: Supplemental blue LED lighting array to improve the signal quality in hyperspectral imaging of plants. Sensors **15**(6), 12834–12840 (2015). https://doi.org/10.3390/s150612834
20. Bock, C.H., Poole, G.H., Parker, P.E., Gottwald, T.R.: Plant disease severity estimated visually, by digital photography and image analysis, and by hyperspectral imaging. CRC: Crit. Rev. Plant Sci. **29**(2), 59–107 (2010). https://doi.org/10.1080/07352681003617285
21. Otsu, N.: A threshold selection method from gray-level histograms. IEEE Trans. Syst. Man Cybern. **SMC-9**(1), 62–66 (1979). https://doi.org/10.1109/TSMC.1979.4310076
22. Maulik, U., Chakraborty, D.: Remote sensing image classification: a survey of support-vector-machine-based advanced techniques. IEEE Geosci. Remote Sens. Mag. **5**(2), 33–52 (2017). ISSN 0274-6638
23. Hsu, C.-W., Chang, C.-C., Lin, C.-J.: A practical guide to support vector classification. BJU Int. **101**(1), 1396–1400 (2008). https://doi.org/10.1177/02632760022050997
24. Budiman, F.: SVM-RBF parameters testing optimization using cross validation and grid search to improve multiclass classification. Sci. Vis. **11**(1), 80–90 (2019). https://doi.org/10.26583/sv.11.1.07
25. Finer, J.J., Vain, P., Jones, M.W., Mcmullen, M.D.: Development of the particle inflow gun for DNA delivery to plant cells. Plant Cell Rep. **11**, 323–328 (1992). https://doi.org/10.1007/BF00233358

Towards the Modeling of the Hot Rolling Industrial Process. Preliminary Results

Maciej Szelążek[1,2], Szymon Bobek[1,2(✉)], Antonio Gonzalez-Pardo[2], and Grzegorz J. Nalepa[1]

[1] Jagiellonian University, AGH University of Science and Technology, Krakow, Poland
{szymon.bobek,grzegorz.j.nalepa}@uj.edu.pl
[2] Computer Science Department, Universidad Rey Juan Carlos, Madrid, Spain

Abstract. In the paper we describe the industrial process of hot rolling of steel. In cooperation with ArcelorMittal Poland we consider a specific fully automated production line. While it is equipped with a number of industrial sensors, the acquired data has only been analyzed on a basic statistical level, mainly for reporting. In the paper we outline opportunities for the use of AI methods in order to improve the process and possibly the quality of the resulting product. We report on preliminary results using selected methods of eXplainable AI.

1 Introduction

Introduction of ICT tools for Smart Industry, or Industry 4.0 is a clearly a gradual process. Even though certain high level guidelines are well defined, their practical implementation in a specific industrial case depends on many factors. Some of them are related to technological constraints, while others depend on the industrial domain, industrial process and machinery considered. Furthermore, the range of applications of industrial AI vary also in the cases where a given factory, or production plant was already build with industrial sensors and computer-driven control and automation tools, or is only converted to a smart factory configuration, from an older design.

This paper is a result of cooperation with the ArcelorMittal Poland (AMP). The company is adopting the latest technologies as lead steel producer on the world, globally controlling sources, logistic, production and sales. In Poland, AMP acquired number of older factories industrial installations related to the steel production. The objective of the paper is to give an introduction to the hot rolling steel production process. Then, we discuss important features of the fully automated steel production line in Krakow. Event though the existing sensor setup is quite robust, so far the acquired data was analyzed only on a very general level, with use of statistical methods. Our objective instead is to explore the opportunities for the use of Industrial AI methods including Explainable AI (XAI). In particular, we aim at using XAI to gain more knowledge about the complex dependencies between the input parameters and the results from

C. Analide et al. (Eds.): IDEAL 2020, LNCS 12489, pp. 385–396, 2020.
https://doi.org/10.1007/978-3-030-62362-3_34

steel production process. This approach is different from the way the machine learning models are utilised usually. Instead of obtaining predictions from the model, we are more interested in getting insight into the process of creating such predictions. This would allow us to better understand the impact of the process settings on the resulting product and possibly improve the process itself.

The rest of the paper is composed as follows. In Sect. 2 we present the industrial process in the steel factory, with the details of the Krakow steel plant in Sect. 3. Then in Sect. 4 we discuss how the industrial process is monitored with the use of sensors and what data is acquired. Based on this, in Sect. 5 we outline different scenarios for the use of AI methods. Preliminary results of selected works are then introduced in Sect. 6. The paper ends with Sect. 7 where we summarize our work and discuss future plans.

2   Overview of the Hot Rolling Industrial Process

Steel production techniques has been evolving for hundreds of years. The development of PLCs (Programmable Logic Controller) and automation systems in the 20th century seemed to have perfected the process, almost excluding human participation. Modern methods of data analysis enable further increase of process precision and cost optimization. The main subject of our interest is one of the main components of the process, i.e. the Hot Rolling Mill. However, to understand how many factors can affect the final product, we briefly describe all stages of steel making. First step of production is performed in blast furnace. In chemical and physical reactions, ore is reduced into liquid hot metal [6]. Depending on technology, coal is processed into coke before addition to blast furnace. Hot metal is an input product in BOF (Basic Oxygen Furnace) converter part. This is next level of steel purification and chemical composition setting due to additions (like chrome or molybdenum). In ~300 tons portion of steel named "heat", oxygen in high pressure increases temperature and mix the additions to obtain an homogeneous material. At the end, low-carbon steel is ready for casting.

Casting is crystallization process when purified liquid steel is transforming into slabs. An average slab is ~22,5 tons solid piece of steel. Its thickness is constant but length and width depend on customers orders. All casted slabs from a single BOF converter process (single "heat") has similar chemical composition. Different steel grades could be made using different forces and cooling settings in the next process – the Hot Rolling Mill.

3   The HRM Process in the Krakow Steel Plant

The Hot Rolling Mill is a completely new installation build in Krakow in 2007. This is the most sophisticated part of the steel production, where final geometrical and chemical properties of the material are determined [5]. The line is fully automated, which made it possible to collect detailed data about the products and the line performance.

Furnace Node. The hot rolling process starts in Walking Beam Furnace. Slabs are heated according to technical data sheet. It is crucial to obtain the plasticity required in rolling processes. Heating quality is directly related to the rolling forces required to achieve end thickness.

Roughing Mill (RM) Node. Roughing Mill is a single stand where thickness reduction is achieved gradually by iterative rolling. After all scheduled passes, the slab thickness is reduced about 5 times. At RM the node, the Edger device sets final strip width of the product. A pair of vertical rollers put pressure on the sides and stabilize geometrical properties of the transfer bar.

Finishing Mill (FM) Node. It is a fully automated combined 6 stand mill. The appropriate configuration of rolling forces and subsequent cooling determine physical and chemical properties of the material. The rolling speed is constantly increasing to compensate cooling of the material, up to 10 m/s, maintaining thickness precision within limits ∼0,03 mm along the strip. Keeping strip directly at the center is important due to shape of the rollers surface. The technology is efficient enough to set end product width in RM node, regardless of over 10 times thickness reduction in further proceeding.

Laminar Cooling and Coiling Node. The next stage is laminar cooling, where water reduces the temperature of the strip before coiling. Cooling strategies affect the chemical and physical properties of steel. Achieving target endurance, ductility or magnetic conductivity needs strict temperature control during cooling. The last stage is coiling. It is performed for logistics reasons on 2 sets of Coilers working alternately. Coil form is universal for rail and road transport, due to suitability for further processing by both customers and inside divisions. Finally, after the complete HRM process, the coil is moved on stock to reach transport safe temperature. Next, product could be transferred for additional refining such as Cold Rolling Mill, galvanizing, painting, cutting into sheets or send directly to customer.

4   The Real Time Sensor Monitoring and Related Data

The HRM uses two levels of automation, therefore the system operates on two kinds of data. Raw measurements (unfiltered) collected by sensors in their nominal working frequency are used by the level one automation as feedback. Data need to be as certain and precise as possible. Second type of data is used for reporting purposes, planning, logistics, maintenance and process control in general. Data resolution is adjusted depending on specific applications. For example, if heat dispersion in steel is changing in essential way on 20 cm interval, using 5 mm measurements for reporting will not bring additional knowledge. The most important parameters collected in data warehouse in the Krakow HRM can be divided into two categories: Static, and Continuous-measured values. Static/single values includes Product IDs, Temperatures targets and limits values, Products dimensions in all nodes, Rollers dimensions, Thickness reductions settings, Chemical composition, Customers information, Logistics information. Continuous-measured values includes: Forces performed on all rollers, Temperatures in all nodes, Rolling and coiling linear speeds, Rollers gap performance, Tension Loopers position, Dimensions in all nodes, Laminar Cooling water flow and Accelerations in all nodes.

Over 3 million tons of steel are produced every year on HRM. The amount of data generated during the process allows to perform multidimensional analyzes and searching for hidden relations between parameters. Process nodes with the most relevant parameters are shown in Fig. 1. The HRM systems collect over 3500 process parameters and ~1 million records for each product.

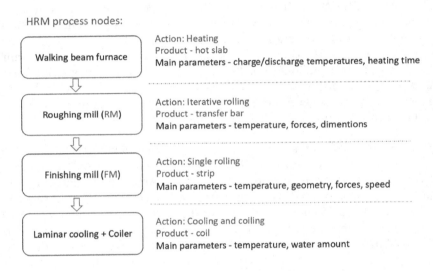

Fig. 1. Hot Rolling Mill process flow

4.1   Furnace Data

The most important parameters in the furnace node are the heating time and the discharge temperature. Heating the slab to ~1200 °C is taking ~120–170 min depending on the steel grade. Its internal structure is determined by temperature and related chemical/grain transformations. Discharge temperature is a single value measurement. More suitable information about heating quality represent transfer bar temperature (RM node). Elongation and thickness reduction makes it possible to look "inside" of the slab.

4.2   Roughing Mill (RM) Data

The most relevant RM node parameters, measured on all passes, are temperatures, thickness reductions, forces, rollers position and transfer bar geometry.

Common strategy is 5 or 7 passes per slab. Rolling is performed in one direction, followed by bypass reverse. In the RM node one can verify the quality of furnace heating by analysing transfer bar temperature. After the RM process, there is one more temperature measurement, just before FM process starts. Additional information is needed, because the transfer bar is cooled during the conveyor transport between the nodes. FM automation systems need precise temperature information for rolling forces calculations. Another significant parameter is the width, used to assess the accuracy of Edger rollers. Minor width deviations could be fixed during FM process.

4.3 Finishing Mill (FM) Data

On th FM node the most significant parameters are related with forces, tensions, temperatures and dimensions. Similar set of data is recorded at all 6 stands separately. Single stand measurements are sharing timestamps (regardless of the parameter), but there is no synchronization between the values gathered on subsequent stands. Due to elongation rollers in all stands had to work with different angular velocity. In addition, to maintain required temperature during the process, speed rolling is constantly increasing (and thus internal friction).

Dynamic strip length and rolling speed across the FM, made measurements comparison between stands ambiguous. For example, at the first stand, at the beginning of the rolling, distance between the measurements is $\sim$0,25 m. At the end of the process, at the same stand, measurements are separated $\sim$0,35 m. These values according to last stand are like 1,6 m at the beginning and 2,2 m at the end of the rolling.

4.4 Laminar Cooling and Coiling Data

The goal of laminar cooling is to control temperature reduction to obtain target internal structure according to grade. Sequence and intensity of the water poured into the strip is set individually per product. Another part of the system is responsible for closed water circuit. Production is continuous, so constant cold water resources and effective management of hot water is crucial. The most important parameters are valves efficiency, water parameters and strip speed.

Last devices in the production line are two Coilers. The most important parameters at this point are coiling temperature, coiling speed and strip tension. Temperature measurements are suitable to verify cooling quality.

5 Opportunities for Data Mining Analysis

Modern production site issues are not about eliminating human errors but to provide efficient cooperation of independent systems, today often based on machine learning. Explainable AI methods [2,9] seem to answer these needs as they allow for complex analysis of the dependencies between process parameters and automated decisions made based on that parameters values. Therefore, in this work we are more concerned about analysis of impact of different parameters on the output of the machine learning model, rather than on a prediction from that model.

Defect Minimization. In the continuous rolling process, despite the high degree of repeatability, there must occur some defects of the product. The main groups are: 1) surface defects, 2) lack of homogeneity along the strip length due to endurance, magnetic or corrosion resistance parameters, 3) internal structure defects, which not only could have an impact on the physical parameters of the steel, but also on further processing e.g. painting. Despite the overall deep understanding of the whole production process, the causes of some defects are not explained. XAI methods have great potential in solving this. A deeper understanding of the relationship between process parameters and product features would improve quality management.

Predictive Maintenance. Another important area with high potential for using XAI is predictive maintenance. The continuous nature of production means the need for constant control of the resources of each element of the production line. Consider phenomena such as the wear in bearings in motors, thermal insulation of furnaces, rollers surface or equipment exposed to high temperatures. Each of them has a specific working time after which it must be regenerated or replaced. The databases of the production line contains information on almost all operational elements of the line. Improvement in the area of predictive maintenance, reduces the risk of unplanned, expensive stoppages.

Energy Savings. Production processes have such a high energy demand that the steel plant needs its own power plant. Unfortunately, current methods of producing energy from renewable sources are too inefficient. Improvement in this aspect means reducing the amount of waste generated in the production of energy using conventional methods. Optimization directions could be related to the analysis in the field of shortening processes times. Creating more accurate models could allow to reduce electricity and heat consumption without quality loss.

6   Preliminary Results

6.1   Data Interpretation Approaches

Data Approaches. The process at the FM node can be considered as 6 independent operations performed on the same product. A similar set of parameters is recorded simultaneously on all frames. These include forces, speeds or distance between rollers (the Gap parameter). We decided to check difference in the results if we analyze the same set of measurements in two aspects named Process and Product approaches.

Product Dataset. Product approach on the data from the FM stands could be considered as horizontal dimension analysis. This method of data preparation allows to capture single snapshot of the parameters in all stands. We compare performance of the line devices during production of the same part of the product. When creating the dataset, we expected results related to products features, such as dimensions and temperatures. Based on timestamp, measurements from each stand were grouped in segments. In the next step, statistical features of each segment were calculated and then we compared the corresponding segments between all stands. The most relevant parameters according to thickness performance were calculated using SHAP [7] explainability algorithm.

Process Dataset. Process approach could be considered as a vertical dimension analysis. The difference is to create a time series using all single parameter measurements instead of analyzing segments. When creating the dataset, we expected results related to process line performance features, such as forces, speeds or accelerations. Products were divided by target thickness values, then we group parameters by distribution similarities. In the next step, statistical features of the parameters as series were calculated, for predicting end product thickness. As in the previous approach, SHAP explainability methods were used to define relation between features.

6.2   Static Parameter Analysis

As it has been previously said, there is a wide number of parameters that affect to the final quality of the product. Some of these parameters are time-dependent whereas the others are static (i.e. its value does not change along the time). This first experiment has two different goals: 1) to predict the final thickness of the products by taking into account only the static parameters, and 2) to determine which are the most important parameters to predict the thickness.

The first step is the data preparation. For this experiment, the features for each product is composed by those parameter that are not time-dependent. This means those parameters whose value is fixed at the beginning of the process and it does not change. We have 9513 products and each product is composed by 294 features (parameters). From this initial dataset, we keep only those features with numerical values. Also if there is any feature whose measurement has failed at any point (i.e. the feature has a null value), it is also removed. Finally, we have a dataset where each product has 70 features, and the goal of the machine learning model is to predict the thickness of the product. We choose Random forest as a model, due to its robustness and good out of the box performance.

The next step is to split the data into training and test. In this case, we use 80% of the data for training (7610 products) and the rest for test (1903 products). Then we build the Random Forest with 50 different decision trees and we have limited the depth for each decision to 15. This last decision prevents decision trees to be really specific for the data. With this configuration, the mean absolute error in the test is 0.1482.

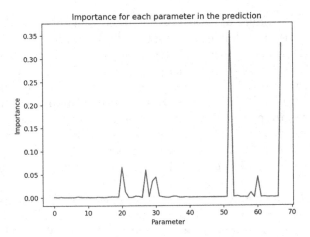

Fig. 2. Importance of each static parameter in the prediction of the final thickness

Table 1. Details of the 5 most important static parameters for predicting the thickness

Param. Numb.	Importance	Description
52	0.3581	Targeted (calculated) width of the transfer bar
67	0.3314	Entrance length of the transfer bar on each pass
20	0.0653	Theoretical slab thickness
27	0.0595	Measured coil weight
60	0.0825	Targeted end rolling temperature

It is important to analyse the relevance, or importance, for each feature in the Random Forest. The importance of a feature is computed as the normalized total reduction of the criterion brought by that feature[1]. This is because it is possible that all the predictions are based only on one or two parameters. Figure 2 shows the relevance for each of the features taken into account in this experiment.

As it can be seen in Fig. 2, not all the features are relevant. In fact, there are few parameters that are used to determine the thickness of the products. Table 1 contains the description for 5 most important parameters. This table shows for each parameters, its importance and its description.

6.3   XAI Results of Continuous-Measured Parameters Analysis

This section describes two complementary approaches that are based on time-series data for various sensors measurements. In **product approach** our goal was to build a predictive model for thickness of the product and then analyze the relationship between sensor measurements and target thickness. In **process**

---

[1] See: https://scikit-learn.org.

approach, we aim at clustering products by their thickness and explain differences between them using sensor measurements to obtain the information of what influences most the differences in the products.

Product Approach. In this approach only time-series parameters were taken into consideration. The aim of the approach was to predict mean thickness of the product based on the measurements and actions taken by automation systems at allof the stands. We transformed time series data into tabular representation by calculating various statistical properties of the series in rolling window of chosen size. There properties included standard deviation, variance, mean, maximal and minimal value. These features were independent parameters, while the dependent variable was mean thickness over the sliding window of the same size. Models were created with XGBoost algorithm, achieving R2 score at the level of 0.90. In the next step, we use TreeShap [7] algorithm that is based on SHAP [8] (SHapley Additive exPlanations) to explained how process parameters impact the target prediction.

In Fig. 3 we can observe the most relevant parameters in thickness predictive model. First on the list is coiling speed (Table 2). Second parameter is strip tension during the coiling process.

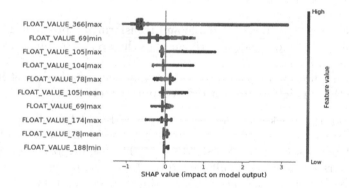

Fig. 3. SHAP values for features with the most impact on model output

Next two parameters are geometrical properties. End product is not completely flat across the width, surface has a slightly lenticular shape. Profile is a measure of that shape. Wedge is a difference between the thickness measured on the sides of the strip. Both of these parameters should not exceed customers specifications. FM automation systems, as a priority, needs to provide uniform, steady thickness. Even if profile or wedge values will be closer to their limits.

In group of the most influence parameters is also rolling speed at stand 1, rolling pressure force and Looper position. All of that parameters are related with strip tension. Importance of forces at only one side of the roller (stand 2) could be associated with wedge parameter performance. Stand 2 may be more important

Table 2. Features with the most impact on end thickness - product dataset

Param. Numb.	Description
366	Coiling speed
69	Strip tension between 6 stand and coiler
105	Strip wedge (after FM)
104	Strip profile (after FM)
78	Roller pressure force DS (FM 2)
174	Rolling speed (stand 1)
188	Loop position (stand 4)

than other from that perspective. The XAI results from the technology angle are consistent (Table 2), after conducting more extensive analyzes, it should be possible to determine detailed relations between these values.

Process Approach. The methodology consisted of three steps:

1. Select products which target thickness was similar and cluster them with respect to this parameter to obtain groups of products which are somehow different to each other.
2. Use cluster labels as target variable and transform the problem into classification task, where dependent variables was time series obtained from sensor measurements.
3. Use XAI methods to explain the classifier, and hence to show most important features that made the product differ from each other.

To cluster the thickness time series we used ROCKET [3] algorithm. To build a classifier we used XGBoost achieving F1 score at the level of 0.98. To create explanations we used TreeSHAP as in the case of previous approaches. All parameters obtained as results are related with strip tension except roller pressure force (Table 3). The mechanism of thickness changes is insufficient compensation of variability in steel plasticity by tensile forces during the process. Rocket results represent both components of this mechanism (Fig. 4).

The main principle of plasticity change over the length of the strip is temperature variation. In this case, rolling temperature at the end of Finishing Mill node has the most impact of all parameters in all considered classes. Next three, the most influential parameters are related to internal stresses. Loop position is information about strip tension. Results suggest the explicit importance of first part of the FM process, where reduction and rolling forces are, compared to final stands, higher. Significant impact of the acceleration values can also be associated with internal stress. Speed or acceleration differences between stands, as mentioned in Sect. 3, may be the cause of various types of deviations.

Among the results is also one side, rolling pressure force parameter at stand 1. The impact of rolling forces in this case may suggest, similarly to the tension

Table 3. Process dataset Rocket algorithm

Param. Numb.	Description
106	Rolling finish temperature
176	Loop position (stand 1)
180	Loop position (stand 2)
184	Loop position (stand 3)
195	Acceleration (stand 6)
179	Acceleration (stand 2)
73	Roller pressure force - DS - (FM 1)
187	Acceleration (stand 4)

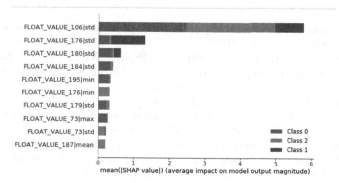

Fig. 4. Process dataset Rocket algorithm

looper parameters, the greater significance of initial finishing rolling stands. It may also be associated with a unique meaning of this parameter. A detailed interpretation requires further analysis.

With such a high level of production accuracy, possible improvements are related to minor adjustments. XAI results show the great potential of detailed process analysis, difficult to achieve by other methods. There are new possibilities to understand the specific relationships even in such well-known technology.

7   Future Works

In the paper we described the steel production process in the hot rolling mill in the Krakow steel plant of ArcelorMittal. We presented preliminary yet promising results of the data analysis we conducted. We transformed the data gathered by the AMP to the format that is acceptable by most of the machine learning algorithms. Based on these dataset we presented three complementary approaches for analysis of the influence of the parameters to the thickness of the product with an usach of machine learning algorithms and eXplainable AI methods.

In the future we are planing to continue the works outlined in Sect. 5. This especially include combining explainations from TreeShap with other methods such as LIME [10], Anchor [11] and others [1] to test them for consistency. We also plann to incorporate more advanced frameworks for time series processing based on Deep Neural Networks to better grasp the characteristic of the series [4].

Acknowledgments. This paper is funded by the National Science Centre, Poland under CHIST-ERA programme, the CHIST-ERA 2017 BDSI PACMEL Project, NCN 2018/27/Z/ST6/03392. We would like to thank ArcelorMittal Poland for cooperation. Special thanks go to the department of Automation, Industrial Informatics and Models in the Krakow steel plant.

References

1. Adadi, A., Berrada, M.: Peeking inside the black-box: a survey on explainable artificial intelligence (XAI). IEEE Access **6**, 52138–52160 (2018)
2. Arrieta, A.B., et al.: Explainable artificial intelligence (XAI): concepts, taxonomies, opportunities and challenges toward responsible AI. Inf. Fusion **58**, 82–115 (2020). https://doi.org/10.1016/j.inffus.2019.12.012
3. Dempster, A., Petitjean, F., Webb, G.I.: ROCKET: exceptionally fast and accurate time series classification using random convolutional kernels. Data Min. Knowl. Discov. **34**, 1454–1495 (2020). https://doi.org/10.1007/s10618-020-00701-z
4. Ismail Fawaz, H., Forestier, G., Weber, J., Idoumghar, L., Muller, P.-A.: Deep learning for time series classification: a review. Data Min. Knowl. Discov. **33**(4), 917–963 (2019). https://doi.org/10.1007/s10618-019-00619-1
5. Jerzy, H.: Walcowanie wyrobów długich, blach i taśm. Wydawnictwo Politechniki Śląskiej (2014)
6. Jerzy, S.: Poradnik odlewnika. T. 1; Materiały. pod red. Jerzego J. Sobczaka. Wydaw. Stowarz. Technicznego Odlewników Polskich (2013)
7. Lundberg, S.M., Erion, G.G., Lee, S.I.: Consistent individualized feature attribution for tree ensembles. CoRR abs/1802.03888 (2018). http://dblp.uni-trier.de/db/journals/corr/corr1802.html#abs-1802-03888
8. Lundberg, S.M., Lee, S.I.: A unified approach to interpreting model predictions. In: Proceedings of the 31st International Conference on Neural Information Processing Systems, NIPS 2017, pp. 4768–4777. Curran Associates Inc., Red Hook (2017)
9. Molnar, C.: Interpretable machine learning (2019). https://christophm.github.io/interpretable-ml-book/
10. Ribeiro, M.T., Singh, S., Guestrin, C.: "Why should i trust you?": explaining the predictions of any classifier. In: Proceedings of the 22nd ACM SIGKDD International Conference on Knowledge Discovery and Data Mining, KDD 2016, pp. 1135–1144. Association for Computing Machinery, New York (2016). https://doi.org/10.1145/2939672.2939778
11. Ribeiro, M.T., Singh, S., Guestrin, C.: Anchors: high-precision model-agnostic explanations. In: McIlraith, S.A., Weinberger, K.Q. (eds.) Proceedings of the Thirty-Second AAAI Conference on Artificial Intelligence (AAAI-2018), the 30th Innovative Applications of Artificial Intelligence (IAAI-2018), and the 8th AAAI Symposium on Educational Advances in Artificial Intelligence (EAAI-2018), New Orleans, Louisiana, USA, 2–7 February 2018, pp. 1527–1535. AAAI Press (2018)

Author Index

Printed in the United States
By Bookmasters